GODFATHER
OF THE KREMLIN

GODFATHER
OF THE KREMLIN

*Boris Berezovsky
and the Looting of Russia*

Paul Klebnikov

Harcourt, Inc.
NEW YORK   SAN DIEGO   LONDON

www.harcourt.com

Library of Congress Cataloging-in-Publication Data
Klebnikov, Paul.
Godfather of the Kremlin: Boris Berezovsky and the looting of Russia/Paul Klebnikov.
p.   cm.
Includes bibliographical references and index.
ISBN 0-15-100621-0
1. Berezovskiæ, B. A. (Boris Abramovich), 1946– 2. Businessmen—Russia
(Federation)—Biography. 3. Capitalists and financiers—Russia (Federation)—
Biography. 4. Russia (Federation)—Economic conditions—1991–
I. Title: Boris Berezovsky and the looting of Russia.   II. Title.
HC340.12.K555 2000
338.092—dc21   99-462183
[B]

Text set in Simoncini Garamond

Printed in the United States of America
First edition
A C E G I K J H F D B

To Jim Michaels for training me as a reporter,
To Forbes *for standing firm,*
To Musa for her support

TABLE OF CONTENTS

Preface		*xi*
Introduction		*1*
Cast of Characters		*7*
1.	The Great Mob War	*11*
2.	The Collapse of the Old Regime	*46*
3.	Trader's Paradise	*77*
4.	Selling the Country for Vouchers	*110*
5.	The Listyev Murder	*144*
6.	Privatizing the Profits of Aeroflot	*170*
7.	The Race for Oil	*188*
8.	The Black Treasury of the Yeltsin Campaign	*212*
9.	Oligarchy	*248*
10.	Vladimir Putin Takes Power	*284*
	Epilogue	*318*
	Appendix I: Boris Berezovsky Appeals to the President	*327*
	Appendix II: The Yeltsin Campaign Covers Up	*337*
	Appendix III: Berezovsky's Empire	*345*
	List of Sources	*346*
	Notes	*361*
	Index	*391*

Things fall apart; the center cannot hold;
Mere anarchy is loosed upon the world,
The blood-dimmed tide is loosed, and everywhere
The ceremony of innocence is drowned;
The best lack all conviction, while the worst
Are full of passionate intensity.

W. B. YEATS

PREFACE

Whenever I met important Russians over the past decade I always taped our conversations. Unless otherwise noted, the assertions made in this book are based on tape-recorded interviews with the businessmen and politicians who ruled Russia in the 1990s.

In Russia the truth has long been fluid, and I found that my interlocutors often made use of this fluidity. Usually, they were telling not so much the whole truth as only part of it. The challenge of this book has been to piece together the fragments of truth, weigh them against one another, and arrive at a complete picture. I found that the task became easier with time. The more a person knows, the less likely he is to be lied to.

In telling this story, I have tried to be conservative in asserting what I believe to be true. Throughout, my sources are noted to allow the reader to judge for himself their solidity or lack thereof. I have left a trail of my research, so that the reader may be in a position to decide for himself what to believe and what not to believe.

So much of contemporary Russian history is conveyed by word of mouth, asserted by people who "were there." Often, these people report things they do not know. That is the minefield any journalist must navigate. But many agreements in Russian business and politics are made with a handshake. Seldom is anything written down. Often these contracts are much more solid than the written ones, so a determined analyst of contemporary Russia must not be discouraged by the lack of Western-style documentation.

Nevertheless, many of the key elements of this story are documented—from publicly available annual reports, registration documents, or investment banking analyses, to more closely held copies of contracts and boardroom minutes. I have also used a number of government documents, both classified and unclassified, to clarify the nature of the relationships among the various protagonists of this story.

In many respects, Russia under Yeltsin was still a police state—the telephones were tapped, confidential documents were exposed, individuals were subject to intense surveillance—albeit a police state that had been privatized. Huge numbers of specialists from the old Soviet security and law-enforcement apparatus (KGB, GRU, MVD, and so on) took jobs in the private sector. Every large financial-industrial group created its own mini-KGB, usually known as the "analytical department," staffed with people who were good at collecting information, eavesdropping on rivals, and stealing documents. The intelligence dossiers compiled by these analytical departments invariably took the form of the standard briefs that went into the files of the old Soviet security services. Often, the data were inaccurate, but I occasionally found these reports to be a useful source of background information. Some of the information collected by these private intelligence agencies would later be confirmed by public events or by law-enforcement agencies.

Among my sources were former members of the Presidential Security Service (SBP). Before it was disbanded in 1996, the SBP was one of the most powerful institutions in Russia. It employed approximately 500 specialists—from commandos trained in special operations, to intelligence analysts equipped with the latest surveillance technology. The SBP's mission was not only to guard Yeltsin but also to investigate allegations of corruption or espionage among high government officials.

During the Yeltsin era, there was a great deal of leakage from the law enforcement agencies. There seemed to be no clear policy about what information could be released to the public. Many individuals took it upon themselves to share documents (videotapes, audiotapes, and printed matter) with me. For the most part, these individuals were frustrated by the inability of the Russian legal system to bring criminals to justice. They believed that when the court system does not work, glasnost—public exposure—can provide some kind of tonic against the epidemic of crime.

Many of the people who talked to me about the events described in this book spoke only on the basis of a prior agreement that they would not be quoted. These interviews served only as background information. If I have included the events these anonymous sources described to me, I have done so only if I have confirmation from an on-the-record source. In the rare cases when an anonymous source was the only one for the material, I have cited him in the notes (as an anonymous source). The most important of these is referred to in the notes as a "RUOP source." This man is a former high-ranking officer in the Moscow RUOP (the police organized-crime squad). I have confidence in the reliability of his information because he has been in a position to know the events he describes. Moreover, I have relied on this man since 1993 and, over the years, he has provided me with a lot of information that would later be proved correct by events. Individuals he identified as gangster bosses, for instance, would later be revealed as such by their involvement in shoot-outs or by their arrest and conviction by Western law-enforcement agencies.

I have tried to avoid relying on newspaper sources as the foundation for my narrative. The times when I do cite newspaper sources, I do so only for their role as a daily chronicle of public events or for their verbatim interviews with one of my protagonists. Most of the times when I felt I had to use an interview conducted by another journalist, the latter did not share the tape of his conversation with the subject. However, I still believe these printed interviews to be accurate, both because of the good reputation of the newspapers in which they appeared and because the interview subjects themselves would often return to the same newspaper to grant another interview months or years later. I am assuming, in other words, that if a person was misquoted in a given newspaper he would not go back to that publication again.

The best sources for this book are the protagonists themselves. In writing this documentary and oral history of the Yeltsin era, I am sure that I have missed a lot. There will almost certainly be books that expose the lives of my protagonists in more revealing detail. But I had the opportunity to catch these men in the early 1990s, in their "age of innocence," when they spoke frankly—often boasting of their criminal exploits—and, equally frankly, lied.

GODFATHER
OF THE KREMLIN

INTRODUCTION

In February 1997 *Forbes* magazine was sued by a man named Boris Berezovsky. This individual had risen out of nowhere to become the richest businessman in Russia and one of the most powerful individuals in the country. I had profiled Berezovsky in a December 1996 article entitled "Godfather of the Kremlin?" He hired English litigators and sued for libel in London High Court. As of the publication of this book, the litigation is still pending. *Forbes* was not intimidated by the prospect of a libel trial and I continued to publish articles on Berezovsky.[1]

I noticed that his shadow loomed over many of the great events that have shaken Russia throughout the past decade. I started going through the tapes of my earlier interviews with the robber barons of Russia's new era, men whose careers intersected with Berezovsky's: commodities traders who briefly controlled the Russian economy; factory bosses lording over sprawling industrial empires; young bankers, unscrupulous and tough, who made fortunes from their political connections. These men were great when Berezovsky was small; they were in the spotlight while he was backstage, waiting.

Many of Russia's business magnates had inherited their wealth from the old Soviet Union—they were authorized millionaires—but Berezovsky built his empire from scratch. While many individuals contributed to Russia's decline in the 1990s, Berezovsky exemplified the spirit of the era. No man was as versatile in negotiating the rapidly changing situation; with each new twist in Russia's agonizing transition to a market economy, Berezovsky reinvented himself to profit from the change. And when he turned to politics, he dominated his peers. Having privatized vast stretches of Russian industry, Berezovsky privatized the state.

Russia's decline from a global superpower to an impoverished country is one of the most curious events in history. The collapse occurred during a period of peace and took only a few years. In its rapidity and magnitude, the collapse is unprecedented in world history.

When Mikhail Gorbachev launched perestroika and when Boris Yeltsin emerged as Russia's first democratic president I expected that Russia would get the same burst of energy that had invigorated China after the reforms of Deng Xiaoping. I expected the same kind of national economic revival that had followed the decollectivization of agriculture by Russian prime minister Pyotr Stolypin almost a century earlier. But soon I understood that Russia was falling apart. When Yeltsin's government freed prices, it unleashed hyperinflation, which instantly impoverished the majority of the population. The introduction of the free market did not produce a more efficient economy— it produced relentless economic decline. Privatization enriched only a small number of insiders. The country was being looted and destroyed by the new property owners.

How could everything have gone so wrong? All trails seemed to lead to Russian organized crime. I wrote articles about the grotesque lifestyles and lurid violence of the new gangsters. While investigating the Russian Mafiya, I was always given the same advice: If you want to write about Russian organized crime, don't focus on the colorful Mafiya chiefs, but on the government. Russia was a gangster state, I was told, and its political system nothing more than a government of organized crime.

The FBI definition of organized crime is "a continuing criminal conspiracy, fed by fear and corruption, and motivated by greed." The definition includes the following: "They [organized crime groups] commit or threaten to commit acts of violence or intimidation; their activities are methodical, systematic, disciplined, or secret; they insulate their leadership from direct involvement in illegal activities through an intricate bureaucracy; they attempt to influence government, politics, and commerce through corruption, graft, and legitimate means; and their primary goal is economic gain, not only from patently illegal enterprises . . . but also from such activities as laundering illegal money . . . and investment in legitimate businesses."[2]

A clear history of criminal activity during the Yeltsin era is difficult to establish. Almost none of the most famous murders have been solved. Even past convictions have been difficult to identify—a major problem for law-enforcement agencies was the ability of well-connected ex-convicts to pull their files, thus erasing any trace of past crimes. The Russian legal code was full of ambiguities and loopholes. Many financial operations that in the West would be classified as crimes (certain types of bribery, fraud, embezzlement, and racketeering, for instance) are not necessarily crimes in Russia.³

Russian gangsters could ignore the police because they had protection at the top. The typical gangster organization in Russia began with street-corner thugs who extorted money from local businesses; these men answered to a citywide criminal organization; this organization, in turn, was subject to mob bosses functioning on the national level. At each level, the gangsters had allies in the government—beginning with the local police precinct or tax inspectorate, proceeding to the mayors and governors, and ending with the entourage of President Yeltsin himself.

Typically, any successful Russian businessman had to deal with these two chains of command. The Russian power structure was a three-sided pyramid, composed of gangsters, businessmen, and government officials.

Behind every historic process stand specific individuals. I wanted to know who Russia's real bosses were. Who had brought the country to such a state? Who stood at the top of the pyramid?

In the autumn of 1996, I began looking into the affairs of Boris Berezovsky. No man stood closer to all three authorities at once: crime, commerce, and government. No man profited more from Russia's slide into the abyss.

It was on a trip to the Volga town of Togliatti, 700 miles east of Moscow, home of Russia's largest auto company, Avtovaz, that I first learned of him. I was writing an article on the Russian car industry and had heard that Avtovaz was somehow connected with an entrepreneur named Berezovsky. (In fact, the tycoon had made his first big fortune at the auto factory.)

When I asked Avtovaz president Alexei Nikolaev about Berezovsky's holding company Logovaz, the auto executive and his aides exchanged worried glances. I saw a flash of fear in the faces across the table. "We no longer have direct links with Logovaz," he mumbled. "They have some sort of business of their own over there [in Moscow]."[4]

Who was this businessman whose very name brought on silence? As I looked into Berezovsky's meteoric career, I discovered a history replete with bankrupt companies and violent deaths. The scale of destruction was extraordinary, even by contemporary Russian standards. He had soaked the cash out of the big companies he dealt with, leaving them effectively bankrupt, kept afloat only by lavish government subsidies. He seemed to gravitate to the most violent corners of Russia—the car-dealership business, the aluminum industry, the ransoming of hostages in Chechnya. Many of his business ventures—from the takeover of Russian Public Television to that of the Omsk Oil Refinery—were marred by the assassination or accidental death of key players. Shortly after his interference in the management of the National Sports Fund, there was an attempt to assassinate the charity's ex-president. There is no evidence that Berezovsky was responsible for any of these deaths. Although he was briefly under suspicion for the Yeltsin era's most famous assassination, in 1995, he has never been charged with any crime relating to these events.

I met Berezovsky in the autumn of 1996 in Moscow. It was immediately evident that he was highly intelligent—he holds a Ph.D. in mathematics. He spoke nervously, articulately, waving a hand scarred from an assassination attempt in the early years of the Yeltsin regime. He was unapologetic about the violence of the Russian business world and quickly took the moral high ground. "The Western press portrays Russia unfairly," he declared. "Russian business is not synonymous with the Mafiya."

I asked why the government was incapable of bringing the mobsters to justice. "Because in the government there are many people who are criminals themselves," he replied. "They are not interested in having these crimes solved."[5]

A month later Berezovsky would be appointed to a key government post: deputy secretary of the Security Council.

The disintegration of Russia allowed Berezovsky a unique chance to realize his designs on a grand scale. And as he grew stronger, Russia grew weaker.

Incredible as it may seem, the foundations of the country's economic and demographic decline were laid by the "young reformers" and "democrats," led by Yegor Gaidar and Anatoly Chubais. First, in 1992, the democrats freed prices and unleashed hyperinflation before they privatized Russia's assets. In just a few weeks the savings of the vast majority of the population were wiped out, instantly destroying the hope of building the new Russia on the foundation of a strong domestic market.

Second, the democrats subsidized the traders—well-connected young men who made fortunes by assuming the role of the old state trading monopolies and arbitraging the vast difference between old domestic prices for Russian commodities and the prices prevailing on the world market.

Third, Chubais's voucher privatization in 1993–94, following the hyperinflation that had destroyed the Russians' savings, was mismanaged. Most Russians simply sold their vouchers for a few dollars to securities traders or unwittingly invested them in pyramid schemes that later collapsed. Instead of creating a broad class of shareholders, Chubais's privatization gave away Russia's industrial assets to corrupt enterprise managers or to the new Moscow banks.

Fourth, Chubais and his allies subsidized these new banks by providing them with Central Bank loans at negative real interest rates, handing them the accounts of government institutions, and rigging the government securities market in their favor.

Finally, in the rigged loans-for-shares auctions of 1995–97, Chubais sold the remaining jewels of Russian industry to a handful of insiders at nominal prices.

The crony capitalism that emerged under Yeltsin was not an accident. The government deliberately enriched Berezovsky and a handful of men in return for their political support. Both the Yeltsin clan and the crony capitalists remained in power, but they presided over a bankrupt state and an impoverished population. The young democrats were supposed to clean up Russia, devise a proper legal system, and foster a market economy. Instead they presided over one of the most corrupt regimes in history.

No one benefited from the arrangement more than Berezovsky. He loved to boast of his accomplishments. He once told the *Financial Times* that he and six other financiers controlled 50 percent of the Russian economy and had arranged Yeltsin's reelection in 1996.[6]

Godfather of the Kremlin is not a biography of Boris Berezovsky. It is not a book about his childhood, upbringing, outlook, or personal life. Neither is it a work of political analysis or a consideration of Russia's changing role in the world. It is an investigation into Berezovsky's business and political career. This book charts two parallel developments—Berezovsky's rise and Russia's decline. It is also a story of the abuse of Western democratic principles: personal liberty, free speech, electoral government, protection of minority rights, market economics, self-reliance, and the opportunity to make an honest living. In the early 1990s, foreign well-wishers thought that by dismantling communism and enshrining these principles, Russians would transform their country into a Western-style democracy. Billions of dollars of aid and enormous amounts of time were spent to help the transition. If it is hard for Westerners to understand how the introduction of democratic principles could have been so poisonous to Russian society, Berezovsky's career holds one of the keys.

Winston Churchill famously characterized the political intrigues of Joseph Stalin's Russia as "bulldogs fighting under a carpet." It was still the same in Yeltsin's Russia. This is a story of corruption so profound that many readers might have trouble believing it. Yet Berezovsky and the other Russian businessmen often did not bother to disguise their depredations. General Aleksandr Lebed accurately described their attitude: "Berezovsky is the apotheosis of sleaziness on the state level: This representative of the small clique in power is not satisfied with stealing—he wants everybody to see that he is stealing with complete impunity."[7]

CAST OF CHARACTERS

This is a select list of names. It does not include such internationally known figures as Boris Yeltsin or Mikhail Gorbachev. Neither does it include lesser figures whose status or role is made clear at each mention.

Roman Abramovich: Berezovsky's partner in the oil business and part owner of the oil company Sibneft.

Vagit Alekperov: deputy minister of fuel and energy in the last Soviet government; later boss of Russia's largest oil company, Lukoil.

Pyotr Aven: minister of foreign trade, 1991–92; president of Alfa Bank, 1992–2000; a friend of Berezovsky's.

General Mikhail Barsukov: chief of the Main Security Administration, 1992–95; head of the FSB, 1995–96.

Shamil Basayev: a Chechen warlord and terrorist commander.

Boris Berezovsky: a businessman.

General Philip Bobkov: first deputy chairman of the KGB until 1991; thereafter, one of the directors of the security service of Vladimir Gusinsky's Most Group.

Oleg Boiko: founder of Natsionalny Kredit Bank and the trading company Olbi; a business partner of Berezovsky's.

Pavel Borodin: head of property management in the Presidential Administration under Yeltsin; currently under investigation in Switzerland and Russia for bribery and money-laundering.

Konstantin Borovoi: chairman of the Commodities and Raw Material Exchange; later a member of parliament.

Viktor Chernomyrdin: boss of the natural gas monopoly Gazprom until 1992; prime minister of Russia, 1992–98.

Anatoly Chubais: chief of the "young reformers"; first deputy prime minister in Yeltsin's government; manager of Yeltsin's election campaign in 1996; architect of Russia's privatization campaign, 1991–97.

Yves Cuendet: a managing director of the Swiss commodities trading company André & Cie.

Oleg Davydov: minister of foreign trade, 1994–97.

Leonid Dyachenko: Yeltsin's son-in-law; a commodities trader doing business with Berezovsky's oil company, Sibneft.

Tatyana (Tanya) Dyachenko: Yeltsin's youngest daughter.

William Ferrero: general director of Berezovsky's Swiss financial company Andava S.A.

Yegor Gaidar: the first of the "young reformers"; acting prime minister, 1991–92; architect of the 1992 price reform and economic "shock therapy."

Nikolai Glushkov: a founding member of Berezovsky's holding company Logovaz; head of finance and foreign relations for the automaker Avtovaz, 1989–95; deputy general director of the national airline, Aeroflot, 1995–97.

Vladimir Gusinsky: founder of Most Bank and NTV; Berezovsky's archrival, 1993–96; his ally, 1996–97; then his rival again, 1998–2000.

Alexei Ilyushenko: acting prosecutor-general, 1994–95; fired in 1995 in connection with the Balkar Trading scandal and the privatization of Sibneft.

Vladimir Kadannikov: chairman of Avtovaz, Russia's largest auto manufacturer; a founding member of Berezovsky's Logovaz.

General Oleg Kalugin: former head of foreign intelligence (First Chief Directorate) of the KGB.

Mikhail Khodorkovsky: former Communist Youth League activist; founder of Menatep Bank; chairman of the oil company Yukos; a business partner of Berezovsky's.

Sergei Kiriyenko: minister of fuel & energy, 1997–98; prime minister, 1998.

Alfred Kokh: one of the "young reformers"; first deputy chairman of the State Property Committee (responsible for privatization), 1995–96; deputy prime minister, 1997.

General Aleksandr Korzhakov: Yeltsin's bodyguard, 1987–91; chief of the Presidential Security Service, 1991–96; member of parliament since 1997.

General Aleksandr Lebed: paratroop commander; presidential candidate, 1996; head of the Security Council 1996.

Sergei Lisovsky: entertainment and advertising magnate; head of advertising at ORT; one of the coordinators of Yeltsin's 1996 election campaign.

Vladislav Listyev: Russia's most successful TV producer and talkshow host; general director of ORT; assassinated in 1995.

Yuri Luzhkov: mayor of Moscow since 1992.

Christian Maret: head of Russian operations for the Swiss commodities firm André & Cie.

Aslan Maskhadov: president of Chechnya since 1997.

Alain Mayor: head of Russian operations for the FINCO department of André & Cie; the primary intermediary between the Swiss company and Berezovsky; since 1997, a director of Berezovsky's Swiss financial companies Forus and Andava.

Boris Nemtsov: one of the "young reformers"; first deputy prime minister, 1997–98.

Valery Okulov: Yeltsin's son-in-law; general director of Aeroflot since 1997.

Badri Patarkatsishvili: a founding member of Logovaz; member of the board of directors of ORT and Sibneft; intermediary in Chechen hostage negotiations 1999; suspected of links with organized crime.

Vladimir Potanin: founder of Onexim Bank and MFK; controlling shareholder of the oil company Sidanco, Norilsk Nickel, and the telecommunications giant Svyazinvest; inventor of the loans-for-shares auctions; since 1997, a rival of Berezovsky's.

Yevgeny Primakov: head of foreign intelligence, 1991–96; foreign minister, 1996–98; prime minister, 1998–99.

General Vladimir Rushailo: head of the Moscow RUOP (police organized-crime squad), 1993–96; minister of internal affairs since 1999.

Yuri Skuratov: prosecutor-general, 1995–2000.

Aleksandr Smolensky: founder of Stolichny Bank (later renamed SBS-Agro); a close business partner of Berezovsky's.

Oleg Soskovets: first deputy prime minister until 1996; a political patron of Berezovsky's.

Colonel Valery Streletsky: Moscow police veteran; head of Department P (anticorruption) of the Presidential Security Service, 1994–96.

Movladi Udugov: first deputy prime minister of Chechnya since 1996; leader of the extremist wing of the Chechen government; Berezovsky's primary negotiating partner.

Grigory Yavlinsky: one of the authors of Gorbachev's 500-Day plan of economic reform in 1990; head of the liberal Yabloko faction in the Russian parliament; presidential candidate in 1996 and 2000.

Valentin Yumashev: deputy editor of the magazine *Ogonyok*; the ghostwriter of Yeltsin's memoirs; Yeltsin's chief of staff, 1997–98.

Glossary of Russian Law-Enforcement and Security Agencies

FSB: Federal Security Service (the name of the domestic arm of the KGB since 1995).

FSK: Federal CounterIntelligence Service (the name of the domestic arm of the KGB in 1993–95 before it was renamed FSB).

KGB: Committee of State Security (the domestic and foreign intelligence service of the U.S.S.R.).

MVD: Ministry of Internal Affairs (responsible, among other things, for the police).

RUOP: Regional Administration for Organized Crime (the organized-crime squad, subordinate both to the MVD and the local government).

SBP: Presidential Security Service (Yeltsin's "mini-KGB," headed by General Korzhakov).

CHAPTER ONE

THE GREAT MOB WAR

The Shoot-out at the Kazakhstan Cinema

The funeral took place on Saturday at noon. It was a typical July day in Moscow: high clouds, noisy traffic, a yellow haze of smog and dust shrouding the city. At the Church of the Archangel Michael on Vernadsky Prospekt, the cars began to pull up, parking on the sidewalks and on the grass—not just the ubiquitous Russian sedans, but also sleek BMWs, Mercedeses, and Volvos. The mourners entering the church in groups of three or four were big guys dressed in black jackets, unbuttoned casual shirts, tracksuits. They had come to pay their respects to Igor Ovchinnikov, a former wrestler who had served as the treasurer and chief lieutenant of a large Moscow crime family that, according to Moscow police, was allied to a powerful organization called the Solntsevo Brotherhood. Inside the church, the mourners filed by the open coffin and gave a respectful nod to Ovchinnikov's boss, nicknamed "Cyclops"—a former boxer whose nom de guerre referred to an eye he had lost in a battle with some Chechen gangsters several years before.[1]

Ovchinnikov had been killed in a gun battle with a Chechen gang outside the auto showroom of Boris Berezovsky. A forty-seven-year-old man who had spent most of his professional life designing computer software, Berezovsky was hardly the type one would expect to succeed in the world of gangsters; yet in just four years, he had parlayed an obscure joint venture into a major fortune in the car dealership business—one of the most criminalized segments of the Russian economy. According to Russian law-enforcement officials, Berezovsky had built his dealership, Logovaz, in conjunction with organized-crime groups from the southern republic of Chechnya. He had used the fearsome Chechens for protection; they were his *krysha*, his "roof" in the auto market. Berezovsky's fortune made him a target for organized-crime groups. He would not survive in this business unless he could protect his gains physically. With the Russian government in disarray, the most effective "security services" available to businessmen was the Mob.[2]

In July 1993, Ovchinnikov's gang began to move onto Berezovsky's turf. One Moscow police detective, charged with tracking Ovchinnikov's movements, described the negotiations as follows. When Ovchinnikov's men approached Logovaz about a partnership, Berezovsky said that he "already had a roof" and that they should "speak to the Chechens." The decisive "conversation" occurred outside the Logovaz showroom in the Kazakhstan Cinema on Lenin Prospekt. Ovchinnikov and his gang pulled up in three cars and began shooting. The Logovaz men shot back. Within minutes three men were dead (including Ovchinnikov) and six had been wounded. It was one of the bloodiest gang battles in Moscow in 1993.[3]

When I asked General Vladimir Rushailo, head of the Moscow RUOP (the organized-crime squad of the police), about the incident, he replied obliquely. "Many believe that businessmen get killed simply because they are businessmen," he said. "Not so. Investigations of cases involving contract killings have shown that the victims were involved in some sort of unclear relationship, to put it diplomatically, with the very people who either ordered or carried out the hits against them. No one kills the law-abiding citizens—those who do not violate the law, who pay their taxes.... The cause of the shoot-out was that the business you and I were discussing [Logovaz] had its own security service and another gang arrived and wanted to collect money from them as well. The result was entirely predictable."[4]

Berezovsky told me that he remembered the shoot-out, but did not know what it was about. He went on to say: "Today we are witnessing a redistribution of property on a scale unprecedented in history. No one is satisfied: neither those who became millionaires overnight—they complain they didn't earn enough millions—nor those who received nothing and are naturally discontented. Therefore, I do not think the extent of crime exceeds the scale of the transformation process."[5]

The Chechens

The Chechen gangsters had arrived in Moscow in the late 1980s. Suddenly, restaurants, hotels, banks, and new private businesses (cooperatives) became victims of Chechen extortionists. If they were lucky, recalcitrant managers received a warning; the second time they resisted, they were murdered. Established Moscow criminal gangs, many of their leaders still in prison in the late 1980s, were summarily pushed aside. The Chechens terrorized their competitors—their thirst for vengeance was frightening, their brutality awesome. Moscow was so traumatized that many unexplained crimes were blamed on "the Chechens." The names of Chechen gang leaders—Ruslan Atlangeriev, Khozha Nukhaev, Lechi the Bearded One, Lechi the Bald, the Talarov brothers, Sultan Daudov, Khozha Suleimanov—were mentioned with dread in the Moscow underworld. Less well known were the names of future Chechen terrorist commanders, such as Shamil Basayev, who were earning money as "businessmen" in Moscow in those years. The number of Chechen bandits in Moscow was, even by generous estimates, less than 1,000 men. Yet, in just a few years, they succeeded in taking the Russian capital by storm.[6]

No one seemed to know how the Chechen "commune" worked or how one Chechen gangster related to another. Russians knew the Chechens mostly from historical lore—as a mountain people, renowned for their blood feuds, their celebration of banditry, and their fierce fighting spirit. One of the strengths of the Chechen community in Moscow was its clannishness—no matter how fierce their internal blood feuds, their quarrels never went beyond the commune. Moscow police officers admitted that they could not handle them; whenever the police closed in on a particular Chechen crime boss, he quickly fled to Chechnya.

Nominally part of the Russian Federation, the Republic of Chechnya had long been striving toward independence. The Yeltsin administration offered little resistance; in fact, it left a large arms depot in place for the local Chechen government when Russian troops withdrew in 1991. For the first three years of the Yeltsin era, Chechnya existed in a legal gray zone. Since it was still part of Russia, it benefited from Russian government subsidies and from participation in the Russian financial system. At the same time, it was beyond the reach of Russian customs control and law enforcement.

One of the Chechen government's first moves after the fall of the Soviet Union was to open the doors of the prisons, releasing an estimated 4,000 professional criminals. Many of the gangster bosses became part of the Chechen government, while maintaining contact with Chechen groups in Moscow and other Russian cities. A good part of the money these groups earned from their rackets was sent back to Chechnya. Grozny Airport became a trans-shipment point for smuggling operations. Chechnya emerged as a hub of the international heroin trade. One smuggling route began in the Golden Triangle (Burma, Thailand, and Laos). The opium and heroin were transported to the Russian naval base at Cam Ranh Bay (where it was handled by Vietnamese drug traders), then shipped to the Far Eastern port of Nakhodka (where it was handled by Chechen and Russian gangs), flown to Grozny, and then distributed (via Russia, Ukraine, and Turkey) to markets in Europe and the United States. Another smuggling route originated in the Golden Crescent (Afghanistan, Pakistan, and Iran); in this case the narcotics passed through the Central Asian republics or Iran and Azerbaijan, before reaching Chechnya, and were then shipped to markets in the West. If the drugs were shipped through Russia, their destination was usually Germany; the contraband was shipped on military transports to Russian army bases in East Germany and then sold to local drug gangs.[7]

In Moscow, the Chechen gangsters quickly took control of the state-owned chain of Beryozka stores, Soviet-era luxury supermarkets that catered to foreigners and the Soviet elite. Next, they set up extortion rackets to cover shops, restaurants, and other businesses around the city. They moved into the wholesale trade when a particularly powerful band of Chechens, led by Khozha Suleimanov, took control of the Southern Port, Moscow's most important river terminal and a big open-air market for automobiles, auto parts, and other big-ticket consumer items. This allowed the Chechens to establish themselves in one of the most important cash businesses in Russia: sales of automobiles (both new and used). They also took over dealerships and repair shops, especially those specializing in foreign cars. Another Chechen gang pushed the Azerbaijani gangsters out of the burgeoning drug trade.[8]

When casinos began appearing in 1992–93, the Chechens quickly took control of many of the best ones. Hotels were also a prime target. The gangster takeover of a hotel usually followed the same pattern: First, control the hotel's prostitution racket, then its shops and restaurants, and, finally, the cash flow of the hotel itself. Subsequently, the Chechens moved into the financial markets, taking dozens of banks under their control.[9]

The seven main Chechen gangs in Moscow (Central, Belgrad, Ukraina, Lazaniya, Ostankino, Salyut, and the Southern Port) fielded a force of approximately 500 fighters in the early 1990s. Many fighters of the Central Group (under the command of Lechi the Bearded One) lived at the giant Rossiya Hotel, across from the Kremlin. From this Chechen-controlled hotel, every destination in the center of Moscow was within easy striking distance. Once their business empire became too large to manage directly, the Chechens began operating through proxies, enlisting gangsters from Georgia, Dagestan, Ingushetia, and Russia itself.[10]

Thieves-Professing-the-Code

As Chechen gangsters moved into Moscow and other Russian cities, they were not entering virgin territory. The criminal world of the

Soviet Union had old traditions, developed over decades in the prisons and gulags. In the slang of the penal colonies, the code of laws governing the professional criminals was known as the "thieves' world" (*vorovsky mir*). The chief of the traditional criminal underworld was a person called the thief-professing-the-code (*vor v zakone*), the closest Russian equivalent to the godfather of the Sicilian and American Mafia. He was typically a hardened criminal, "crowned" by other thieves-professing-the-code while in prison; one of his duties was to mediate between competing gangs. There were several hundred thieves-professing-the-code in the former Soviet Union. Most were members of one of Russia's ethnic minorities. Out of the sixty or so thieves-professing-the-code in Moscow in 1993, for instance, more than half were from the Republic of Georgia, a country which in Soviet times was famous for its people's penchant for good living and black-market trading. Another dozen came from neighboring countries in the Caucasus Mountains.[11]

Berezovsky's primary emissary to the traditional underworld was his partner Badri Patarkatsishvili, one of the cofounders of Logovaz. Whenever Berezovsky's business career led him into particularly violent terrain, he would turn to his partner to sort things out. Badri was a Georgian and had good ties with many important Georgians in Moscow. According to the Presidential Security Service, the FSB–KGB, and at least one major private security agency, Badri also had close contacts with organized-crime groups from the Caucasus.[12]

"The official position of Badri at that time [1993–94] was deputy chairman of the board of directors of Logovaz," recalls General Aleksandr Korzhakov, former head of the Presidential Security Service. "In fact, his responsibility was to ensure the repayment of debts and to provide protection against gangsters. In Soviet times Badri Patarkatsishvili was one of the directors of an auto-repair network in Georgia. One of his brothers, Merab, is a thief-professing-the-code; another brother, Levan, is an 'authority' of a Georgian organized crime group. Badri also has an alias, like any gangster. In the criminal underworld he is known as 'Badar.'"[13]

In the 1970s, the most notorious thief-professing-the-code was a

Russian ex-convict named Vyacheslav Ivankov, also known as "the Jap" (*Yaponchik*). Having been crowned as a thief-professing-the-code during a brief stay in prison in the early 1970s, the Jap developed an extensive operation extorting money from black-market entrepreneurs and corrupt officials; he also smuggled narcotics, jewelry, icons, and antiques. He had a reputation for flamboyant brutality, often taking recalcitrant black marketeers into the woods outside town and torturing them. According to one legend, when the Jap was crossed by a Moscow restaurant manager, he had the man buried alive and a road paved over him. "Killing someone is as easy as lighting up a cigarette," he apparently liked to say. From Riga to Sverdlovsk, from Kazan to Moscow, he left a trail of fear. Yet, despite his loud criminal career, the Jap did only two short stints in prison—one for carrying false documents and one for carrying a gun. Only in 1981 did a combined KGB–MVD operation succeed in netting him. He was convicted of banditry and sentenced to fourteen years in prison.[14]

The Mafiya chief had friends in high places. In November 1991—as the Soviet Union was falling apart—the Supreme Court of the Russian Federation freed the Jap, officially because of his bad health. Among those who allegedly petitioned for his release were the famous nightclub singer Iosef Kobzon, the world-renowned eye surgeon Svyatoslav Fyodorov, and the venerable campaigner for human rights (and old colleague of Andrei Sakharov's) Sergei Kovalyov.[15]

"[The Jap's] wife came and petitioned me," Fyodorov recalled laconically. "[The Jap] was in bad health and an extra four years [in prison] wouldn't do him any good."

Fyodorov perhaps had weightier reasons for dabbling in such affairs. In addition to his famous hospital in Moscow and the clinics he had opened abroad, he also had equity stakes in two big Moscow hotels and the Casino Royale (a rival of the Chechen-controlled Cherry Casino). One theory about the Jap's premature release was that Russian law-enforcement authorities, including the former KGB, were anxious to produce a counterweight to the Chechen gangs dominating Russian streets in 1991.[16]

The Jap did not remain in Russia for long. In 1992 he went to New York to establish a criminal network there, preying on the burgeoning Russian émigré community. But he retained strong links with the gangsters back home, mostly through the international narcotics trade.[17]

One of the custodians of the Jap's interests in Russia was Otari Kvantrishvili ("Otarik"). This dark, bearlike man was not a thief-professing-the-code; he was what the police termed "an authority" in the underworld. "[The Jap] is one of the most honest people around," Otarik gushed to me in 1993. "Even if he was a criminal, he was only a private criminal. And private criminals today are nothing in comparison with the state criminals who have bankrupted Russia and are . . . buying themselves homes in Florida."

As a young man, Otarik had been a talented wrestler, with a good shot at making the Olympic team. But in 1966, when he was eighteen years old, he participated in a gang rape. He was treated leniently by the authorities. After four years in prison, he was diagnosed as a schizophrenic, transferred to a psychiatric hospital, and freed soon after. In the early 1980s, he got a job as a wrestling coach for the prestigious Dinamo sports complex in Moscow. He trained many of the boxers, wrestlers, martial-arts specialists, and weight lifters who would later join the various crime gangs of Moscow.

By the early 1990s, he had become a highly visible entrepreneur and philanthropist. His official position was chairman of the Lev Yashin Charitable Foundation for the Social Rehabilitation of Athletes. Funded largely by lucrative customs-tax exemptions, this foundation was a job-placement program for Russian athletes. Otarik claimed that he channeled his athletes into clean professions. "I do not let my lads go out and start robbing and killing people," he declared.

His office was a row of suites on the top floor of the Intourist Hotel, where the KGB used to have its offices. I managed to speak to him in August 1993. He was remarkably candid about his gangster connections. "I do not pay money to any Mafiya organizations," he said. "On the contrary, they give me money."

Otarik ran several commercial operations. His main company was called 21st Century, co-owned with another entrepreneur and the singer Iosef Kobzon. Whether through this company or other commercial entities, Otarik owned equity stakes in some shady enterprises, including the Gabriella Casino and the U Lis`sa Discotheque (founded by advertising magnate Sergei Lisovsky). He was also believed to have interests in automobile dealerships and oil trading firms such as Hermes. He was reported in the newspapers to have been involved in a bloody battle to take over an oil refinery in the Volga town of Samara.[18]

Otarik claimed that the profits from his commercial operations funded stadiums, sports programs, and training centers. "Our country is being overrun with drug addicts and homosexuals," he declared. "Sport is the only means of saving the nation. That's why I am building sports schools and encouraging the love of sports—to divert young people away from drugs and from screwing each other up the ass."

Moscow police considered Otarik the epitome of a Mafiya godfather, networking between different criminal groups, receiving a cut from their operations, and arbitrating their disputes. "I have a large organization of friends and comrades, some of whom are strong and some of whom are weak," he told me. "I burden the strong with taking care of the weak."

He was quick to attack Russia's political leaders, whom he called "state criminals"; the politicians enriched themselves, but did not give a penny to help the children or the elderly, he fumed. The gangsters, on the other hand, could be relied upon to contribute to worthy charitable causes. "For you they are gangsters," he stated. "For me they are decent, honorable people."[19]

Both Otarik and the Jap were intent on rolling back the Chechens. But neither they nor any other traditional organized-crime bosses had much influence with the Chechen gangs. (The Chechens spurned the established rules of the Russian underworld and only a few thieves-professing-the-code were Chechen.) When the Chechen gangsters began arriving in Moscow, the city's established mob bosses tried to put the newcomers in their place. The first showdown occurred in 1988, when a dozen thieves-professing-the-code summoned top Chechen leaders to Café Aist for a dressing-down. The Chechens, despite being outnumbered, fell upon their hosts; two thieves received serious knife wounds (this was before the Moscow underworld had obtained firearms) and the rest fled. Ultimately, the thieves-professing-the-code in Moscow would split over the Chechens; some would join the Chechen cause whereas others would join their rivals.[20]

The leading role in the emerging anti-Chechen coalition was taken not by the Jap, Otarik, or any other traditional gangster boss, but by a new organization: the Solntsevo Brotherhood. Named after a grim, sprawling suburb of Moscow, the Solntsevo Brotherhood was established in the 1980s as a series of "sports clubs" by a former waiter, black marketeer, and cardshark named Sergei Mikhailov (nickname: "Mikhas"). During the perestroika years, according to Moscow police, Mikhailov turned his sports clubs into a powerful criminal empire, controlling protection rackets, prostitution rings, narcotics smuggling, and auto-theft rings. A member of the younger generation, Mikhas was not a thief-professing-the-code; he called himself a businessman. In 1989, Mikhas and other top Solntsevo bosses were arrested on suspicion of "banditry" and imprisoned. Several Chechen gangs seized the opportunity to take control of parts of the old Solntsevo territory. But after the Soviet Union collapsed, Mikhas was freed from prison; he was determined to reassert his authority in the city.[21]

In 1992, according to Moscow police sources, Berezovsky approached Mikhas to buy the Orbita, a Solntsevo-controlled supermarket on Smolensk Square, next to the Foreign Ministry building; he wanted this prestigious site for one of his car dealerships. The price Mikhas demanded was supposedly $1 million; Berezovsky apparently found the price too high. Whether on the advice of his partner, Badri, or on his own initiative, Berezovsky ultimately teamed up with the enemies of the Solntsevo Brotherhood: the Chechens.

Meanwhile, Berezovsky's rivals in the auto market, beholden to other organized-crime groups, were jealous of his success. They resented his inside deal with Russia's largest auto company, Avtovaz, and his successful lobbying of the Ministry of Foreign Trade to impose high customs duties on imported automobiles.[22]

Thus Berezovsky found himself in the middle of a war between Moscow's predominant gangster families. The shoot-out at the Kazakhstan Cinema was just the beginning. Throughout the next year, Logovaz would be subjected to violent attacks. Berezovsky himself would come within two feet of losing his life.

The War Begins

Gangsters had been fighting and killing each other throughout the Gorbachev and Yeltsin years, but the bloodshed that was unleashed in 1993–94 was special. The "Great Mob War" was fought mostly in Moscow, but its echo was heard as far as Vladivostok, Krasnoyarsk, Sverdlovsk (Yekaterinburg), Samara, St. Petersburg, Tbilisi, Grozny, London, and New York. The root of the conflict was economic. Dozens of top gangster bosses had come out of prison after Communism fell, and they found the nation's prime economic assets up for grabs. Huge industrial companies, mines, and oil fields were being privatized. Anyone ruthless enough could attain unimaginable wealth almost overnight. One popular analogy of what was happening in Russia at this time was that of a car wreck—a vehicle stuffed with dollar bills crashes, the money is scattered on the ground, and the bystanders push one another away, trying to grab the biggest bundle. Both the older mob bosses (the thieves-professing-the-code) and the younger "gangster-businessmen" engaged in a savage struggle with one another to stake their claim.

Though the mob war was largely a free-for-all, the main organized-crime groups generally adhered to one of two rival alliances. On one side were the Chechens and the thieves-professing-the-code allied with them. On the other side—loosely termed the Slavic alliance—was the Solntsevo Brotherhood and its allies. The Jap, from New York, lent his support to the anti-Chechen cause, as did Otarik. Another key ally of the Solntsevo Brotherhood was a young mobster recently freed from prison: Sergei Timofeyev (nicknamed "Sylvester" because of his resemblance to Sylvester Stallone).[23]

The war began with the assassination of a mob boss named "Globus." His real name was Valery Dlugach; he was a thief-professing-the-code, controlling the Bauman organized-crime group. In Moscow, Globus also represented the interests of the Kazan Tatar gangs. The precise reason for Globus's assassination is unclear. Some say he was exploring an alliance with some of the Chechen gangs; with their support, he was moving into the Solntsevo Brotherhood's narcotics and automobile markets. Globus was also apparently struggling with Solntsevo-connected mob boss Sylvester over control of the Arlechino nightclub. Others say that Globus was not yet committed to a confrontation with the Solntsevo Brotherhood and was merely playing a complicated game, maneuvering between the Chechens and the Slavic gangs.[24]

On the night of April 10, 1993, Globus went to a disco called U Lis`sa. This establishment supposedly belonged to advertising magnate and future Yeltsin campaign manager Sergei Lisovsky; in fact, it was owned, at least partly, by Otarik and the Solntsevo Brotherhood. As Globus left U Lis`sa and walked to his Chevrolet Caprice in the parking lot, he was killed by a sniper.[25]

Two days later, the chief enforcer of the Bauman Group, a man known as "Rambo," was killed in front of his apartment house. The next day, another key member, Viktor Kogan (nicknamed "the Kike"), was ambushed in his car in central Moscow and riddled with bullets. Nine months later, the new boss of the Bauman Group, Vladislav Vanner, was assassinated as well.

As the Bauman Group was being obliterated, the Solntsevo coalition began to move in on a prime Chechen protectorate, Berezovsky's Logovaz. The shoot-out at the Kazakhstan Cinema was followed several months later by other attacks. On at least two occasions, Logovaz dealership lots in Moscow were attacked by men with grenades. Logovaz refused to cooperate with police detectives investigating the attacks. One detective told the press that the attacks were "a continuation of the war among organized-crime groups over the auto market."[26]

Berezovsky chose to spend most of that winter in the West. In November he went to Israel and obtained Israeli citizenship; he resided in a Tel Aviv suburb with his then wife, Galina, and his two youngest children.[27]

"In 1993 I was experiencing tremendous pressure from people whom I don't want to name," Berezovsky would later tell a Russian newspaper. "I was not the only one who was under pressure. Many were. And that is when I went to Israel for several days . . . and asked for Israeli citizenship, which was granted to me."[28]

But exactly who was pressuring him and why remains unclear. According to some Russian reports, Berezovsky also traveled to the United States at this time and obtained a "Green Card."[29]

In Lausanne, Switzerland, his business partner, the big commodities trading firm André & Cie., was alarmed at the gangster war raging in Moscow and the travails of its Russian partner. When André & Cie. teamed up with Berezovsky, it had not reckoned on getting involved in a wave of bombings and assassinations. But André's point man for Russia, Alain Mayor, managed to soothe the nerves of his colleagues.

"It was something that was happening in Russia; it wasn't something that was happening in Switzerland," he explained. "We established the facts and . . . my superiors at the company accepted the facts. That was all. There was no alternative. Obviously the alternative at that time could have been to say: We stop all relations. But that didn't happen.

"At that time, there were very strong rivalries in the automotive milieu in Moscow, that's for sure," he continued. "People used very violent methods. It wasn't very reassuring. In fact, it was painful, disagreeable to see, of course."[30]

André & Cie. decided to stick by Berezovsky.

When I arrived in Moscow in the summer of 1993 to report a story on organized crime, I unwittingly found myself in the middle of the mob war. Virtually every day the newspapers carried stories of a shoot-out in the city. One morning *Nezavisimaya Gazeta* carried a photograph of an unidentified man strung up on a lamppost, with a crowd of passersby looking up in astonishment.

I often found my protagonists being killed before I could interview them. I tried to meet Valery Vlasov, a crime boss who was part of the Solntsevo Brotherhood. Vlasov's base was the Casino Valery, a rather grim establishment in southwestern Moscow. I telephoned and introduced myself.

"I am an American journalist writing an article about Russia's new entrepreneurs and I would like to interview Mr. Vlasov."

"He's not here. He will be in tomorrow afternoon. Call then."

The next day I called again and asked to speak to Vlasov.

"Who are you?" I was asked. Again I introduced myself— American journalist, covering new entrepreneurs, I was promised an interview, etc. I heard frantic whispers. Then another voice came on the line:

"I'm sorry, Mr. Vlasov will not be able to do the interview. He died last night."

I later discovered that Valery Vlasov had been shot down by a sniper as he was leaving his casino the night before.

I had much the same experience with Otarik. I had an interview scheduled with him one day in August, but the Mafiya boss was absent. He was arranging the funeral of his older brother, who had just been killed in a gunfight with the Chechens. A month later, I asked Otarik who had been responsible for his brother's death. "Don't touch that subject!" he shouted. "Never ask those kinds of questions of a member of the family!"

Otarik's older brother, Amiran, had gone to talk to a Chechen gang at the offices of a trading company five minutes from the Kremlin. He was accompanied by an allied crime boss named Fyodor Ishin ("Fedya the Possessed") and three members of the Lyubertsy organized-crime group. As Amiran was leaving, gunfire erupted; he and his four friends were killed.

The Cherry Casino

This was an exciting time for Russia's gangsters. They featured prominently in the pages of the newspapers, appeared in photographs with government ministers and big-city mayors. Russian citizens began to buy dictionaries of prison slang, laborious analytical works about the criminal world, and cheap novels about the exploits of mob heroes. Russians were curious about the new authorities in their lives. Hundreds of Western-style cafés, overpriced restaurants, and glittering nightclubs had opened throughout Moscow. Members of the new ruling class—with their Rolex watches, Italian shoes, gold bracelets, cellular phones, and wives or girlfriends draped in Versace—sulked over their drinks. The most beautiful girls of provincial Russia were available for a relatively low price—the best of them became Mafiya molls.

The first and most important consumer purchase for any mobster was the automobile. The streets of Moscow sprouted an impressive array of expensive imported cars. In 1993, the preferred model was a Mercedes S-600 sedan (retailing for around $100,000 in the United States, double that in Russia); the Chechens and the Georgians seemed to like BMWs and such large American cars as Lincoln Continentals and Buicks. A few years later the vogue became a predatory-looking sport-utility vehicle: a Toyota Land Cruiser, Jeep Grand Cherokee, Land-Rover, Mitsubishi Montero, or Isuzu Trooper. The gangsters drove their cars fast and hard, often on the wrong side of the road, ignoring traffic signals and policemen. The typical gangster caravan was a big black Mercedes with tinted windows and a Land Cruiser "doing the bodyguard drive." Many expensive cars roared around the streets of the capital without license plates. No policeman stopped them.

The most visible architectural symbol of the new Russia was the casino. Within two years after the fall of Communism, scores of casinos had opened around Moscow; the Russian capital began to resemble a tawdry version of Las Vegas. Some casinos were garish gambling palaces, others were squalid little operations hiding behind neon signs.

One summer night in 1993, I visited the Cherry Casino, on Moscow's main shopping avenue, Novy Arbat. It had opened in July and was the hottest place in town; it was home to young Russian businessmen, a few Americans and Europeans, dozens of high-class prostitutes, and lots of professional criminals. The gangster bosses were middle-aged men dressed in black—one had his jacket draped debonairly over his shoulders—each surrounded by half a dozen subordinates. By their looks, almost all the bosses seemed to be from the Caucasus.

The gaming tables were crowded, with the players gleefully slapping down piles of chips. Many players were nonchalantly betting more than $1,000 a shot. These men enjoyed showing that such colossal sums meant nothing to them. Commenting on this strange phenomenon, the casino's British-born floor manager, Dave Sayer, told me: "Most of these people have little confidence in tomorrow, so they spend everything they earn."

The casino itself was making out very well. "Heavy profits," gloated Sayer. "If we keep going like this, we will have earned back our original investment [$5 million] within four months."[31]

Upstairs, in the disco, music boomed and strobe lights flashed. Gorgeous women twisted and turned, their faces flushed from dancing and drugs. To the side of the dance floor I noticed a delicate man in his thirties surrounded by thugs. His hair was painted red and he wore a bright orange jacket; he looked like Alex, the protagonist of *A Clockwork Orange.* I noticed that this man seemed to know many people in the discotheque; an endless number of thugs approached him and respectfully whispered in his ear. Later, when the fellow returned to his side table with only two aides, I went up and introduced myself.

He identified himself as "Sergei" and he said he was a doctor specializing in neuropathology. Considering that doctors were among the worst-paid members of the impoverished Russian educated elite, how did he have the money to come to a club that charged a $30 door fee and $10 drink prices? "I manage my own business," he said. "You meet a lot of friends in a place like this."

Perhaps a businessman like Sergei could afford it, but where do all the big boys in leather jackets get such cash? "They steal it," Sergei said, grinning.

Sergei was contemptuous of the new Russian banks because they had been started with Communist Party funds and had participated in the "Great Chechen Bank Fraud." This was a famous scandal. In 1992–93, using a network of corrupted officials inside the Russian Central Bank, several organized-crime groups and gangster-connected banks carried out the biggest bank fraud in Russian history. Incredibly, it was a simple check-kiting scheme. There was so much chaos in the Central Bank's payments department and the bank's employees were so susceptible to bribery that the criminal organizations were able to make a fortune.

The typical scheme involved two shell companies, usually registered banks. Using codes purloined from the Central Bank, the first bank would send a fake wire transfer to the second. The second bank would then present the wire transfer to one of the 1,400 Central Bank payment centers and receive the money in cash. By the time government officials realized what was happening, both banks had disappeared along with the Central Bank's money.

This bank fraud was one of the biggest disasters of the "reformist" government of Acting Prime Minister Yegor Gaidar. According to the Russian government, the amount successfully embezzled in 1992–93 was $500 million (a third of the credit line the International Monetary Fund—IMF—granted to Russia that year). Many analysts estimated the losses in the billions. Chechen organized-crime groups, acting partly for themselves and partly on behalf of the treasury of the self-styled Chechen Republic, took a lot of the proceeds. (Chechnya was the ideal place to register fake banks.) The bank fraud, however, would not have been possible without the participation of many of Russia's leading commercial banks and trading companies.[32]

I asked Sergei, my acquaintance at the Cherry Casino, about foreign organized-crime groups. Was it true, as some people were saying, that foreign gangsters were moving into Russia? "The Italian Mafia sniffed around here, but then they went back home," Sergei said. "Our boys were too tough for them. Now, the Colombians are a different matter."

Moscow was full of Colombian narco-dollars invested in real estate and various businesses, Sergei claimed. The Russians who arranged the purchases had received big payments in the West. "A lot of people here are living off those dividend payments," he noted with pride.

Who owned the Cherry Casino? "The Chechens," said Sergei blithely. Later, a RUOP officer confirmed that a large portion of the take from the casino went to one of the Chechen Mafiya groups. When asked about the casino's ownership, the manager, Dave Sayer, mentioned four equity partners: the city of Moscow (the landlord), a closely held Swedish company specializing in casino operations, a Russian foreign-car dealership called Trinity Motors, and a private Russian holding company called Olbi.[33]

Of the four, Trinity Motors, a dealer in Chrysler and other imports, had the most notorious reputation as a gangster-connected business. Trinity's primary showroom was in one of the best locations in Moscow: across the street from the Bolshoi Theater. One of the official cofounders of the dealership, and a frequent patron of the Cherry, was Vladislav Vanner, Globus's successor as boss of the Chechen-controlled Bauman organized-crime group.[34]

Trinity Motors's connection with the Cherry Casino was perhaps predictable, but Olbi was a surprising shareholder. The company belonged to a thirty-year-old entrepreneur named Oleg Boiko—the name Olbi stood for Oleg Boiko Inc. He was one of Russia's most famous new millionaires. He owned a chain of upmarket electronics stores called Olbi-Diplomat, and one of Russia's largest banks, Natsionalny Kredit. He was the main outside investor in Russia's leading newspaper—*Izvestia*. He was also the primary financier of the political party of his friend Yegor Gaidar, the former prime minister, hero of the Western media, and leader of the "young reformers." Boiko was the chairman of Gaidar's party, Russia's Choice.

Years later, I asked Gaidar why he chose to work with an entrepreneur who, among other things, was a major investor in a Chechen-controlled casino.

"I knew to a certain extent about his business," Gaidar answered. "At that time he was a pretty large businessman and was considered a solid one."

Did Gaidar know about Boiko's dealings with the underworld?

"No."

"Would you choose him again to work for your party?"

"Of course not. This was still at a time when we had many more illusions about the new Russian business world, about the sense of social responsibility [of these people]."[35]

Superpower of Crime

Barely two years into Russia's democratic experiment, most Russians understood that the country had fallen into the hands of a criminal caste. A foreign term was used to describe the new authorities: Mafiya. The popular view was that Russia was run by killers and crooks, and everyone in government was an accomplice. In the spring of 1993, Russian president Boris Yeltsin claimed in a speech that two thirds of all commercial structures in Russia were linked to organized crime and that organized crime represented the largest threat to Russia's national security. A year later, Yeltsin would again air his concerns, declaring that Russia was becoming the "superpower of crime." The Ministry of Internal Affairs (MVD) tried to be more precise, estimating that 40 percent of all private businesses, 60 percent of all state-owned companies, and up to 85 percent of all banks had links to organized crime.[36]

In 1993, the MVD identified about 3,000 organized-crime groups in Russia. A year later, a special United Nations conference on international crime counted 5,700, with 3 million members. But the frightening thing about Russian organized crime was not the rapid increase in the number of gangster organizations, but the degree to which the gangsters dominated Russia's economy.[37]

Economic liberalization had led, not to black-market operations evolving into legitimate businesses, but to newly incorporated businesses being sucked into the black market. Russia's new businesses were pushed into the world of organized crime by the corruption of the government apparatus, which meant that commercial success was overwhelmingly dependent on political connections. Businessmen were hampered by a crushing tax code—which impelled enterprises to conduct business off the books—and the absence of an effective legal system, which meant that contracts were unenforceable and debts uncollectible without the use of mob enforcers.

At the dawn of Yeltsin's Russia, thousands of new banks were established. Sometimes these institutions boasted sparkling marble offices, sometimes they resembled shabby grocery stores with pistol-toting guards. Banking was one of the most lucrative activities in the early years of the Yeltsin regime. The key was access to government funds. If bankers had the right connections, they could make huge profits without knowing anything about finance or credit policy. The biggest source of income for Russian banks was loans from the Central Bank at negative real interest rates. In February 1993, for instance, a well-connected commercial bank could borrow money from the Central Bank at a monthly rate of 7 percent. That same month, however, the consumer price index rose 25 percent. The bank simply had to turn the Central Bank's rubles into something that would hold its value (marketable commodities, for example, or dollars) and reconvert it into cash at the end of the month, netting a real profit of 15 percent or so in just a few weeks. Well-connected banks earned equally large profits by handling the deposits of central government ministries, local government departments, and the big state-owned oil companies and arms exporters. Here again, they would pay the government an interest rate far below the inflation rate, while using the government funds to make enormous profits on commodities trading or foreign exchange operations.[38]

Not surprisingly, banks were among the most frequent targets of organized-crime groups. Dozens of top bank executives were assassinated. The state-owned Rosselkhoz Bank, Russia's third-largest, was the biggest institution to be affected; its chairman was killed in December 1993.[39]

In 1993, I visited one of the young Russian bankers in his luxurious offices on Novy Arbat in central Moscow. A large man with lizard-like eyes, Vladimir Sipachev was the chairman of Aeroflot Bank (40 percent owned by the airline, 60 percent by six individuals). Sipachev, thirty-five, told me that he had made his career as the "financial manager" of an engineering firm called Atommash in the southern city of Rostov-on-Don. In 1989, he began building a business empire. By 1993, he owned Aeroflot Bank, a big marble quarry, a small aircraft-production company, a metal trading company, and a network of radio stations. Sipachev boasted of his skills as a financier, noting that his businesses would produce about $100 million in revenues that year. I decided to pitch him some standard questions.

"What do you think of the IMF austerity program Russia recently adopted?"

He was genuinely surprised: "The IMF? What's that?"

I tried again. "As a prominent Russian banker, are you worried about all those bank managers being murdered?"

"Why should I be?" he answered. "There is nothing abnormal in this. Bank managers get killed in the West all the time."[40]

One of his business partners—Aleksandr Petrov, chairman of Moscow Profbank—was murdered outside his apartment in December 1991. But nothing, it seemed, could shake Sipachev's cavalier attitude. One British businessman related how Sipachev gave him a ride in his unmarked Mercedes S-600. Instead of driving on the street—the busy shopping avenue of Novy Arbat—the banker's chauffeur drove the car briskly along the sidewalk, through the afternoon crowds, scattering pedestrians and street vendors.[41]

Russia's lucrative export industries were also targeted by organized crime. In 1993, directors of state enterprises undergoing privatization—oil refineries, aluminum smelters, timber companies—were being assassinated en masse. One of Russia's most respected businessmen, Ivan Kivelidi, a major chemicals industrialist, chairman of Rosbusiness Bank, and founder of the Russian Business Roundtable, was disposed of in particularly gruesome fashion. Someone rubbed an obscure nerve toxin onto the telephone receiver in his office; he collapsed, frothing at the mouth, and died three days later.[42]

Murder had become a form of dealing with business competition. "Businessmen, instead of deciding their differences in the market or in court, are hiring professional killers and deciding their differences with guns," lamented Georgi Khatsenkov, a former Tass news agency employee who had started a chain of jewelry stores.[43]

Hundreds of shoot-outs were occurring in Moscow, often in broad daylight. The battles (razborki) included handguns, Kalashnikov automatic rifles, car bombs, even rocket-propelled grenades. "We are witnessing a veritable flood of sophisticated weaponry," noted General Rushailo.[44]

The 29,200 murders committed in 1993 represented an official murder rate twice that of the United States at the height of its own murder wave that year. In Moscow the number of murders rose eightfold between 1987 and 1993. As horrifying as these figures seem, they represented only a fraction of the actual number of people being killed in Russia. Many murder victims were listed in the official statistics under other cause-of-death categories: suicides, accidents, disappearances. The "disappeared" category was particularly voluminous. Andrei Pashkevich, spokesman for the Moscow RUOP, observed that, in addition to the 30,000 people murdered annually, another 40,000 people had "disappeared." The majority of those, he said, were almost certainly murder victims. Hence, while official statistics put Russia's murder rate at 20 per 100,000, or twice the level of the United States, the real rate was three or four times higher than in the United States. Well-known and important people were being assassinated in Russia, and law-enforcement authorities apparently found it impossible to catch the killers.[45]

Though the police were incapable of stemming the violence, many policemen died trying. In 1994, 185 policemen were killed and 572 wounded in shoot-outs with the organized-crime gangs. The police did not have a sufficiently powerful investigative unit to combat the mob. The primary force, RUOP, was understaffed, badly equipped, and insufficiently insulated from political pressures.[46]

The judicial system was worse. Russia had no judges specializing in organized crime; there were no provisions for plea bargaining; there was no witness-protection program. Moreover, Russia's judges were notoriously malleable—trained to follow the dictates of political leaders and vulnerable to bribes and threats. Police officers complained that even when they arrested the worst mobsters, key witnesses usually withdrew their testimony and judges dropped the case.[47]

There were statutes against extortion and banditry, but these were useful only against criminals caught red-handed. There was no statute against racketeering, making it illegal to belong to an organized-crime group. Laws against such white-collar crimes as bribery, embezzlement, money laundering, and fraud were inadequate; to get a conviction on fraud charges, for instance, prosecutors had to prove that the accused knew that he was committing fraud (ignorance of the law, in other words, was an excuse). Russia had nothing to match the RICO (Racketeering-Influenced and Corrupt Organizations) statute that had broken the back of the Mafia in the United States. In 1995–96, the lower house of parliament finally passed a bill on organized crime, only to have it voted down by the upper house, the Federation Council, which contained Russia's regional bosses and members selected by President Yeltsin; it was not until 1997 that the organized-crime bill was finally passed and signed by the president. A money-laundering bill was passed by the parliament only in the summer of 1999, but President Yeltsin vetoed it.[48]

In 1996, CIA director John Deutch informed the U.S. Congress that Russian organized crime had strong links to the country's parliament.

Congressional committee chairman Benjamin Gilman speculated that these links were the reason for the delay in the passage of effective crime legislation. The greatest resistance to passing effective anti-Mafiya provisions came not from the parliament but from the Kremlin itself—from the group of men (and women) advising Boris Yeltsin.[49]

"I feel like a person sitting in a pile of shit with his hands and legs tied—he can smell everything, he can see everything, but he can't do a damn thing," grumbled one Moscow detective.[50]

Men with scarred faces, broken noses, thick necks, and bulging biceps were seemingly everywhere in Moscow. These were the enforcers, graduates of the formidable Soviet athletics apparatus or unemployed members of the KGB or the Soviet army special forces— victims of the drastic reductions in military spending, professional killers with nowhere to go. They had trained as weight lifters, gymnasts, wrestlers, hockey players, boxers, karate experts, professional snipers, or explosives experts; some were even members of the old Soviet Olympic Team. The vision of socialist glory and Olympic gold medals had vanished. More often than not, the only possible employment for many of these soldiers and athletes was as a security guard or mob enforcer (not always separate vocations). The total number of privately employed "security guards" in Russia was estimated at 800,000—these were not old guys, keeping watch over some parking lot, but highly trained professionals.[51]

The army proved a particularly fertile recruiting ground for the mob. When Russian units returned from Eastern Europe, more than 100,000 officers found themselves without housing; they and their families often lived in tents and other makeshift shelters. The officers' meager salaries were routinely paid several months late. As the Russian army shrank, nearly 1 million professional soldiers were sent home without pensions. "To be an officer used to be a post of the highest prestige," Grigory Ivanchenko, a recently demobilized lieutenant colonel, complained to me. "Now what are we supposed to do, become janitors or night watchmen?"

It was not surprising that so many professional soldiers found employment with organized-crime groups. Aleksandr Lebed, the former paratroop general who had become one of Russia's leading politicians, described the process. "[The demobilized soldier] says: 'I served for twenty years, ruined my health, sacrificed myself for my country.... Why did they just throw me out like garbage?' He goes to the criminal organization and there he is received with open arms: 'My dear comrade colonel, look at what a career you had in the Spetsnaz [special forces]! Why have you been orphaned with a one-time payment of 900,000 rubles [$150]? Here, take $3,000 and let's roll.' "[52]

To ensure the loyalty of the military and to guard against the possibility of a coup d'état, President Yeltsin closed his eyes to the corruption of the army high command. "The top generals are corrupt," General Lebed observed. "[Defense Minister] Pavel Grachev both gave them huge cash payments and allowed them to use military resources to build private castles for themselves. Then they were told: Remember why you were given these privileges—now, obey."[53]

As corruption and anarchy spread, Russia's enormous nuclear stockpile was left dangerously exposed. Over the course of four decades, the Soviet Union had produced more than 50,000 nuclear warheads. Even after most of the U.S.S.R.'s nuclear weapons were decommissioned, the key components—the plutonium or enriched-uranium pits—remained. One Western inspector saw 23,000 plutonium pits stored in two sheds in a formerly secret city; the sheds were secured by a barbed-wire perimeter and ordinary padlocks on the door. Pieces of Russia's nuclear arsenal soon began appearing on the international black market. In 1993–94, alone, German and Czech police made at least half a dozen busts of Russian organized-crime groups exporting nuclear weapons components to the West.[54]

Even the former KGB was incapable of preventing the infiltration of organized crime into the most sensitive parts of the military-industrial complex. The Russian security service itself had been corrupted. When General Aleksandr Korzhakov, Yeltsin's security chief, inspected the KGB's elite Alfa antiterrorist squad in 1993, he found that the unit had gone freelance. "There was no discipline in the group at all," Korzhakov noted. "The officers earned money on the side, sometimes by working for the mob. There were cases when a shopkeeper would be approached by Alfa men from two different detachments, each group demanding money for his protection. Protection against whom? Against their own comrades. One group came in the morning with threats, the second group came in the evening promising protection."[55]

When I asked Konstantin Borovoi, chairman of the Russian Commodities and Raw Materials Exchange, how he would define the Mafiya, he replied: "The Mafiya is an attempt to imitate the government. It has its own tax system, its own security service and its own administrative system. Any entrepreneur, in addition to paying taxes to the government, has to pay taxes to this shadow government: bribes to the health inspectors, local police, tax inspectors, government landlords, and, of course, protection money to the gangsters guaranteeing his security."[56]

"The power of the government has completely collapsed, but individual bureaucrats retain control over the nation's resources," explained Georgi Khatsenkov, who ran a small publishing and gem-trading business. "They want to profit from their position, but to do that on a big scale they need an organization—allies in the bureaucracy, commercial structures to handle the money, and professional [hitmen] to enforce their will. So they ally themselves with the criminal groups."[57]

The main reason for the explosion of organized crime in Russia, in other words, was neither poverty nor unemployment. It went back to the earliest days of Communism. Lenin and his heirs understood the gangster mentality, and their secret police had used gangster methods to intimidate or eliminate political opponents. With privatization, Russia's political bosses and industrial managers found themselves in control of the nation's prime industrial enterprises. To operate these companies in their own interests (rather than the interests of the Party or the state), they needed protection.

"Under the old system, the different power centers [the local Party organizations, the KGB, big state-owned enterprises, and so on] protected their particular rackets by jailing any whistle-blowers or nonconformists," Borovoi observed. "Today, these power centers do not have such an ability to protect themselves, so they use professional criminal groups instead."[58]

I asked the liberal parliamentary leader Grigory Yavlinsky why virtually all small businesses in Russia had to pay off one gangster organization or another. "Because the government promotes that," he replied. "They work together, after all. It's an oligarchy. In Russia there are two groups of people who are especially contemptuous of the law: the very top of society and the very bottom. When the top and the bottom unite, you get a blanket that covers everything."[59]

When I asked Berezovsky about the reasons for Russia's crime epidemic, he, too, pointed to the collusion of top government officials with the gangsters.[60]

Two years after the establishment of the Yeltsin regime, the racket was operating at the highest levels of the state apparatus. James Moody, head of the organized-crime division of the FBI, observed that a corrupt government was the single most important factor allowing organized crime to flourish. "Organized crime always tries to penetrate the government," Moody noted. "If the government is corrupt, who is going to stop organized crime?"[61]

The Bombing of Logovaz

It was in this lawless environment that Berezovsky was building his empire. Alone among the big businessmen who would come to be known as "the oligarchs," Berezovsky found himself personally involved in the war between Chechen and Slavic gangs. After his car dealerships were attacked on three different occasions by heavily armed gunmen, he spent much of the winter of 1993–94 abroad. Refreshed from his travels, he returned to Moscow to find the mob war still raging. He was targeted almost immediately. The door of his apartment was booby-trapped with a grenade. Fortunately, it failed to explode.[62]

The Slavic gangs, meanwhile, were being subjected to much worse. On the afternoon of April 5, 1994, Otarik left his office near the

Krasnaya Presnya Bathhouses, not far from Russian government head-quarters. Before he got to his car, he was shot three times by a sniper in an apartment house across the street. He died within minutes.

From his perch in New York, the Jap was watching the mob war in Russia with increasing alarm. Otarik, one of his best friends and the central figure in the war against the Chechens, had been killed. Russia was turning into a chaotic battleground, with too many criminal bosses dying and too much bad publicity being shed on the gangster organizations. After Otarik's death, the Jap reportedly called an assembly of Mafiya bosses in Vienna to discuss who would lead the fight against the Chechens. The choice fell on the mob boss Sylvester.[63]

In the late afternoon of June 7, 1994, Berezovsky walked out of the new Logovaz headquarters in central Moscow and got into the back seat of his Mercedes. His bodyguard sat in front, next to the chauffeur. As the car pulled out, the neighborhood was shaken by a massive explosion. Someone had parked an Opel sedan filled with explosives on the narrow street and detonated the car bomb by remote control as soon as Berezovsky's car passed by. Berezovsky saw his driver decapitated by the blast. The bodyguard was severely wounded—he would lose an eye. Half a dozen passersby were injured. Berezovsky stumbled out of the car, his clothes smoldering; his burns were severe, requiring months of treatment in a Swiss clinic. A few days later, the headquarters of his Obyedinyonny Bank was also bombed.

Like the other famous contract killings in those years, the Logovaz car bombing remained unsolved. Berezovsky blamed his competitors in the auto industry. One of his colleagues at Logovaz told the newspaper *Kommersant* that the assassination attempt was linked to the aggressive pricing policy pursued by Logovaz. Months later, in private conversations with Yeltsin, Korzhakov, and other authorities in the Kremlin, Berezovsky implicated TV magnate Vladimir Gusinsky (and his patron, Moscow mayor Yuri Luzhkov). When I asked him in late 1996 who was responsible for the bombing, Berezovsky replied: "This organization [that commissioned the bombing] continues to operate today."[64]

Wasn't he concerned that the assassins would try again? "No," he answered. "They understand I know who stood behind this deed and this gives them pause."

Then, after a menacing silence, he added: "But I am not one of those people who seek vengeance."[65]

Investigating the Logovaz bombing, Moscow police focused on the business dealings of Sylvester. The police discovered that he and Berezovsky had had the kind of "unclear relations" that General Rushailo had mentioned to me a year earlier. In March 1994, a Berezovsky-led investment fund called Avva had placed money with Mostorg Bank, buying two 500-million-ruble short-term promissory notes. Mostorg Bank was controlled by Sylvester; the bank did not pay back the obligation on time, having transferred the money abroad.[66]

The assassination attempt put Berezovsky into the national spotlight for the first time. The police declared that it was "the most powerful explosion in Moscow that year." President Yeltsin ordered his security forces to cleanse the country of its "criminal filth." Within a few days, Mostorg Bank finally made its payment (with interest) on the promissory notes it had sold to Avva. Berezovsky went to Switzerland and left his old associate at Logovaz, Badri Patarkatsishvili, to sort out relations with the gangsters.[67]

The mob war continued. In August, the Slavic gangs struck a fatal blow against one of the strongest Chechen gangs—the Lazaniya Group—by ambushing and killing Gennady Lobzhanidze (known as "Gena the Scar"). Within a year, one of the other leaders of the Lazaniya would be arrested by the police, while a third fled to Turkey.

But the Slavic gangs could not exploit their victories. In the early evening of September 13, there was a powerful explosion just off Tverskaya Street (Moscow's most elegant shopping district). Police found a Mercedes S-600 sedan that had been destroyed by a remote-controlled bomb attached to the underside of the car. A charred corpse was pulled from the wreckage. It was Sylvester.

Berezovsky, who by then had returned to Moscow from Switzerland, was briefly considered a suspect, but the case was never solved.[68] With Sylvester's death, the gangster assaults on Logovaz ceased.

The Invasion of Chechnya

That autumn, the mob war in Moscow was overshadowed by a bloodier conflict in the breakaway republic of Chechnya. The two conflicts were related; the Chechen War was a gangster turf war writ large.

Chechen organized-crime groups in Moscow and other Russian cities maintained subsidiaries in their ancestral homeland. Chechnya was a key transit point in the Russian narcotics trade, and the Moscow-based gangsters sent a large part of their profits back to the homeland. The same Russian officials and security officers who patronized Chechen organized-crime groups in Moscow also patronized the Chechen government of President Jokhar Dudayev by allowing the Chechen government to appropriate millions of tons of Russian oil at little or no cost. Chechnya was an important nexus for Russia's oil pipeline system, serving as a transshipment point for both Caspian Sea oil from Baku and Western Siberian oil on its way to the port of Novorossiisk. Most of this oil, whether refined or crude, was exported.[69]

"How did they export it? Who exported it? Where did this oil disappear to? No one ever tried to investigate it," recalls former Foreign Trade Minister Oleg Davydov. "At least during my term, there was never a single appeal by the Chechens nor a single decree [by the Russian government] to give them a certain quota, a certain amount of pipeline capacity, or whatever. They [the Chechens] operated through frontmen, through some kind of influential individuals, who helped them and were paid for their services."[70]

President Dudayev and the warlords and gangsters who controlled the Chechen government were able to sell the Russian oil on foreign markets and earn hundreds of millions of dollars, both for the republic's treasury and for themselves personally. Meanwhile, the

Chechen militias received their armaments from corrupt Russian army commanders, based as far away as the Far East. The Russian security services contributed to building up the Chechen army as well. When the Black Sea province of Abkhazia decided to declare independence from the Republic of Georgia, unleashing a brief civil war in 1993, the Russian security services secretly aided the Abkhazian separatists. These fighters were reinforced by Chechen militias, who received Russian weapons and training. A number of the most deadly Chechen warlords, such as Shamil Basayev, cut their teeth fighting in Abkhazia.[71]

"Then Jokhar Dudayev decided he had become big and strong and stopped sharing the booty with his Moscow sponsors," General Lebed says. "So [the Russian government] decided to punish him with the army."

Lebed cited another reason for the decision to invade Chechnya: to cover up massive corruption in the military high command. As the Russian army was withdrawing from Germany in 1991–94, equipment was sold on the black market by the commanders of the Western Group of Forces. Lebed claims that more than 1,000 armored vehicles were sold on the black market, mostly to Serbia and Croatia during the Yugoslav civil war. The looting of the arms depots of the Western Group of Forces, according to Lebed, was done with the connivance of the group commander, General Matvei Burlakov, and the minister of defense, General Pavel Grachev. The journalist Dimitri Kholodov of the newspaper *Moskovsky Komsomolets* began investigating these charges. In October 1994, Kholodov picked up a briefcase that supposedly contained key documentary evidence for his case. The briefcase contained a bomb and Kholodov was killed instantly when it exploded in his office. In any case, Lebed says, the Western Group of Forces scandal provided another incentive to launch the Chechen War. "These so-called generals needed a big war to break out somewhere, so that a large number of armored vehicles could be destroyed and written off to this war."[72]

Though it was well known that there was an unusual amount of corruption in the Western Group of Forces and Russian newspapers

frequently alleged that Defense Minister Grachev himself was corrupt (his nickname was "Pasha Mercedes"), I never managed to confirm that the decision to invade Chechnya was guided by a desire to
hide black-market sales of armaments. One thing is clear: In the
same way that, as RUOP chief Rushailo explained, the businessmen
targeted by gangsters almost always turned out to be gangster-linked
themselves, Russia's invasion of Chechnya was the result of some
kind of corrupt relationship gone sour. The gangster organizations
that had taken power in the Republic of Chechnya were linked both
to such prominent Moscow businessmen as Berezovsky and to numerous top Kremlin officials.

The Russian army invaded Chechnya on December 11, 1994.
The decision was made by Yeltsin and a few of his closest cronies:
Grachev, Soskovets, Lobov, Korzhakov, Yerin, Stepashin—the so-
called Party of War. "This was not the party of war," snarls Lebed.
"This was the party of business."[73]

Shortly before the invasion began, Defense Minister Grachev
boasted that "a battalion of paratroopers can take Grozny [the
Chechen capital] in two hours." In fact, it took two months and the
most horrific bloodshed. It was not Russia's elite paratroopers who
did the fighting, but the military units closest to Chechnya—mostly
underequipped units composed of bewildered eighteen-year-old
provincial boys. They were sent into the center of the city in long armored columns, without infantry support or close air support. The
generals evidently had learned nothing from World War II—both
the Battle of Stalingrad and the Battle of Berlin had demonstrated
that tanks were useless in a big city. "A tank in a city is like an elephant
in a pit," notes General Lebed. "Any kid can throw gasoline down on
it, throw down a cigarette and the tank burns."

From the windows and balconies of the apartment buildings, the
Chechens attacked the Russian armored columns with antitank rockets and grenade launchers; the tactic was first to disable the lead
tank, then the last one, then destroy the remaining tanks in the
middle. The Russians were burned alive in their armored vehicles or

picked off by snipers as they tried to climb out. The Russians later returned and blasted the Chechens out of the apartment buildings.

By the end of January 1995, Grozny was taken and the fighting moved to the mountains in the south. The Russian war effort was a mess. Some commanders refused to carry out orders to attack; others refused to honor cease-fires. Many commanders accepted bribes to let encircled Chechen units escape, while others sold weapons to their Chechen opponents. The Russians bombed civilian settlements indiscriminately. For their part, the Chechens wallowed in blood lust: They made videotapes of their slaughter of Russian soldiers and publicly cut the throats of Russian prisoners.

On June 14, 1995, the Chechens achieved a devastating victory. Shamil Basayev, the former Moscow "businessman" who had become the Chechens' top field commander, launched a terrorist attack in the Russian heartland. He led several dozen elite fighters concealed beneath the canvas covers of dump trucks through seventy miles of Russian-held territory; the convoy passed through numerous checkpoints, bribing the guards not to inspect the vehicles. Finally, Basayev and his fighters were stopped at the Cossack town of Budyonnovsk. They stormed the town hall and the police station, rounded up 1,500 hostages, and hunkered down in the Budyonnovsk Hospital. The building was soon surrounded by Russian police and special forces; several attempts to storm it failed miserably. Russian newspapers ran a photograph of one of Basayev's men holding his gun out the window from behind a terrified Russian woman—the human shield he had found in the hospital corridors. Basayev's men executed several Russian prisoners and threw them out the windows.

After a five-day standoff, Russian TV watchers saw a videotape of their prime minister, Viktor Chernomyrdin, talking humbly with Basayev. Chernomyrdin agreed to let Basayev go. He was provided with buses and fuel to facilitate his return to Chechnya. He was also allowed to take several dozen Russian hostages back with him. Horrified by the bloodshed at Budyonnovsk (at least 120 people died in the raid), the Russian government agreed to a cease-fire—which was to last half a year before the war began again.

The End of the Old-timers

By this time, the mob war in Moscow had burned itself out. Neither the Slavic gangs nor the Chechens emerged as the victors. They had largely destroyed one another. Dozens of thieves-professing-the-code had been killed. The surviving Chechen bosses tactfully withdrew into the shadows. Many Slavic bosses went abroad. The Jap was jailed in New York, to serve a nine-and-one-half-year term for extortion. Mikhas, the boss of the Solntsevo Brotherhood, was imprisoned in Switzerland, serving two years before he was acquitted of criminal charges. After having captured the attention of the entire nation in the early Yeltsin years, the traditional gangster bosses retreated back to the margins of society.

The winners were Russia's new businessmen. Many of these men had worked closely with the Chechens, the Solntsevo Brotherhood, or the thieves-professing-the-code. They had paid the gangsters the underworld tax, had used them to eliminate rivals, had even put mob representatives on their boards of directors. But now the businessmen were powerful enough to ignore the old gangsters.

"Our oligarchs passed relatively quickly through the Mafiya shoot-out era," remarked First Deputy Prime Minister Anatoly Chubais three years later in an interview with the newspaper *Kommersant*. The businessmen played a significant role in ending the gangster violence. Chubais cited one incident in 1994 or 1995, when Berezovsky and other leading businessmen apparently got together to "adopt certain principles of coexistence," including "abstaining from the use of contract killings."[74]

Berezovsky, by this time a fierce political rival of Chubais, felt compelled to reply in the same newspaper. He issued no direct denial of the charges. "Do you remember that passage where [Chubais] says that we gathered in 1994 and agreed to refrain from contract killings?" Berezovsky asked. "This is beyond hypocrisy. After all, Chubais is accusing himself first of all. The conclusion we must draw is that all this time he [Chubais] has been dealing with gangsters, right? And the president [Yeltsin] was the protector of the gangsters? That's totally out of line."[75]

Chubais was remarkably frank about his decision to deal with gangsters. "We did not have a choice between an ideal transition to a market economy and a criminalized transition," he told me in early 1998. "Our choice was between a criminalized transition and civil war."[76]

The civil war materialized anyway—in Chechnya. But in Moscow, at least, things had calmed down. In the last days of December 1994, one of the most powerful Chechen gangsters, Khozha Suleimanov, head of the Southern Port group, was assassinated. It was the last big killing of the great mob war.

It had been an extraordinary two years. Bombings, assassinations, wealthy businessmen murdering their rivals, gangster bosses fighting over the nation's prime industrial assets, top government officials making deals with hardened criminals—rarely has any nation gone through such anarchy in peacetime. What had brought Russia to such a pass? Ironically, it was Mikhail Gorbachev's liberal reforms that gave free rein to Russia's gangsters. Along with the political prisoners released from the gulags, thousands of professional criminals were freed as well. More important, Russia's main gangster-business empires were established under Gorbachev. Unbeknownst to the Soviet leader, many of them were sponsored by such pillars of the Soviet establishment as the KGB and the Central Committee of the Communist Party.

CHAPTER TWO

THE COLLAPSE OF THE OLD REGIME

Gorbachev's Reform

In 1986 I attended Revolution Day celebrations in Moscow with a group of Italian Communists. It was still the old Soviet Union. November 7—the supreme celebration of Soviet power. We were assigned a place in front of the National Hotel to watch the parade. It was a bitterly cold morning. Armored vehicles were parked along Gorky Street: tanks, armored personnel carriers, half-tracks, amphibious landing craft—the paint glistening on their metal hides. The tank drivers, in their leather helmets, shifted on their feet to stay warm as the winds swept up the boulevard. The troops had been in place a long time.

The Soviet army had paraded here on Revolution Day every year since the first anniversary of the Bolshevik seizure of power in 1917. The most famous Revolution Day parade, of course, was during a snowstorm in 1941; Hitler's panzers were poised on the outskirts of Moscow, and as the Soviet troops finished marching past Red Square they went directly to the front.

Forty-five years later, the parade was still impressive. At the appointed hour, the massed troops snapped to attention. The crowds grew silent. An open limousine swept into Revolution Square carrying a bemedaled Soviet marshal. The car stopped in front of each formation and the marshal's deep voice boomed over the public address system: "Comrades, I congratulate you on the Holiday of the Great October Socialist Revolution!"

"Hurrah!" the troops responded in a mighty cheer.

Then the marshal drove away, toward the towers of Red Square. There was a deafening roar as the armored vehicles' engines were started. The tanks rumbled past, wave after wave, shaking the pavement, emitting plumes of exhaust smoke over Moscow's Stalinist boulevards. Then the foot soldiers came on, marching in broad columns, faces granite-still, their step perfectly timed, their boots polished to a winter sheen.

The Italians were impressed.

But the Soviet Union was less vigorous and powerful than it appeared. That year a new type of Soviet leader was in power. Mikhail Gorbachev was the first of the post–World War II generation. He began reforming the Soviet system with an energy not seen in half a century. He abandoned the foreign policy of Communist aggressiveness and reached an understanding with President Ronald Reagan to end the Cold War. At home, he opened the gates of the gulag and launched a series of reforms to reinvigorate the country.

When I met Gorbachev in May 1992, the deposed Soviet president impressed me as energetic and straightforward. He had a rich, deep voice, embroidered with the accent of the Cossack territories of southern Russia. His warm brown eyes immediately convinced you that he was well-meaning and honorable.

"At that time everything was already fermenting in the Party, in the country, and in society," Gorbachev recalled. "At first we went the way all reformers had gone before us: We began with the idea of accelerating economic growth. We began the modernization of our production facilities and the introduction of new technologies. We made large investments in microelectronics and machine tools."[1]

Almost immediately, things began to go wrong. In December 1985 the Saudi oil minister Sheik Ahmed Yamani, the leading figure in the OPEC oil cartel, announced that Saudi Arabia would no longer support oil prices by limiting its production. World oil prices fell 69 percent within eight months. With its other industries struggling, the

Soviet Union was dependent on its vast production of oil and gas. The U.S.S.R. was a petroeconomy—oil and gas typically accounted for more than half of the country's cash exports.[2]

Dragging the country down further was the endless war in Afghanistan. The Soviets' defeat in this war would have a domino effect: Within a decade, the Russian army would be fighting professional Islamic guerrillas, not in the remote Afghan mountains, but within the borders of the former Soviet Union—in Tajikistan, Dagestan, and Chechnya. In 1986, Gorbachev's regime was rocked by another ominous event: the accident at the Chernobyl nuclear power plant. The Soviet Union—a massive, muscle-bound structure, armed with the world's most dangerous technology—was self-destructing. Soviet consumers, meanwhile, received few rewards from Gorbachev's reforms. New housing developments were even shoddier than before. Durable consumer goods, from refrigerators to televisions to automobiles, were still hard to obtain. Soviet economic growth rates continued to decline and around 1988 the economy stopped growing altogether.[3]

With hopes for an economic revival fading rapidly, Gorbachev turned his attention to political reform. He launched glasnost (openness) and perestroika (restructuring). For the first time in seven decades, freedom of speech was allowed. The press, universities, and government institutions became arenas of spirited public debate. The Communist Party abandoned its official monopoly on power and allowed popular elections to the nation's parliaments. Insofar as Russia can be considered a democracy today, the liberties defining that term were established under glasnost and perestroika.

But the reforms produced results quite different from those Gorbachev had intended. Once the controls of the central government were relaxed, provincial political bosses and enterprise directors did as they pleased. Their first impulse was to enrich themselves.

Vodka

The clearest example of a Gorbachev reform measure having unexpectedly malignant consequences was the anti-alcohol campaign. Almost immediately after taking office, Gorbachev decided to tackle Russia's most intractable social problem: vodka. The only other time prohibition had been attempted was under Czar Nicholas II, who abolished vodka during the First World War. That measure helped bankrupt the government and made the monarchy extremely unpopular. The czar's prohibition remains one of the most underappreciated causes of the Russian Revolution of 1917.

Gorbachev's anti-alcohol policy was not an absolute prohibition—it merely made vodka very scarce and very expensive. The Soviet president was cursed on street corners and at kitchen tables across the country. In provincial cities and towns, the anti-alcohol campaign was the most visible symbol of Gorbachev's rule. Instead of solving Russia's alcohol problem, prohibition merely brought it into the streets—in the form of horrendous lines and grubby riots of beggars.

Most vodka production, meanwhile, moved to the black market. Rogue distilleries sprang up across the country. Bootleg vodka was produced at collective farms, in food-processing plants—almost everywhere—with the connivance of local political bosses, and then sold on the street or under the counter in state-owned grocery stores. Some of the vodka was produced from cheap perfume, window cleaner, or shoe polish. Every year, thousands of Russians died of poisoning after drinking these noxious cocktails.

Prohibition also poisoned the Soviet economy. The state alcohol monopoly had long been a pillar of Soviet government finance, typically accounting for nearly 25 percent of budget revenues. With prohibition, the profits from vodka no longer went to the state, but to bootleggers, laying the foundation for Russia's first criminal capital.[4]

This bootleg revenue was often invested in "cooperatives"—the new private businesses (trading companies, banks, restaurants, shops) permitted by the reforms. As the vodka Mafiya evolved into a national network, it relentlessly corrupted the government apparatus—starting with the local police, then the courts, mayors, and regional Communist Party bosses, and ending with ministers of the central government. Soon, the government—the only institution capable of fighting organized crime—began to crumble.[5]

After a few years, the antialcohol campaign was abandoned. But it was too late. The damage to government finances, Gorbachev's popularity, and the fight against organized crime had already been done.

The Ruble Overhang

Within a few years after Gorbachev's accession to power, such government watchdogs as the KGB and grass-roots Party organizations, noticed an alarming rise in public discontent. "People's expectations are much higher than our actual real possibilities," admitted Prime Minister Nikolai Ryzhkov in 1990.[6]

Seeking to address this discontent, Gorbachev allowed wages to rise steeply. In 1989, for instance, average incomes rose 12 percent, even though the economy remained stagnant. As government revenues fell because of the decline in oil prices and the loss of vodka revenues, social expenditures grew. The government financed its ballooning deficit by printing money. There was little evident inflation because the prices of most products were still rigidly controlled. The inflation revealed itself in growing shortages of consumer goods and longer waiting lines.[7]

The rapid accumulation of money in the hands of the population, combined with the growing scarcity of consumer goods, became known as the "ruble overhang." Its size was estimated at between 460 billion and 500 billion rubles (about $800 billion at the official exchange rates)—half the size of the U.S.S.R.'s gross domestic product. The money was kept either in bank accounts or in cash at home. Ideally, this money would have constituted long-term savings, but in fact it represented the involuntary savings of people waiting to buy something. One way of eliminating the overhang would have been to free prices, turning the latent inflation into real inflation. But this path would lead to hyperinflation, a social catastrophe. Fortunately, there were other ways to solve the problem.[8]

In October 1989, Federal Reserve chairman Alan Greenspan visited Moscow to meet with Gorbachev and other top Soviet officials.

He argued that the Soviets should not free prices until they tackled the weakness of the ruble. "I think [the Soviets] have delayed, if not abandoned, the notion that price reform comes first," Greenspan told me shortly after his return. "They recognize that they need a change in the monetary structure [first]. They do need to get their so-called monetary overhang out of the market. It's got to be absorbed and eliminated before the fixed-price structure is unwound. Probably the easiest way to get rid of the monetary overhang is to float some bonds—ruble-denominated bonds with all of the principal and [perhaps] part of the interest being guaranteed in convertible currencies or gold."[9]

Greenspan also spoke of the necessity of privatization as a means of addressing the ruble overhang and cutting the Soviet budget deficit. Only after these two steps—issuing government bonds and privatization—were taken should prices be freed.

The best Soviet economists had come to a similar conclusion. In 1990, Gorbachev's adviser Stanislav Shatalin, and a young economist, Grigory Yavlinsky, developed an emergency series of measures called the 500-Day Plan. The plan promised to neutralize the ruble overhang, stabilize the currency, free prices, and transform the country into a market economy. The centerpiece of the 500-Day Plan was privatization. Since the state owned all the assets and the population had all the money, the 500-Day Plan envisioned a swap. The privatization program would start with such small assets as apartments, plots of land, shops, trucks, and small workshops, and move gradually to large factories, mines, and oil fields.[10]

Partly because of the opposition of the conservative wing of the Communist Party and partly because of his own doubts as a Marxist, Gorbachev never implemented the 500-Day Plan. This would prove to be a fatal mistake.

Incapable of helping themselves, Gorbachev and his colleagues turned to the West. They wanted to borrow $30 billion to buy Western consumer goods and resell them to Soviet consumers at ten times the cost. This, they argued, would take care of the excess cash of the population and stabilize the ruble, allowing market reforms to proceed. For a few months, policy makers in the West debated whether to develop a new Marshall Plan to help the Soviet Union make the transition to democracy, free markets, and prosperity. No such aid package materialized. Western governments, perhaps distracted by other foreign policy concerns, such as Iraq's invasion of Kuwait, sent economists, lawyers, and consultants to advise the Soviets, but no money.

"For whatever reason, the U.S. government did not come to the Soviet Union's aid with either bank loans or with debt rescheduling," Gorbachev would later note bitterly. "It waited until the country and the ruble had fallen apart to step in."[11]

Boris Berezovsky Goes into Business

While the crisis worsened in the Soviet Union, Boris Berezovsky made his first foray into business. Up until 1989 he had been a member of the Soviet scientific elite. (In 1991 he would be elected correspondent-member of the highly prestigious Academy of Sciences.) As one of the Soviet Union's top scientists, Berezovsky would have received a salary of somewhere around 500 rubles a month ($800 according to official exchange rates). But today he downplays the financial incentives that propelled him into business. "It's simply that I am better adapted, genetically speaking, for business than I am for science," he says. "I was very happy working in science, but in my opinion, science is less dynamic than business."[12]

Born on January 23, 1946, into a Jewish family that was part of Moscow's educated elite, Berezovsky grew up in the Soviet capital and received the best education the U.S.S.R. had to offer. He studied at the electronics and computer science faculty of the Forestry Institute. This was one of the Soviet Union's secret scientific institutions—Berezovsky's faculty was involved not in forestry, but in the Soviet space program. He proceeded to graduate school at the famous mechanical-mathematics department of Moscow State University. Next, it was on to the prestigious Academy of Sciences, where he spent twenty-five years researching decision-making theory.[13]

Apparently he was a good scientist. He earned a master's degree (equivalent to an American Ph.D.) in applied mathematics in the 1970s and then, in 1983, a Ph.D. (a higher qualification than the American degree). At the Academy of Sciences, he became director of one of the "laboratories" of the Institute of Management, which specialized in developing automation and computer systems for industrial enterprises.[14]

Like many people, Berezovsky recognized that Gorbachev's reforms were failing; instead of reinvigorating the Soviet Union, they were precipitating its disintegration. The anti-alcohol campaign had drained the state of revenue and produced a generation of bootlegger millionaires. The black market was flourishing and crime was increasing dramatically. The ruble overhang was becoming larger each month. The lines were becoming longer, the shops emptier.

As a member of the scientific-industrial elite, Berezovsky did not need to follow the steps of most other budding Soviet entrepreneurs at this time—starting such small enterprises as grocery stores, restaurants, or construction companies. He needed only to find a big, established Soviet industrial enterprise and attach himself to it. He found his target 700 miles east of Moscow, in the provincial town of Togliatti—home of Russia's largest auto manufacturer, Avtovaz. At the Institute of Management, Berezovsky had worked with the auto giant on its automation systems. Now he would approach it with a commercial proposition.

Avtovaz had been established in the late 1960s as a showpiece of Leonid Brezhnev's program to provide Soviet citizens with the kind of consumer goods the West was enjoying. Italian automaker Fiat provided the equipment and the blueprints for the first models. When I first visited Avtovaz, in the summer of 1996, the auto plant had changed little from its original Italian design. The town of Togliatti, named after the 1960s Italian Communist leader, had few distinguishing characteristics—there were just rutted roads and shabby low-rise apartment blocks. Nearby, gently rising fields stretched to the horizon and the Volga flowed majestic and sealike, approaching its Caspian estuary.

The factory was on the outskirts of town. It was a monster auto plant: huge, vertically integrated, and badly run. The technology was hopelessly outdated. The factory produced 2,000 cars a day, but the vehicles—principally Lada sedans and Niva off-roaders—were primitive jalopies (essentially 1960s-era Fiats). It took Avtovaz thirty times the man-hours to produce a lousy car that it took a U.S. or Japanese factory to produce a good one.[15]

The world's top automakers preferred to operate relatively small assembly plants, employing flexible manufacturing techniques and using large amounts of outsourced parts, usually installed as modules directly on the assembly line. But the Avtovaz assembly line was quite different. It stretched a mile and a quarter, workstation after workstation; there were no modules being slapped in—the cars were assembled piece by piece. Evidently, these pieces did not fit properly. The hammering was incessant: hammering down the gaskets, hammering in the doors, hammering on the bumpers. On the engine line I saw a man screwing in pistons by hand and whopping them with a hammer. If there was a robot on the line, I didn't see it.

It was in this dinosaur that Boris Berezovsky saw his first commercial opportunity. In 1989 he approached top Avtovaz managers and proposed a private company to provide computer programming for the factory. Most private companies in the Soviet Union were either cooperatives or joint ventures with foreign partners. Berezovsky was up-to-date on the latest changes to the Soviet commercial code—he knew that a joint venture was easily the best route, since it entailed significant tax advantages and the right to transfer half the profits abroad. At this early stage in his business career, Berezovsky already had an international perspective—he wanted to make money in Russia while keeping at least some of his earnings abroad. As a joint-venture partner, Vladimir Kadannikov, the chairman of Avtovaz, suggested the Italian factory-automation specialists Logo System; this Turin-based outfit had been working with Avtovaz for years and they promised to be a docile partner.[16]

In May 1989, Logovaz was established. The company was a collection of individuals—Berezovsky, Kadannikov, and a handful of the most commercially minded managers of Avtovaz. Kadannikov was chairman of the new company, Berezovsky was general director. The stated mission of Logovaz was to provide Avtovaz with automation software. Raising productivity at such a key manufacturing enterprise resonated well with Gorbachev's intention to modernize Soviet industry. But, almost immediately, Logovaz dropped the business of upgrading industrial computer systems and began selling Avtovaz's automobiles.[17]

The KGB's New Mission

That a top Soviet scientist should become a car dealer was not unusual in the late Gorbachev era. Even unlikely institutions such as the Central Committee of the Communist Party were going into business. Housed in a complex of large Art Nouveau buildings on Old Square, not far from the Kremlin, the Central Committee was the traditional seat of power in the U.S.S.R. The parking lot was usually filled with black and gray Volga sedans, immaculately maintained, their drivers waiting obediently nearby. The Central Committee buildings contained more than a thousand offices. They were linked by private pedestrian overpasses and tunnels.

The intrigues that went on in the red-carpeted halls of the Central Committee building complex are legendary. It has always been difficult for outsiders to know exactly which faction was opposing another. Yet the outcome of these mysterious struggles determined the fate of the nation. It was in the Central Committee in the 1920s that Stalin emerged victorious and proceeded to have his rivals executed. It was in the Central Committee in 1953 that Nikita Khrushchev succeeded in ousting Stalin's secret police chief Lavrenty Beria and having him executed. It was in the Central Committee in 1964 that Khrushchev lost his hold on power and was forced to yield to Leonid Brezhnev. It was in the Central Committee in 1985 that Mikhail Gorbachev was designated leader of the U.S.S.R.

The International Department of the Central Committee was in charge of organizing overseas-financing schemes for foreign Communist parties. Its staff became adept at using offshore tax havens, setting up shell games, and laundering money. The International Department's projects in this field were usually given to the KGB for implementation. In the 1980s, the KGB set up many fake banks and commercial enterprises offshore, especially in Greece, Cyprus, Italy, and Portugal. Billions of dollars were transferred to these institutions through VE Bank, the Soviet foreign trade bank. The typical mechanism involved the sale of a shipment of oil, metal, or timber to the KGB-sponsored enterprise abroad at a fraction of the world market price. The enterprise would resell the commodity at the market price and keep the trading profit for itself. Now, with the Soviet Union's demise imminent, the International Department and the KGB employed the same scheme for the benefit of the Soviet Communist Party itself.[18]

Apart from a few departments of the Central Committee, the KGB was the only Soviet institution responding decisively to the deteriorating situation. It was given the task of preserving the power of the Soviet Union's ruling caste—the Communist nomenklatura—even if Communism itself should end. Together with the International Department of the Central Committee, the KGB developed a strategy to transfer billions of dollars of Communist Party capital to captive companies in the Soviet private sector and abroad. It was an extraordinary operation.[19]

Russians have always been good policemen. In the days of the czars, the secret police were not just run-of-the-mill snoops, but part of a highly inventive organization, conceiving extraordinary projects on a grand scale. In order to divert Russian workers from revolutionary socialism, for instance, the czarist secret police created a series of monarchist trade unions. These organizations were highly successful until they spun out of control and precipitated the very event the authorities wished to avoid: a revolution.

The Soviet secret police were far more zealous than the czar's secret police. Until 1991, the KGB was involved in industry, transportation, telecommunications, the army, the police, and culture. It also had a long history of infiltrating opposition groups. The terms "agent provocateur" and "provocation" were an omnipresent reality in the Soviet Union (and anywhere else around the world where the KGB spread its tentacles). A textbook provocation involved infiltrating the ranks of the opposition to guide your enemies onto the path of self-destruction. The Soviet secret police specialized in such operations and surreptitiously established themselves in such potentially threatening institutions as monarchist societies, dissident groups, churches, Zionist organizations, art circles, and ethnic separatist groups.

In the fall of 1990, I met Oleg Kalugin, a major general of the KGB and former head of the agency's foreign counterintelligence division. Kalugin described his agency this way: "The KGB was characterized simply by greater flexibility. If the Party administration considered that rock music should be banned, for example, the KGB argued that rock should be permitted, but kept firmly under its control."[20]

When I asked about whether the KGB had infiltrated the nascent democratic movement as well, General Kalugin, who was normally oblique in his answers, reacted unusually firmly. "Nothing of the kind," he snapped. "There never was such a policy."

Nine years later, when Kalugin and I met in Washington, D.C., where the former spymaster was living comfortably in exile, he revealed the true role of the KGB in the birth of "democratic" Russia.

In the landmark elections of 1990—the first free elections for a national parliament, as well as republic-level and regional parliaments—the KGB had fielded several thousand candidates; most were elected. The KGB helped create the first non-Communist political party: Vladimir Zhirinovsky's absurdly named Liberal Democratic Party. Zhirinovsky's mission was to talk big, serving as a lightning rod for nationalist sentiment, but to undertake no radical action of any kind. The nationalist group Pamyat, whose extremist rhetoric evinced great cries of alarm in the West in the late 1980s, was also a KGB creation. Some KGB officers assumed prominent roles in the democratic movement; Vladimir Putin, for instance, became a top aide to the most eloquent champion of liberty and democracy, Anatoly Sobchak.[21]

The man in charge of the KGB's surveillance of the domestic political scene was General Philip Bobkov. A veteran KGB man, already in his late fifties by the time of perestroika, Bobkov was the head of the KGB's Fifth Chief Directorate. "The mission of the Fifth Chief Directorate was very broad: It was to look after the political hygiene of the Soviet regime," recalls General Kalugin. "That means controlling, first of all, the intelligentsia, as the most infectious part of the population. That means controlling the church, as a representative of a hostile ideology. That means controlling the scientific community, and so on."[22]

Bobkov's Fifth Chief Directorate was the arm of the KGB responsible for persecuting Soviet dissidents, from Aleksandr Solzhenitsyn to Andrei Sakharov. It had perfected the art of planting spies and informers among potentially threatening civic groups, and inserting KGB agents in leadership positions. With the arrival of Gorbachev, however, the Soviet government renounced political surveillance. Bobkov's Fifth Chief Directorate soon began to infiltrate the growing democratic movement and the new private enterprises. Bobkov himself was promoted to deputy chairman of the KGB. He remained responsible for the Fifth Chief Directorate, but also took control of the Sixth Chief Directorate, which had formerly investigated economic crimes and was now responsible for overseeing the new cooperatives (private businesses).[23]

In 1990 a certain Colonel Leonid Veselovsky of the First Chief Directorate of the KGB outlined the new strategy in a secret memorandum to a Communist Party functionary named Nikolai Kruchina. Kruchina was in charge of Communist Party property at the Party's Central Committee. The KGB memo sought Kruchina's support in creating a network of captive banks and trading companies, both in Russia and abroad, to collect billions of dollars of government funds during the "period of emergency" and keep them for the Communist nomenklatura until a more propitious time arrived.

"The earnings which are accumulated in the Party treasury and are not reflected in the financial reports can be used to purchase the shares of various companies, enterprises, and banks," the Veselovsky memo states. "On the one hand, this will create a stable source of revenue, irrespective of what may happen to the Party. On the other hand, these shares can be sold on the securities exchanges at any time and the capital transferred to other spheres, allowing the Party to keep its participation anonymous and still retain control.... In order to avoid mistakes in the course of this operation during the 'period of emergency,' it is essential to organize, both in the U.S.S.R. and abroad, special rapid response groups, staffed by specially trained instructors from the active reserve of the KGB of the U.S.S.R., as well as by trusted individuals volunteering their cooperation and by individuals who, for one reason or another, have lost their job in the field units or administrative departments of the KGB of the U.S.S.R."[24]

The strategy was implemented. The ruling Communist Party had concluded that since it could not crush the black market, it must join it. The new entrepreneurs chosen by the Central Committee and the KGB to serve as repositories for this Communist "black treasury" included freelance secret-service agents and gangster businessmen arising out of the vodka economy and the cooperatives. Many of Russia's future billionaires obtained their first real capital in this manner.[25]

Berezovsky apparently was not one of the recipients of the KGB's secret funds. He was not one of the Communist Party's authorized millionaires. Among other reasons, he may simply have been too old—the entrepreneurs chosen to handle the Communist Party funds tended to be members of the Communist Youth League, who were in their twenties, while Berezovsky was in his late forties at this time. But he could not fail to notice that commercial success in Russia depended on official patronage.

"As the KGB was leaving the scene, it didn't simply disappear—it left behind various political and commercial blocs, using people whom the KGB had helped in the past," General Kalugin would later recall. "Don't forget that the Party retained a significant fortune and huge amounts of property. The KGB didn't have this money itself—it was the distributor. As soon as privatization began, these funds began to disappear. They didn't end up in the government—since, legally speaking, these were not government funds, but Party funds. The money ended up in the black market."[26]

Much of this KGB-distributed money went to foreign bank accounts. General Kalugin himself, while head of foreign counterintelligence, helped set up some of the vehicles for this money. In 1978, Kalugin's First Chief Directorate started an economic-espionage agency called the Eighth Department to undertake foreign banking operations. "We would introduce our people, KGB men who were specialists in finance, into both Soviet banks and joint-venture banks in London and elsewhere," recalls Kalugin. "On this foundation, we developed the ability to manipulate the gold market, for example." (As the world's second-largest gold producer, the Soviet Union had the ability to influence the price of gold.)[27]

Western intelligence agencies were aware of the KGB money-laundering program, but they decided not to do anything, even when it became apparent that a large portion of the KGB money was being embezzled by freelance operatives, rogue businessmen, and gangsters. When the CIA, for instance, received an indirect request from the Russian government in 1992 to help track down the missing billions, it decided against it, for fear of jeopardizing its own agent network.[28]

It was impossible to keep such large-scale financial operations completely secret, however. In February 1991 (the last year of the Gorbachev regime), a curious story appeared in Russia's newspapers. Gennady Filshin, the deputy minister of foreign trade in Boris Yeltsin's new Russian government (as distinct from the Soviet government of Mikhail Gorbachev), had apparently been working on a deal to buy $7.5 billion in exchange for 150 billion rubles. The counterparty was an unknown British company called Dove Trading International. The president of this company was a British fugitive living in South Africa. Under the terms of the Filshin deal, Dove Trading would sell the $7.5 billion to the Russian government and use the rubles to buy Russian commodities and export-oriented enterprises. It was unclear whether the billions of dollars the Russian government received would stay offshore or be repatriated to Russia. It was also unclear where an insignificant company like Dove Trading could get such a huge sum of dollars (later, European law-enforcement agencies suggested that the British firm might have been fronting for a Colombian cocaine cartel). Before the Filshin deal could be implemented, however, the operation was leaked to the Russian press. The Soviet parliament began an inquiry and the deal was aborted. Filshin was forced to resign, but the matter was never fully investigated.[29]

A similar operation was set up in November 1990 in the offshore tax haven of Jersey (an island off the British coast) by the Soviet Central Bank and a Paris-based affiliate often used for KGB financial operations abroad. This company was known as Financial Management Co., or Fimaco. Over a five-year period in the 1990s, Fimaco would secretly handle immense sums for the Russian Central Bank ($50 billion according to the estimates of Russian prosecutor-general Yuri Skuratov) and would accumulate hundreds of millions, perhaps billions, of dollars in profits offshore. There was no sign that these profits were being repatriated to Russia; instead they were distributed to a number of private banks, consulting companies, and non-profit foundations.[30]

The massive funds the Central Committee and KGB were hiding abroad were rarely sent via bank transfer. They had to be laundered first. The easiest way of disguising a large movement of funds offshore was by rigging foreign trade contracts. Toward the end of the Gorbachev era—1990 and 1991—Soviet foreign trade practices began assuming a distinctly unorthodox character.

No one better exemplified the Soviet Union's increasing tendency to conduct its foreign trade through dubious middlemen than the notorious international commodities trader Marc Rich. A wunderkind oil trader in the 1970s, Rich had fled the United States in

1983, when he was indicted for racketeering, fraud, tax evasion, and trading with the enemy (Iran). Now a billionaire in his mid-fifties, Rich lived in Zug, Switzerland, where he combined the glamorous life of big-time international deal-making with the furtive existence of an outlaw. Switzerland refused to extradite him, but the United States had strong extradition treaties with most other countries in Europe. Interpol had a "red notice" on him, giving the highest possible priority to his apprehension.[31]

Marc Rich was a businessman who thrived on the misfortune of others. In 1990 he found himself in a tough situation. Metals prices had fallen. Two of his big oil trading partners—Iraq and Kuwait— had dropped out of the world trading system. Many of his traditional markets were drying up. When I asked Rich's trading rivals what was happening with his business, they painted an unremittingly gloomy picture.

"When there was an embargo on South Africa, Rich provided them with oil, and they paid handsomely for that," noted Dieter Boettcher, director of a London-based metals trading subsidiary of Metallgesellschaft A.G.: "Now that the embargo is being lifted, the margins are slimmer."[32]

Latin America? "Twenty years ago Latin American countries like Chile and Venezuela were very much traders' country," said Jonathan Platts-Mills, a director of the British conglomerate Lonrho, Plc. "But now they've become more sophisticated, dealing directly with clients and suppliers."[33]

The Middle East? "The days of wild commodity trading, of the incredible deals when you could use your friendship with some prince in the Middle East to get oil at $4 a barrel, are long gone," said Sveinung Medaas, head of the Moscow office of Salomon Inc.'s oil trading subsidiary, Phibro Energy Inc.[34]

With many of his traditional markets drying up, Rich found another way to make money: Shortchange the desperate Soviets. Throughout the former Soviet Union there was unprecedented turmoil—profit margins on both oil and metals trading were at least two or three times the average trading margins in other parts of the world. The Soviet business became a key part of Rich's $30 billion global trading volume.[35]

Though Rich maintained very close relations with various Communist Party chiefs, it is unclear what role he played in the KGB money-laundering program. He certainly played a part in facilitating capital flight out of the Soviet Union. He was also a pioneer in teaching former KGB officers how to go freelance and appropriate international trading profits for themselves.[36]

Rich traded all kinds of products with the Soviet Union. He sold grain, sugar, zinc concentrate, and alumina (a bauxite extract that is the main component of aluminum) to the Soviets, and they paid him back with oil and aluminum, as well as nickel, copper, and other metals. The Russian commodities gave Rich tremendous weight on some of the most important world commodity markets. Thanks largely to his Soviet purchases, for instance, Rich handled 2 million tons of aluminum annually, controlling one third of the world spot market in this metal.[37]

"[We are providing] Russian companies with investment, know-how, and help in entering the world market at a time when other Western firms are either turning away from Russia or are making intolerable commercial demands," Rich declared to the Russian press several years later, after his operations had come under scrutiny.[38]

In fact, Rich was gouging the Russians, buying commodities at insider prices, selling them abroad, and registering his profits in the tax haven of Zug, Switzerland. Russian traders would perform the same operation repeatedly in the 1990s, but Rich had the distinction of doing it first and doing it big. Much of what he was doing was technically illegal according to Soviet laws, but he had astute collaborators on the inside. His deals usually involved secret arrangements with the directors of oil and aluminum enterprises and complicated payment schemes that spanned the globe.[39]

One of Rich's key counterparties was a forty-year-old entrepreneur named Artyom Tarasov, a man who emerged as one of the pioneers of Russian robber capitalism. Tarasov, who was half-Georgian, had grown up on the Black Sea coast, had studied at the Mining Institute and the state planning agency Gosplan's Higher Economic School, and had taken a job with the Moscow municipal government. As soon as it became legal in 1987 to establish private businesses ("cooperatives"), he started a cooperative called Tekhnika, to trade Russian raw materials for personal computers. He made a small fortune. His next company, Istok, grew into an export-oriented empire, including leased railcars, depots, port facilities, ships, and warehouses.[40]

In the summer of 1990, Tarasov took an important role in financing the government by participating in the "Harvest 90" campaign. This program was devised by the cash-strapped Russian government to pay collective farms with promissory notes, which could later be exchanged for imported consumer goods. Tarasov was chosen to redeem 10 percent of the outstanding Harvest 1990 promissory notes. Boris Yeltsin's republican government granted him a license to export Soviet heating oil and permitted him to keep his earnings abroad—an unprecedented privilege for a private trader—with the stipulation that he use the profit to redeem the promissory notes. He sold the oil abroad, but Soviet farmers never received their consumer goods. It was a famous scandal.[41]

"Yes, Tarasov certainly was our teacher," Oleg Davydov, a veteran official at the Ministry of Trade, would later note ruefully. "He bought heating oil domestically at $36 a ton and sold it for $80 a ton abroad. And all the while he appeared on TV teaching us that people have to do these kinds of deals, how much money it's possible to earn, and that the Ministry of Foreign Trade is such a bad organization, because it's preventing people from doing this. Well, the Ministry of Foreign Trade was doing exactly the same thing, only it put the difference [between domestic and world prices] not into the pocket of Tarasov, but into the government budget."[42]

There is no evidence that Berezovsky ever did business with Marc Rich or even met him, though the two men briefly shared an interest in the aluminum export market. It is a fact, however, that Berezovsky adopted many of the capital-flight strategies pioneered by Rich. In this way, the Switzerland-based trader served as a mentor to Berezovsky, much as he did to many other Russian traders and financiers.[43]

In 1991, another dubious foreign trade operation was exposed. This one involved an outfit called ANT, a closely held company established by several military-industrial managers with the aid of Soviet prime minister Nikolai Ryzhkov and several of his ministers. The president of the company was a KGB general. Throughout 1990–91 ANT secretly handled a number of large commodity export deals, completely bypassing the Ministry of Foreign Trade. The sales were made at prices far below world market levels, with the rest of the purchase price being appropriated by unknown counterparties abroad. ANT was eventually busted when it tried to export a shipment of tanks, artillery, and other military equipment to buyers abroad. The matter was never fully investigated, but Prime Minister Ryzhkov was forced to resign.[44]

I would later talk to the first post-Soviet prime minister, Yegor Gaidar, about capital flight in the last years of the Soviet Union. "In the structure of Soviet foreign trade contracts there were lots of mysteries," Gaidar said. "We bought all kinds of equipment at abnormally high prices and with cash up front, while much of our production was sold at very low prices."

Whatever the nature of these clandestine operations—Fimaco, ANT, the Filshin affair—the Soviet Union's gold and foreign exchange reserves disappeared sometime around 1990. I asked Gaidar what happened. Did the Communist bosses and the KGB smuggle this wealth out of the country? "You can't prove that with statistics on foreign trade," Gaidar responded. "All the money went to pay for imports or foreign debt service. Whether these imports were necessary and at what prices the goods were bought—that's another question. But you can't check that in the statistics."

In the early 1980s, the Soviet gold reserve stood at 1,300 tons (about $30 billion in those days). Over a period of just two years, between 1989 and 1991, most of this gold reserve (approximately 1,000 tons) was sold. At the same time, the Soviet Union's foreign exchange reserves had fallen from $15 billion at the start of Gorbachev's rule to a paltry $1 billion in 1991. Though the true picture of the U.S.S.R.'s balance of payments at this time is almost impossible to ascertain, it was clear that the Soviet Union lost about $20 billion in capital flight in 1990–91.[45]

As its coffers emptied, the Soviet Union had no choice but to accept a reduced status on the world stage. The Berlin Wall fell at the end of 1989 and the Soviet client states in Europe broke free. Half a year later, after Iraq (a traditional Soviet ally) invaded Kuwait and the United States prepared to go to war, Gorbachev supported the Western alliance against Saddam Hussein. Inside the Soviet Union, Gorbachev's hold on power was steadily weakening.

In November 1990, I was again in Moscow on Revolution Day. This time I did not have an insider's post. I watched the beginning of the parade on television and heard the rest of it on the public address system still functioning throughout the center of Moscow. I decided to go to Red Square.

The streets leading up to the Kremlin were empty, blocked by heavily manned checkpoints. I tried the back end, by the giant Hotel Rossiya. Only authorized personnel were permitted through, but with my press credentials and a lot of cajoling, I emerged onto Red Square by St. Basil's Cathedral. Officials and Communist activists passing me on the way back from the review were tense, somewhat bewildered. (A man with a shotgun marching past Gorbachev in Red Square had shot at the Soviet president and had been arrested.)

Under the walls of St. Basil's stood a small group of angry middle-aged activists holding banners and yelling slogans. GORBACHEV AND YAKOVLEV [Gorbachev's closest adviser] ARE ZIONIST SPIES! read one banner with a large star of David. STOP THE BETRAYAL OF THE FATHERLAND! read another. It was Pamyat—the anti-Semitic group that had emerged so visibly in the late stages of the Gorbachev regime. Pamyat would later be revealed as a KGB-sponsored provocation, designed to channel Russian nationalist sentiments into a politically reliable entity and to frighten Western public opinion into supporting the moderate Gorbachev as the "only alternative" to extremist forces.[46]

Hearing about a plan for a counterdemonstration of democrats, led by Boris Yeltsin and the current Moscow mayor, Gavril Popov, I walked around the Kremlin to Revolution Square. Several thousand people—mostly young, but including elderly intellectual types—were there. It was a motley crowd: anarchists, *babushkas* (older ladies), bespectacled young intellectuals, students, as well as ubiquitous gray little men worming their way into every group and conversation—KGB *agents provocateurs.* In the front row was Yeltsin, defiant, his huge chest thrust forward in a white shirt, linking arms with Popov, a diminutive, swarthy man, who looked around nervously.

A line of policemen prevented the column from pushing its way into Red Square. Some crafty old *babushkas* emerged from the ranks of the democrats. "Please, dear, don't hurt an old lady, let us through," they implored the servicemen. "We have a right to go to Red Square," the men shouted. "We have Yeltsin with us and Popov."

The police broke their line and the democrats swept noisily up the cobbled street toward St. Basil's and Lenin's Mausoleum.

"Resign! Resign!" the crowd shouted to the vacant marble terrace of the Lenin Mausoleum, where the Politburo had stood just an hour before. "Democracy!" they shouted to the Kremlin parapets and the government buildings beyond.

The participants in the official Communist demonstration had long since gone. The only ones who remained were the street sweepers and a regiment of Communist thugs with their backs against the Kremlin wall—burly young men, scowling with that brutal glare cultivated by all good Communists. They seemed to be either plainclothes members of the KGB or martial-arts specialists of the Communist Youth League. A few years later I would see the same faces in the organized-crime groups of Moscow. But on that day they were guarding the old order against Yeltsin. They glared and shifted on their feet, itching to break free and pummel the bohemians and intellectuals parading under the democratic banners. The democrats marched past them and dispersed by the statue of Minin and Pozharsky.

The Lausanne Connection

At this time, Berezovsky was making good money selling Avtovaz cars, but, having watched Marc Rich and other commodities traders from the sidelines, he wanted to get into the lucrative business of exporting Russian raw materials. For this he needed to find a partner experienced in trading commodities around the world and in discreetly managing international money flows. In early 1991, accompanied by Nikolai Glushkov, chief financial officer of Avtovaz and a Logovaz founding member, Berezovsky traveled to Lausanne, Switzerland.

Lausanne is one of the financial capitals of Switzerland—not nearly as big as Geneva (in terms of private banking) or Zurich (in terms of corporate banking), but well located and boasting low levels of taxation and regulation. It is one of those remarkably prosperous Swiss towns—a village that has sprouted skyscrapers. Expensive restaurants, boutiques, and shopping centers grace its cobbled streets. The most well-known corporate resident is the International Olympic Committee, which twice every four years holds the nations of the world in thrall as it chooses the site for the winter and summer games. A more discreet resident of Lausanne, but only slightly less international, is the big commodities trading company André & Cie.

Established in 1877 and still owned by members of the André family, the firm trades principally grain and sugar, but also a wide selection of other goods and services around the world. The company's headquarters are in the center of town: a large glass building, set among lawns and pine trees, with a spectacular view of Lake Geneva. In the marble lobby, the security officer sits properly at his desk; across from him are paintings and models of oil tankers and other bulk carriers. Upstairs, on five or six well-lit floors, ambitious young traders work the phones and computers.

André began doing business in Russia in 1978, when a number of big Western corporations were trying to exploit détente. By the time Berezovsky approached it, in 1991, André was selling the Russians grain and sugar, as well as Swiss industrial machinery, and wanted to expand its operations.[47]

One of the company's point men was Alain Mayor, a middle-aged Swiss businessman who had been working on Russian projects since 1981. Based in Lausanne, Mayor was head of the Russian desk of an outfit called Finco (Financial Compensation), a semiautonomous André subsidiary charged with procuring payment from countries without convertible currencies. The Finco outfit specialized in a variety of countertrade deals, barter arrangements, and complicated currency-exchange schemes in the darker corners of the global market. "It is a very opportunistic operation—this compensation department at André—trying to make business out of any opportunity which arises," Mayor recalls. "The basis of making it work is getting into contact with partners with whom you feel a common interest."[48]

Alain Mayor quickly realized the implications of the changes unleashed by Gorbachev's perestroika. During the Soviet era, business in Russia for Western companies was relatively straightforward. It was important to have the right contacts and introductions, but basically Soviet government officials overseeing the trade deals acted according to precise economic principles. Typically the officials were informed by the central planners at Gosplan that the U.S.S.R. needed a particular quantity of a particular type of good; then the only question was which Western firm offered the best deal in terms of price, quality, and delivery schedules. The Soviet Union had an excellent reputation as a commercial partner: The Soviet trading establishment was an honest counterparty, did not engage in bait-and-switch shenanigans, and did not try to renege on the bill. In the late 1980s, once the central government began losing its authority and more business was handled by semi-independent commercial organizations, the market became corrupted—an inside deal among cronies.

"For the success of deals in Russia, I considered it very important to have a partner in Russia," says Mayor. "It wasn't the case before.

But since 1990 it became indispensable to have someone who was on the inside of the market. And future events would prove that it is on the inside of the market that business is interesting, so to speak."[49]

Boris Berezovsky was the partner Mayor had been looking for— the man who could get André into the "interesting" part of the Russian market. "I really met [Berezovsky] the first time in Lausanne," Mayor recalls. "He asked for a rendezvous. [I said] if you have a proposition, I'm game. At that time I was very interested in finding a partner in Russia. He presented what he was doing, what Logovaz was doing in cars. I don't remember what we talked about, but I found the person interesting and agreeable. They [Berezovsky and Glushkov] said: 'Couldn't we create something together?' We simply replied: 'Well, why not?' "[50]

One of the André staffers involved in the early negotiations with Berezovsky was Christian Maret, who would later head up the Swiss grain trader's Moscow office. Maret was impressed with Berezovsky. "In contrast to the stereotypical Russian businessman, he gave the impression of already being much more Westernized," Maret recalls. "As a man, I would say, he has lots of ideas, but the problem is to follow his ideas, to put them into practice. Glushkov, on the other hand, was more matter-of-fact."[51]

The traders from André & Cie. were eager to get in on the emerging Russian gold rush. "We were interested in buying automobiles," says Mayor. "We knew that at the time there was already a lot of reexport of Avtovaz cars. We thought it might be interesting to have access to this kind of activity, since we were not involved in these kinds of things at that time."[52]

Alain Mayor convinced André & Cie. to go into business with Berezovsky not only as a counterparty but also as an equity partner in a series of different companies, beginning with Logovaz. Why would a century-old trading house, jealous of its reputation, want to get into bed with a little-known entrepreneur in a market already notorious for its corruption and criminality? "The choice at that time in Russia wasn't exceptional," said Mayor. "It would have been difficult to find someone with a great reputation in 1991 in Russia, someone with a fantastic track record. There is a lot of intuition when you judge the qualities of a potential partner. You meet someone, you start something, and then you see what happens next."[53]

For his part, Berezovsky wanted to reincorporate Logovaz and retain a foreign-registered company as half-owner, but he was no longer satisfied with Logo System as a partner. Automated control systems were no longer interesting—he needed an outfit specialized in buying and selling. He wanted Logovaz to be co-owned by a Swiss company, but not controlled by anyone but the Russians. He turned to André for ideas.

André found the perfect vehicle for Logovaz, a Swiss company called Anros S.A. It had been established by André in 1977 for business in Southeast Asia. Anros was owned on the basis of bearer shares—which means the owners are not identified in a register, they are simply individuals who physically possess the shareownership certificates. This practice used to be common in the United States until the stock market crash of 1929. When Anros was founded in 1977, 100 percent of the bearer shares were held by André. When Logovaz was reincorporated on May 28, 1991, Anros was reincorporated as well—first in June, then in September—with 99 percent of the shares eventually belonging to the Russian partners.[54]

Formally speaking, Logovaz was still a 50/50 Russian-Swiss joint venture, entitled to various tax breaks and to keep part of its profit abroad, but in fact all but a small percentage of the company belonged to Berezovsky and his Russian partners.[55]

In the autumn of 1996 I asked Berezovsky who owned Logovaz. "The main shareholders are individuals," he answered. "The number of shareholders in [Logovaz] can be counted on the fingers of one hand. These are the people who founded the company. Two of them have no business other than Logovaz and they are my principal partners." He added that the owners of Logovaz were generally the same people who held the top management positions in the company.[56]

One of the extraordinary things about the shareholder structure of Logovaz, considering the company's later role in nearly bankrupting Avtovaz, was the presence of several of the top managers of the aker, including president Vladimir Kadannikov, finance chief ai Glushkov, marketing chief Aleksandr Zibarev, and Kadannikov's special assistant on financial matters, Samat Zhaboev. "There were people who were on both sides, that's for sure," André director Yves Cuendet would later admit, "people who were both on the side of Logovaz and on the side of Avtovaz."[57] In other words, Avtovaz was selling its cars on special terms to an independent dealership called Logovaz and engaging Logovaz to handle its finances at the same time that top Avtovaz executives benefited personally from the relationship as Logovaz shareholders.

For Berezovsky the reincorporation of Logovaz represented an extraordinary achievement. The Soviet Union had not yet fallen, and here was a Russian businessman, operating without the advice of the KGB or other internationally minded parts of the Soviet establishment, setting up a sophisticated international financial structure, complete with reputable foreign partners, shell companies, and tax shelters. The joint venture with André & Cie. confirmed Berezovsky as one of the pioneers of Russia's peculiar version of capitalism. The car dealer had established the two key elements of the strategy that would make him immensely wealthy: a personal connection with the managers of a major Russian enterprise and the international financial network to bleed that enterprise of cash flow.

The Putsch

Dissatisfaction continued to grow throughout the Soviet Union. Cigarettes were scarce and tobacco riots broke out in many cities; mobs of deprived smokers trashed shops, kiosks, and bus stations.[58]

The 1991 grain harvest forecast was down 23 percent. Soviet farmers began distress slaughters of their livestock. Food stores were emptier than usual. There were reports of desperate villagers hijacking freight trains on the Trans-Siberian Railroad. As the government's deficit continued to grow, Gorbachev was forced to reduce social spending. For the first time in the Soviet Union's history, there were coal miner strikes and mass labor unrest. "Gorbachev has only months left in power," Oleg Voronin, a coal miner leader, told me in early 1991. Meanwhile, the policy of covering the deficit by printing money meant that a monetary explosion was imminent. Something had to be done quickly. But the Gorbachev government was paralyzed by indecision.[59]

My friends talked about an approaching military coup. The Soviet army was unhappy, having lost its East European client states and having been forced to reduce spending and retire 500,000 army officers. Army units were now confronting rebellious crowds with live bullets in several cities: Vilnius, Baku, Tbilisi. Two Soviet republics—Armenia and Azerbaijan—were engaged in open warfare for the disputed territory of Nagorno Karabakh. General Boris Gromov, former commander of the Soviet forces in Afghanistan, was often mentioned as a candidate for the role of Napoleon to carry out the coup d'état and revive the country.

Then, on August 19, 1991, an astonishing spectacle unfolded before the world. On CNN, in yellow, fuzzy pictures, filmed at night, there were tanks on the streets of Moscow. Tank and paratroop units had seized control of the capital. Gorbachev was being held captive at his Black Sea villa. An entity called the State Committee for the State of Emergency (GKCHP) had taken power. Unhappy army officers and disgruntled Communist Party members were carrying out a Putsch against Mikhail Gorbachev and his reforms.

The new parliament of the Russian Federation, based in a white skyscraper on the Moscow River—nicknamed the White House—was holding out, however. The elected president of the Russian Federation, Boris Yeltsin, had at first been stranded by the coup, waiting helplessly in his dacha outside Moscow. By means of an audacious sleight-of-hand trick, Yeltsin's loyal bodyguard, Aleksandr Korzhakov, managed to smuggle him into the White House. On the way, Korzhakov and Yeltsin passed through a cordon of KGB special troops who had apparently been ordered to prevent the Russian president from reaching his destination at all costs; surprisingly, the KGB troops failed to fire a single shot.[60]

The armed forces loyal to the Putsch surrounded the White

House, but democratic activists held them off during the perilous midnight hours. A tense two-day standoff ensued. Several army units, including the crack Tula Paratrooper Division of General Pavel Grachev and General Aleksandr Lebed, refused direct orders to storm the democrats' bastion. Instead the paratroopers formed a cordon of armored vehicles to guard the White House against "disorder and vandalism." It was a key turning point in the siege (both Grachev and Lebed would be rewarded with high government posts in the Yeltsin government: Grachev as minister of defense, Lebed as secretary of the Security Council).[61]

On the third day, the army units retreated. The leaders of the coup fled and were soon arrested. Boris Yeltsin made his dramatic speech from the White House balcony, and hundreds of thousands of joyful citizens celebrated democracy in Red Square with a gigantic Russian (non-Communist) flag. On Lubyanka Square, in front of the KGB headquarters, the KGB-sponsored nationalist organization Pamyat led a mob in removing the statue of Felix Dzerzhinsky, the founder of the Soviet secret police.

Throughout the attempted coup the majority of the Soviet population remained neutral. Their destiny had been decided by a surprisingly small group of politicians, soldiers, and activists.

With the failure of the August Putsch, the KGB, which had served as the "sword and shield" of the Communist Party, fell apart. Its departments were distributed among at least four different government agencies. A huge number of KGB officers, often the brighter ones, left the service altogether and formed "mini-KGBs" at private companies. The KGB's vast network of agents and informants dispersed. Among these individuals were many of the new businessmen who had gotten their start with Communist Party funds.[62]

Most of the KGB-designated entrepreneurs wanted to keep the funds for themselves. Many of the most famous murders that swept the Russian business world in the next few years resulted from the struggle over these funds. Since the KGB had been an organ not of the Soviet Union but of the Communist Party of the Soviet Union, its operatives acknowledged neither the call of the new Yeltsin government nor that of the reconstituted rump of the Russian Communist Party. The latter was particularly bitter about the success of the KGB-sponsored entrepreneurs and kept pointing out that, among other things, the entrepreneurs had stolen the Party's funds.

Parking Communist Party funds in commercial enterprises was not the only KGB-sponsored perestroika project that would ultimately harm Russia. The KGB also infiltrated and sponsored leading organized-crime groups. The FBI, for instance, had long noted that the Soviet secret service was behind the rise of many leading Russian gangsters, almost certainly including the Jap, Otarik, leading members of the Solntsevo Brotherhood, as well as several Chechen gangs. It would not be long before the "agents" would break from the control of the "masters" and transform the country in their own image.[63]

Here was the root of Russia's downfall: The KGB and the Communist Party were incapable of playing it straight. They continually concocted some kind of double game. They made the same mistake that the czarist secret police had made a century earlier—they thought they were infiltrating the revolutionary movement and controlling it, but actually they were feeding the revolution. The KGB had created the gangsters and the capitalists who went on to destroy the Soviet Union. It was an outcome neither a genuine Communist nor a Russian patriot could have desired. In retrospect, the Central Committee–KGB plan was a colossal mistake: Dr. Frankenstein creating the monster, only to see him break his chains and run amok.

For Berezovsky the fall of Communism was good news. He had never been particularly close to the power brokers of the Soviet Communist Party, but he was on very good terms with new authorities—Yegor Gaidar, Anatoly Sobchak, Anatoly Chubais, and the other "young reformers" who constituted Yeltsin's kitchen cabinet. Some of them, like Pyotr Aven, the young economist who was Yeltsin's new minister of foreign trade, were old family friends; others, such as Mikhail Khodorkovsky, the Deputy Minister of Fuel & Energy, would soon be close business associates; still others were social acquaintances made in the 1980s. On September 6, 1991, just three weeks after the August Putsch, Logovaz received a special export license from the Ministry of Foreign Trade. This gave the car dealer permission to export oil, aluminum, and a variety of strategic raw materials. Berezovsky could now proceed with his plan to tap into Russia's prime sources of hard currency.[64]

Death in the Courtyard

The fall of Soviet Communism was largely a bloodless affair. Officially only three people died during the Putsch—three democratic activists crushed by tanks. There was another notable casualty, however: Nikolai Kruchina, the man who controlled Communist Party property at the Central Committee headquarters at Old Square. Several days after the Putsch, Kruchina "jumped" out of his office window. With this death, Russia's new entrepreneurs could breathe easier. The man who knew where all the Central Committee and KGB money was parked would not be able to divulge his secrets to the world.[65]

CHAPTER THREE

TRADER'S PARADISE

The Decision to Dissolve the Soviet Union

Yeltsin's first task after emerging victorious from the August 1991 Putsch was to get rid of the other man with a claim to the leadership of the nation: Soviet president Mikhail Gorbachev. Since Gorbachev had assumed his post legally, in accordance with the Soviet constitution, there was only one way to go.

In early December, Yeltsin flew to a Belarussian hunting resort called Belovezhskaya Pushcha to meet with the leaders of the two other big Slavic republics—Ukraine's Leonid Kravchuk and Belarus's Stanislav Shushkevich. Together, on December 8, they decided to abolish the Soviet Union and declare independence. This decision, made just nine months after a nationwide referendum in which 76 percent of Soviet citizens voted to keep the union intact, was unconstitutional and antidemocratic.[1]

The once rigidly centralized Soviet state dissolved into fifteen nationalistic and largely anticapitalistic republican governments. In the Republic of Georgia, for example, the new president was a former dissident named Zviad Gamsakhurdia. Upon taking power, Gamsakhurdia revealed himself to be an old-fashioned tyrant, turning his militia goons on political rivals; he was eventually ousted in a coup and died in a shoot-out with government troops in a mountain village. The Ukrainian government was taken over by the republic's former Communist bosses, who combined extreme nationalist symbolism with an absolute refusal to change their nation's social or economic order. In Kazakhstan, the largest net exporter of grain among the former Soviet republics, the government prohibited Kazakh farmers from selling their produce beyond the boundaries of the republic. Throughout the former Soviet empire, new border posts appeared and venal customs agents haggled over burdensome trade duties. Neighboring villages and family members suddenly found themselves on opposite sides of international frontiers.

For Russia the geographic change was devastating. Its borders were rolled back roughly to those of 1613. At the stroke of a pen it lost perhaps 50 million people who considered Russian their native language. One person who was particularly outraged by this development was the dissident writer Aleskandr Solzhenitsyn.

"Imagine that one not-very-fine day, two or three of your states in the Southwest, in the space of twenty-four hours, declare themselves to be a fully sovereign nation, independent of the U.S., with Spanish as the only language," Solzhenitsyn fumed. "All English-speaking residents, even if their ancestors have lived there for 200 years, have to take a test in the Spanish language within one or two years and to swear allegiance to the new nation. Otherwise they will not receive citizenship and will be deprived of civic, property, and employment rights."[2]

The gross domestic product in all the former Soviet republics (except for the tiny Baltic states) plummeted. Many years later, the liberal parliamentarian Grigory Yavlinsky would note: "There was no doubt that the Soviet Union was politically doomed. But it was absolutely indispensable to retain a [unified] free market and a [unified] distribution system. They [Yeltsin's ministers] not only failed to do this—they did everything they could to push the other republics away."[3]

Gaidar's Reform

Thirty-three-year-old economist Yegor Gaidar answered questions in a soft, high-pitched voice; he spoke quickly, as if the answer were obvious to everyone. Boris Yeltsin was impressed by him and put the fate of Russia in his hands.

"The socialist economic system is a complete system," Gaidar explained. "You cannot pull out one little element, such as controlled prices, and expect it to work. For it to work you need...an effective Gosplan, a system of orders which are obeyed, the ability to jail the factory head who does not deliver his goods to the place he is supposed to, to fire the head of the local government if he doesn't deliver the grain he's ordered to, the ability to seize the grain from a collective farm that doesn't want to surrender it. Under these conditions, the system works more or less.

"This system was disintegrating fast in 1989–90. By the autumn of 1991, when we formed our government, it was gone altogether. In Gosplan, people were still working, but once no one was jailed anymore for disobeying orders, the orders were no longer obeyed.

"Imagine you are the director of a state farm. You are told to surrender your grain for worthless money, which is incapable of buying anything. Previously, you knew that if you didn't obey, in the best of cases you would be dismissed from your job. In the worst of cases, you would be thrown in jail. Today, you know that the law says that you can't be fired and you won't be jailed. Are you going to surrender your grain? Of course not.

"This situation is very similar to the one that existed in Russia in 1918. There are only two ways out of the situation. The first is to start shooting people again, to requisition the grain and to throw people into prison if they don't agree. The other alternative is immediately and without any hesitation to create the necessary conditions for money to become valuable again."

In the winter of 1991–92, the shops were empty. People were hoarding supplies. "I knew what was going on with my grain supplies,"

Gaidar says. "I knew how many railcars of grain I had, how many reserves I had. I knew that at best, assuming complete freedom of movement and lowered consumption rates, we had enough grain to last us until the middle of February."

Starving to death and freezing in winter—the two great phantoms of the Russian imagination. Perhaps Gaidar could not be blamed for panicking. "There is no time for discussions," he remembers thinking. "People will begin to die of hunger."[4]

Gaidar knew that freeing prices with the ruble overhang unsolved would unleash hyperinflation. But on January 2, 1992, all prices save for those of a few strategic goods were freed, and they immediately skyrocketed. Shopkeepers scrambled to add zeros to price tags. Shoppers wandered around in a daze. By the end of the year, the price of eggs had increased 1,900 percent, soap—3,100 percent, tobacco—3,600 percent, bread—4,300 percent, milk—4,800 percent. Savings accounts, meanwhile, yielded only a few percent interest, and salaries increased only slightly. The great mass of savings accumulated by Russians over a generation was wiped out.[5]

Gaidar's "shock therapy," as the wags would soon note, was "all shock and no therapy." Russia's gross domestic product plummeted 19 percent in 1992, another 9 percent in 1993, a further 13 percent in 1994, and so on for most of the 1990s. By the end, a huge superpower had been reduced to the status of an impoverished Third World country.

One of the clearest measures of the extent to which Russia's market transition was failing was the relentless decline of the ruble/dollar exchange rate. The days when one ruble in Gorbachev's Russia was equivalent to roughly one dollar was a distant memory—by the end of 1992 it took 415 rubles to buy a single dollar. By the end of the Yeltsin era, it would take roughly 28,000 rubles to buy one dollar. The Yeltsin government did not entertain the idea of temporary currency controls as China had done so successfully throughout its long economic boom in the 1980s and 1990s. As American supply-side economist Jude Wanniski observed at the time, "Money is the government's non-interest-bearing debt to its own people and the government should honor that debt."[6]

Grigory Yavlinsky argues that Gaidar's government made a huge mistake by freeing prices so quickly. There was no burning need to capitulate to hyperinflation, Yavlinsky notes. "It is true that there was a certain hysteria, but from my point of view there was no threat of famine at that time," he says. "The shops had empty shelves, but they had been empty in all the past Soviet years as well."[7]

The hyperinflation of 1992, combined with the lack of measures to protect the elderly, the poor, and the sick from rising prices, caused a dramatic decline in Russian life expectancy, but Gaidar saw no reasonable alternative to his policy. He would later tell me that he could not privatize apartments or garden plots for cash because the Communist-dominated parliament was against it; it did not make sense to privatize shops, he argued, before prices were freed; and government bonds were impossible because the population did not trust the guarantees of the government.[8]

While price reform was wreaking havoc, other aspects of democratic reform were being ignored. "They paid no attention to the other aspects of building a civilized society: creating an effective government, a workable constitution, enforcing the rule of law, creating a viable parliament," says Yavlinsky. "In other words, they completely failed to create a civil society to serve as the framework for economic activity. Even questions of economic policy that did not directly stem from the need to free prices were ignored: industrial policy, trade policy, de-monopolization, enforcing competition, rationalizing the tax system, establishing a viable banking sector, creating clear rules of the game."[9]

The lack of institutional reform helped doom Gaidar's program. The Soviets had built their industry in a "rational" fashion, avoiding rivalry and duplication among industrial enterprises. When enterprises were owned by the state and subject to the dictates of Gosplan, it made sense not to create competing outfits. For many goods—passenger cars, tampons, jet engines, laundry detergent—there were only one or two large factories serving the Soviet market. After they were freed from government regulation, these enterprises became predatory monopolies, dictating terms to both customers and suppliers.

"The key question of 1991 was which path were we going to choose: Were we going to liberate the old Soviet monopolies or were we going to liberate society from the old Soviet monopolies?" says Yavlinsky. "Are we going to free the Communist nomenklatura from all control, are we going to tell the Communist directors and the Communist nomenklatura: You're free, do whatever you want?"[10]

Alisa

When Yegor Gaidar broke apart the planned economy, freed prices, and abolished Gosplan, chaos ensued. Industrial enterprises did not know whom to ship their products to, how to get paid, where to get their supplies. In this chaos, trading companies like Berezovsky's Logovaz assumed a central role in Russian wholesale and export trade.

In March 1992, I visited the headquarters of Alisa, one of Russia's top new trading companies. It was a large, nondescript building on 45 Lenin Prospekt. Alisa's entrance was a bare steel door in the corner of the building. Inside, two scowling armed men in camouflage checked my documents. After placing a brief phone call, one of the guards led me down in the basement. We passed a series of dingy secretarial offices; cheaply lit corridors and linoleum floors stretched out in an underground labyrinth.

We walked into a windowless den. A green, tasseled lamp hung over a round table surrounded by old armchairs and sofas. An ashtray overflowing with cigarette butts lay on the table. In the corner was a hunting rifle. The room looked more like a poker parlor than a conference room.

Presently my host arrived: German Sterligov, twenty-five years old, a thin, scraggly fellow with matted hair and dirty fingernails. As cofounder of Alisa, he was one of Russia's richest men. He boasted of his exploits and delighted in his scandalous reputation. He pointed to two bullet holes in the wall, explaining that some gunmen had shot up the place a few months earlier.

"We later caught up with them," said Sterligov. "It's like Chicago in the 1920s." Then he lit a cigarette and put on a fedora—a parody of James Cagney. "Do you want to see how my security service works?"

He pressed a button under the table. Silence. Then there were loud noises in the corridor, shouts, running feet. After a painful moment of anticipation, the door exploded as three muscular men in camouflage leapt into the room. They screamed something incomprehensible, pointing the barrels of their automatic pistols at my heart. I froze. Then they looked around. "Everything okay?" they inquired.

Sterligov nodded and the gunmen left, shutting the door behind them. "If it had been a combat alarm, you would have been thrown on the ground first and asked questions later," Sterligov said. "Because you might have had a pistol under the table."

Sterligov explained that his security agency was called Alisa X and contained sixty men, including KGB commandos who had stormed the Kabul presidential palace at the start of the Afghanistan War.[11]

Alisa carved out a niche in the new market by selling strategic materials from the stockpiles of specialty metals that the Soviet Union had hoarded in case of nuclear war. Other Moscow commodities traders pointed out that the two nominal owners of Alisa, German Sterligov and his older brother, Dimitri, were nephews of Aleksandr Sterligov, a major general of the KGB, and now notorious for his fierce, Stalinist public declarations. I asked German Sterligov about the family connection. "Malicious gossip," he snorted. (Over the course of a year, he repeatedly denied any connection to the KGB general, but would later admit to having worked for his uncle all along.)[12]

Sterligov had dropped out of high school, done his military service, and then spent the perestroika years bumming around. He worked at a Moscow auto factory, got bored, went to the Far East to work on the Baikal–Amur Railroad, then floated down to Kazakhstan, where he herded horses for a Chechen entrepreneur. In 1989 he traveled to Nicaragua, Cuba, and the Dominican Republic (where, he claimed, he won and lost $28,000 in a Santo Domingo casino in a single night).[13]

After Sterligov returned to Moscow in early 1990, he and some old friends went to work for Artyom Tarasov, the urbane entrepreneur who had made a fortune by trading Russian raw materials for personal computers; Sterligov was also a partner of the fugitive American commodities trader Marc Rich.

I had first met Tarasov and Sterligov in a small office Alisa was renting on the corner of Wall Street and Broadway in New York. Tarasov had no official role in Alisa—he claimed the company was owned exclusively by the two Sterligov brothers—though it was clear he was in charge. He presided over our interview like a wise uncle, answering the difficult questions himself and softly correcting Sterligov's more impulsive declarations.

Tarasov explained that he first met the Sterligov brothers in the summer of 1990, when they approached him to set up a trading company for selling construction materials. The new company, called Alisa after German Sterligov's sheepdog, was founded in November 1990 with $3 million in capital borrowed from a recently established commercial bank, Stolichny Bank. "My role here was very small—I recommended to the bank that it issue the loan," said Tarasov modestly. "Aleksandr Smolensky [president of Stolichny Bank] had worked a long time with Istok [Tarasov's trading company], he trusted Istok, he had provided Istok with lines of credit."[14]

Stolichny Bank, which Tarasov described at the time as a "half-private, half-state bank," had been established in February 1990 by thirty-six-year-old Aleksandr Smolensky, one of Berezovsky's key business associates. In the 1970s and 1980s, Smolensky, like many of Russia's future multimillionaires, had been an entrepreneur in the black market. According to Russian law-enforcement agencies, he had been convicted several times of economic crimes, including theft and illegal commercial activity, and had spent time in prison. In 1988, at the height of perestroika, Smolensky started a successful construction-materials cooperative in Moscow. In 1990 he opened his bank, which would grow into a financial powerhouse, playing a key role in the 1995 loans-for-shares auctions and Yeltsin's 1996 reelection campaign. Stolichny remained a closely held institution, revealing few operational details to the outside world, but at some point Berezovsky acquired a big stake in Stolichny—25 percent according to some government estimates.[15]

The $3 million invested in the Alisa project was one of Stolichny Bank's first loans. It seemed reasonable to lend money to a company backed by a man like Artyom Tarasov. He was efficient, intelligent, and bursting with good ideas. Since Russia had lost most of its good ports with the collapse of the U.S.S.R., Tarasov was trying to borrow $200 million to establish three new commercial ports: on the Black Sea, on the Pacific, and at the old naval base of Murmansk. He also planned to create a new private airline on the ruins of the former state monopoly, Aeroflot. Yet another project was to establish a new Russian export-import bank in Switzerland, registered in the Cayman Islands. Tarasov also spoke of his desire to buy Angola's old debts to the Soviet Union and then sell them back to the Angolan government in exchange for tourist properties on the African coastline. These were sensible ideas and would almost all be implemented by someone else within a few years.[16]

Tarasov's most important counterparty abroad remained Marc Rich. Since 1990, Rich had accounted for a significant portion of Russia's exports of oil, aluminum, and other commodities. But one year into the Yeltsin regime, his success turned into a political liability for Tarasov. The Russian government had finally begun to examine the fine print of the trading deals that Rich had signed with Russian exporters. In 1992, when I asked Berezovsky's old friend Pyotr Aven, the minister of foreign trade, whether it was true that Rich had bribed enterprise directors and government officials to get his deals done, he said that Rich probably had.

"We have a significant amount of information that such long-term relationships lead to informal and often corrupt relationships and this is not always good," Aven replied evasively. "These are companies that have been working here a long time and they know whom to pay and how much, and consequently they buy their goods cheaper than it costs to produce them. This is no secret. We approach such companies carefully."[17]

Though it was clear that Tarasov was the boss of Alisa, German Sterligov represented himself as the front man. The youngster relished the role of the newly rich Russian businessman: In Moscow he established a "Club of Millionaires" and gave frequent interviews to the newspapers. He boasted that he could not begin to count his wealth. Besides cash stored in one of Alisa's five offshore companies (in the Isle of Mann, Panama, the British Virgin Islands, and elsewhere), Sterligov said that he had bought thousands of tons of strategic metals, 75,000 acres of farmland in central Russia, office space in cities across the country, even a valuable art collection, including paintings by Kandinsky, Malevich, and Chagall.[18]

His wealth quickly evaporated. When I spoke to Sterligov in the summer of 1992, he was tight-lipped, nervous, gloomy. Only once did he conjure up his old bluster—when he boasted of having recently sold nuclear-power equipment to a buyer in Iran. "I don't know what kind of equipment it was. I am not a specialist. It had twelve letters and seventeen numbers." Otherwise, he was full of gloom. "Everything is at a standstill. No one has any rubles. No one is buying or selling anything. There is no trade turnover. This is a catastrophe for the country!"

The real reason for Sterligov's worries was that the authorities were cracking down on his windfall trading profits. I asked Sterligov which government agency I could contact about his affairs. "Go to the KGB," he replied. "The Sixth Directorate [Philip Bobkov's old fiefdom—the department responsible for fighting against organized crime and illegal economic activity]. They have the best information on our affairs."[19]

Within a few months, Alisa was disbanded and its Wall Street office closed. German Sterligov sank into the shadows. Artyom Tarasov abandoned his ambitious projects of building commercial ports and setting up investment banks. Marc Rich, meanwhile, was accused in the Russian press (based on information leaked from the government authorities) of bribery, illegal export of raw materials, aiding in capital flight, even laundering drug dollars. Parliament began an inquiry and Rich found his most ambitious business plans rolled back. Tarasov was never prosecuted for "Harvest 90" or any of the other alleged embezzlement operations; in December 1993, he was elected to the Russian parliament and received legal immunity.[20]

Some of the most famous traders of the early Yeltsin era were not so lucky. Many were assassinated. After surviving at least two murder attempts (a bomb exploding on his doorstep and an ambush by machine gunners on a country road), Konstantin Borovoi, the founder of the Russian Commodities and Raw Materials Exchange, got out of business and entered politics; he, too, was elected to parliament.[21]

"Today there are no entrepreneurs in Russia, in the sense of an honest and normal understanding of the term," declared the jewelry trader Georgi Khatsenkov. "I would compare our economy with Nigeria's: a state-mafioso economy, when state institutions merge with criminal structures, when there are no normal laws. There is no market. There are only corrupt officials, bribes, total anarchy."[22]

Whatever deals commodities traders had struck with the authorities in, say, 1990 collapsed with the fall of the Soviet Union. A new group of men had taken power in the Kremlin—one part of the Communist establishment (around Yeltsin) was pushing another group (the old Soviet government) away from the trough. In the end, the early commodities traders (Marc Rich, Tarasov, the Sterligov brothers, and others) lost out, but not because they were finally hemmed in by the law. They were simply pushed out of the market by their rivals. One of the survivors of the free-for-all that characterized the first years of post-Communist Russia was Logovaz.

Dealing in Ladas

Not long after Boris Yeltsin took power, an American entrepreneur named Page Thompson went to Russia in search of deals with Russian automakers. Thompson was a retired treasurer of Atlantic Richfield

(Arco), the U.S. oil company. With a taste for adventure, he had begun a second career selling auto parts to the former Soviet Union. Success came quickly. He landed a $4 million contract to sell Goodyear tires to Avtovaz, the auto company that was serving as Berezovsky's cash cow. In the course of the negotiations, he asked Avtovaz for a letter of credit from a Western bank guaranteeing its payment. He was told that the French bank Credit Lyonnais would provide it. Credit Lyonnais would later collapse in a famous fraud and embezzlement scandal, but at the time it seemed a strong and respectable institution. Its guarantee was good enough for Thompson.

Then the deal turned strange. Thompson was told that he could pick up the letter of credit not in Paris (where Credit Lyonnais was based), but in Lausanne, Switzerland, where he was supposed to make contact with an outfit called Forus Services S.A. He arrived at the Forus office in Lausanne to find two Russian men lounging around. "They did business in an absolutely empty office, a beautiful office, but it didn't have anything in it, except furniture—just empty desks and empty chairs," Thompson recalls. "There were three or four rooms, a big bottle of Jack Daniel's, and no people except these two guys and a secretary who didn't speak any language I understood."

The letter of credit had not yet arrived. After two days, Thompson was finally presented with the document. It contained the names of Credit Lyonnais, Avtovaz, and Thompson's company, but there was no mention of Forus Services S.A.

"It made quite an impression on me," Thompson says. "Rather than having the financial VP of Avtovaz call up Credit Lyonnais or Chase Manhattan Bank or Deutsche Bank and say, 'We'd like to have a letter of credit for a few million bucks,' they ran that letter of credit through Forus Financial. When I was treasurer of Atlantic Richfield and we wanted to borrow money, I'd pick up the phone and I'd call Chase and I'd say, 'We're interested in borrowing a couple of hundred million dollars,' and they'd say, 'We're interested in lending it to you and we'll come out and talk about it.' I didn't have to call Joe, Mike, and Moe at, you know, the Wilshire Financial Company or something, to do my borrowing for me at Chase."

Thompson concluded that this "strange, unpopulated office" was the site of a kickback scheme. "It was a company that someone powerful at Avtovaz had set up to run these financing transactions, charge Avtovaz a fee, and then split the fee among the owners. I never knew who the people were...but Forus was basically a cat's-paw for some people in Avtovaz."[23]

In fact, Forus Services had been established on February 13, 1992, by Boris Berezovsky (representing Logovaz), Nikolai Glushkov (representing Avtovaz), and the Swiss commodities trader André & Cie. Though all three founders almost certainly owned a large share of the company, the actual ownership structure of Forus was deliberately opaque. The Swiss company Forus Services S.A. was owned by Forus Holding (Luxembourg), which in turn was owned by the holders of bearer shares (including Anros S.A.). The owners of Forus, in other words, were insulated by at least two front companies—in vintage espionage terms these would be called "cutouts" (the intermediaries insulating the spymaster from the spies he runs).[24]

Officially, Forus was a financial company trading currencies and organizing lines of credit and other financial operations for Russian companies abroad. It remained, however, very much an insiders' club. Its main significance was not as a financial enterprise but as a holding company, owning the shares of some of the most important entities of Berezovsky's growing empire. The first institution Forus helped establish was Obyedinyonny Bank, registered in Moscow in 1992. Obyedinyonny would eventually become the primary bank for both Avtovaz and Aeroflot. Despite such large corporate accounts and the impressive political support Berezovsky could muster, however, the bank remained a small, private affair.[25]

Berezovsky and his Swiss partners, André & Cie., went on to create numerous other financial companies, the most important being Avva, Andava, AFK, and FOK. They also established subsidiaries in such tax havens as Cyprus and the Cayman Islands. They became involved with at least one shell company registered in Dublin, Ireland, which in turn was partly owned by shell companies in Panama. It was a sophisticated financial network, geared toward getting the money out of Russia, maximizing financial flows around the world, minimizing taxes, and avoiding scrutiny.[26]

But the key to Berezovsky's empire remained his link to Avtovaz. Amid the disintegrating Russian economy, the auto industry remained remarkably healthy, producing the one domestically made product Russians were still eager to buy. Few foreign automakers could match the low prices charged by the Russian producers. While the destruction of Russians' savings accounts in 1992 meant that there was no longer a ten-year waiting period for Ladas, demand consistently outstripped supply. Blessed with low raw materials costs and incredibly cheap labor (the average worker received $250 a month, usually paid several months late), Avtovaz should have been an enormously profitable enterprise. In fact, it was bleeding cash and piling on debt.[27]

Hundreds of small companies had been created to trade Ladas and spare parts; they were independent, but they were financed by Avtovaz Bank and they were linked to various members of Avtovaz's top management. The giant auto factory became dependent on a dealership network that was widely known as one of the most criminalized segments of the Russian economy.[28]

When I asked Avtovaz president Alexei Nikolaev in the summer of 1996 about the dealership problem, he admitted that the automaker was selling its cars to the dealers at a loss. "On average, we get $3,500 [per car]," Nikolaev explained. "That's the price of an automobile. In reality, this car has a much higher production cost—approximately 30 percent higher [$4,700]."[29]

Since most dealers sold their Ladas for $7,000 or more, they were making 100 percent gross trading profits. Whether they were Moscow gangster organizations or local racketeers outside the Avtovaz factory gates, dealers were taking the cars as they rolled off the assembly line and walking off with half of Avtovaz's sales revenues. If you were an independent dealer trying to purchase cars from Avtovaz without going through the established gangster channels, you were unlikely to get any cars, or, if you did, you would receive your vehicles with windshields smashed, wiring pulled out, and tires slashed. Or you would get shot.[30]

"You can't just go over and become a dealer in Ladas," Page Thompson recalls. "If the other people who deal in Ladas allow you to become one, you're going to pay for the privilege. Some guy's going to show up and tell you that you have a partner."

Thompson cites the example of one of the biggest Avtovaz distributors in Moscow, an outfit called Lada Strong. "He had his cars stored in two lots and he had to pay one criminal gang tribute for Lot A and another criminal gang tribute for Lot B," Thompson says. "Through some gaffe, one of his employees moved fifty cars from Lot A to Lot B, whereupon the guys who were collecting a tariff from Lot A kidnapped him and held him hostage in a cellar until the distributor came up with $50,000 for the insult."

One dealership was a big outfit run by a young Russian businessman right outside Avtovaz's gates in Togliatti. "He had a whole empire over there in Togliatti—he was a huge dealer in Ladas and parts and all sorts of hard-to-get stuff. To visit his lair, you came in and there was a whole bunch of tough-looking guys lying around watching Woody Woodpecker cartoons on TV and bristling with firearms. And everywhere he went, there was a car with four armed guys in back of him."

Thompson went into business with this man, selling him secondhand American cars for shipment to Russia. "He was just a nothing kind of guy," Thompson says, laughing. "He was getting cars directly from Avtovaz in Togliatti, basically on credit, selling them instantly, and making $100,000 a month. He borrowed a million dollars from Avtovaz Bank for spurious purposes and fled the country. He just defrauded and robbed everybody. This is just what the guy told us. He boasted about it. But there was someone inside Avtovaz that fed him cars that should have gone somewhere else."[31]

Thompson says the problem lay in the venality of Avtovaz managers. If you wanted to get a shipment of spare parts, for instance, you had to bribe the manager responsible for spare-parts sales. "I knew the guy who used to take the money," says Thompson.[32]

I asked Alain Mayor, Berezovsky's primary partner at André & Cie., about the rampant corruption at Avtovaz. "I think that is the case with a lot of Russian companies, not just Avtovaz," he said. "It's a question of mentality: Collective property is property that belongs to everyone."[33]

André & Cie. had plenty of opportunities to witness this corruption firsthand. In 1993–94, André worked on arranging a $100 million credit for Avtovaz from the Italian foreign trade bank. They were working with Glushkov, Avtovaz's finance director. The money was supposed to be paid back over seven years with the sale of Ladas in foreign markets like Africa. "It didn't work out," noted Christian Maret, head of André's Moscow office. "Because each manager of Avtovaz had his own pet distribution network."[34]

Page Thompson continued trying to do business with Avtovaz; he did a few successful deals, lost money on others, and finally decided to stop. "I lost a couple partners," he says. "One of my Avtovaz partners in Bishkek, Kirghizstan, was assassinated in his office. My partner in Togliatti was certainly a major criminal who got fired both by Avtovaz, and then by Avtovaz Bank, where they put him for a while to dry out. I've just decided that it ain't a game worth playing over there."[35]

Law-enforcement agents who tried to break the cycle of criminality paralyzing Avtovaz met grisly ends. In 1994, when Radik Yakutian, head of the investigative department of the Samara Region prosecutor's office, began looking into organized-crime activities related to Avtovaz, he was assassinated. The auto company gained the reputation as the most gangster-ridden of any large Russian company. When police finally decided to sweep the automaker in 1997, they identified no fewer than sixty-five contract murders involving Avtovaz managers and dealers.[36]

Berezovsky was subjected to his share of gangland violence (the shoot-out at the Kazakhstan Cinema, the attacks on the dealership lots, the Logovaz car bombing), but he persevered, and emerged as the largest dealer of Avtovaz cars. In 1991, he told the Russian business newspaper *Kommersant* that Logovaz had received its first working capital from a $20 million syndicated loan from six Russian banks. That year he was already selling 10,000 Avtovaz cars annually, many of them for foreign currency, rather than rubles. Within three years, Logovaz would be selling 45,000 Avtovaz cars a year and grossing revenues of nearly $300 million on this operation alone.[37]

When I asked Avtovaz president Alexei Nikolaev in the summer of 1996 about the problem of gangsters controlling his dealer network, he said simply: "The problem exists."

Nikolaev explained how the problem had developed. After the collapse of Communism, Avtovaz had a network of several hundred dealerships and service stations. But these were still government-owned enterprises, forced by law to sell their vehicles at low, fixed prices. The new dealers (first and foremost Berezovsky's Logovaz) faced no such restrictions. "The government effectively drained us of all our working capital," Nikolaev complained. "But this small group of people [the independent dealers] bought our cars at one price and resold them two or three times higher. That way they quickly accumulated their own working capital. That's how Avtovaz financed the creation of an alternative dealer network."[38]

The independent dealers also made money by taking out loans from Avtovaz. The automaker typically sold only 10 percent of its cars for cash; the rest were sold for barter or on credit. The key to making money in Russia in those days was to have capital, which almost no one did. In the car-dealership business, the game was to demand cash up front from car buyers and delay paying the car company for six months or more. This was an incredibly lucrative operation. The dealer would sell his Ladas for hard cash and simply sit on the money; with inflation running at 20 percent a month, the dealer was almost guaranteed a large profit. By delaying payment to Avtovaz for, say, three months, the dealer ended up paying half price for his cars. After the ruble began to stabilize in 1995, dealers could invest their cash in three-month Russian Treasury bills, which had annualized dollar yields of 100 percent or more. By 1995, dealers like Logovaz would owe the automaker $1.2 billion—one third of the company's total sales.[39]

Why did Avtovaz continue to sell to the gangster-dealers who were bankrupting the company? Probably a combination of the stick and the carrot. The stick was the fear of getting murdered; the carrot was payment under the table for Avtovaz managers. "It is not so easy to take back the provisions of a contract," Nikolaev mumbled.[40]

Berezovsky's particular scheme was called reexport. Export contracts typically stipulated an even lower price for Ladas than domestic dealership contracts, and granted an even longer grace period for dealership payments (up to one year). Berezovsky actually sold his cars in Russia, but their "export" status allowed him to receive foreign currency. The cars remained in the country, but their documentation showed them to be exported and then imported back into Russia.[41]

"No, listen, Monsieur, 90 percent of cars in Russia are sold in the same manner," Alain Mayor protested. "There is not much of a difference in favor of Logovaz. There are very different conditions existing as a result of personal relations between people, people who were friends with the managers at Avtovaz."[42]

Berezovsky had the best "personal relationship" imaginable, since Avtovaz's chairman, head of finance, and head of after-sale service owned a substantial portion of Logovaz stock. The relationship went beyond car sales. Berezovsky helped set up several companies in Russia and abroad to rationalize Avtovaz's finances. One of these was the Automobile Finance Corporation (AFK). This outfit was supposedly majority-owned by Avtovaz itself, though it had several important minority shareholders, including Berezovsky companies such as Forus Holding; AFK's general director simultaneously managed the affairs of other Berezovsky financial companies in Moscow. Owning a controlling stake of Avtovaz, AFK served as a vehicle ensuring that the auto company would be controlled only by insiders. Theoretically, it had been established to rationalize Avtovaz's financial flows (half of Avtovaz's cash revenues passed through AFK) and to collect investment capital for new automobile projects. In fact, according to Russian tax authorities, AFK was the lynchpin of an elaborate system allowing Avtovaz to avoid paying taxes. Avtovaz's cash situation continued to deteriorate, no new investment projects materialized, and the car company soon had the largest tax arrears in Russia.[43]

While Avtovaz dealers like Berezovsky were becoming immensely wealthy, the automaker sank deeper into debt. Avtovaz maintained high production levels and was blessed with a good pricing environment (continued strong demand for its cars), but its managers were conniving with crooked car dealers and the factory was hemorrhaging money as a result. A large portion of its sales was made on the basis of bad quality receivables or cumbersome barter arrangements. The company was so short of cash that it could not pay its taxes, electrical bills, or workers' salaries. The only thing preventing the Yeltsin government from declaring the auto plant bankrupt was the stigma of having to admit the failure of such a key Russian industrial enterprise.[44]

"We Dismantled Everything"

One of the most egregious mistakes of the Gaidar team was the hasty dismantling of the state monopoly on foreign trade. Here was the origin of Russia's great robber-capitalist fortunes. Gaidar's price reform laid the foundations for the big private fortunes by destroying the old planned economy, but Russia's new tycoons would never have gotten so rich if they hadn't been able to exploit the breakdown of the country's foreign trade system.

Gaidar, Pyotr Aven, Anatoly Chubais, and the other "young reformers" were acquainted with the reforms of czarist Russia at the beginning of the twentieth century. Back then, the most gifted czarist ministers, Sergei Witte and Pyotr Stolypin, also tried to establish a capitalist economy and the rule of law under very difficult circumstances. They succeeded in achieving tempestuous economic growth. By contrast, the policies of Yeltsin's team of young reformers were both primitive and destructive.

The old foreign trade network was one aspect of the Soviet

system that was not broken. It was a highly efficient system, pouring hard currency into Moscow's coffers. A large portion of the Soviet government's revenues came from capturing the difference between the domestic price of commodities and the price at which they were exported; the Ministry of Foreign Trade was in charge of procuring the funds. "[During the Soviet era] the Ministry of Foreign Trade was simply a big corporation," veteran foreign trade officer Oleg Davydov would later tell me. "It was an effective business organization, with representative offices and trading companies all over the world. They worked on commission—0.5 percent of sales."[45]

The idea of the state's withdrawing from the export-import business and abolishing customs duties may seem sensible, but for Russia in 1991–92 this measure was disastrous. The country's official exports dropped 40 percent in two years; the government suffered an even sharper drop in revenues it traditionally received from export and import tariffs. Pyotr Aven, the thirty-something economist who presided over Russia's trade policy in 1992, was quick to blame such external factors as falling oil prices in the late 1980s and collapsing metals prices in the early 1990s.[46]

In fact, the problem was that Gaidar's price reform was not a complete liberalization—a shortcoming that his defenders say lay at the root of Russia's subsequent economic decline. The government continued to control the prices of Russia's most exportable commodities; the domestic Russian price for oil and gas, as well as aluminum and other metals, timber, coal, and fertilizer was set at a fraction of the level prevailing in world commodities markets. Gaidar says the failure to liberalize commodities prices was a "mistake" and was forced upon him by political pressure from the conservatives. The country functioned on a double price system with respect to its key export commodities—one price on the world markets and another price (much lower) on the domestic market. This was a license for private traders like Berezovsky's Logovaz to make enormous profits.

Yet, urged on by officials of the International Monetary Fund and other Western government advisers, Yeltsin's young reformers decided that the state should get out of foreign trade altogether.

They eliminated the barriers preventing private traders from buying commodities at domestic prices and selling them abroad. Within months, 30 percent of Russia's oil exports and more than 70 percent of the metal exports were bypassing state trading organizations. By 1994, the majority of Russia's foreign trade was being handled by private trading companies.[47]

"We dismantled everything," remembers Davydov. "We began liberalization in the absence of any control at all. And when foreign entrepreneurs, mostly crooked ones, arrived and taught our entrepreneurs how to do it, you were no longer talking about a 1 percent [commission on the trade], but dozens of percentage points."[48]

The first to benefit from the new regime were the directors of the main export-oriented enterprises. They simply established trading companies abroad and sold them their produce for a fraction of the world price; the profit rarely returned to either the enterprise or the Russian state.

Typically, the scheme worked like this. Ivan is the director of an oil company called National Oil. Ivan also personally controls an oil-trading company called Volga Trading. Volga buys a shipment of crude oil from National Oil; it pays with an IOU. Volga sells the oil on the market for cash. But instead of using the cash to pay National Oil, Volga deposits the money in Ivan's personal bank account in Switzerland. Putting on his best conscientious face, Ivan then goes to Moscow and asks the Ministry of Finance or the Central Bank to give National Oil soft credits to alleviate its cash-flow shortage.

Immediately after the fall of Communism, Berezovsky emerged as a big trader in a wide variety of commodities. In the first half of 1992, according to a petroleum-export yearbook based on Russian trade registrations, Logovaz exported 236,000 tons of crude oil (sold to counterparties in Switzerland and the United States), 95,000 cubic meters of sawn timber (sold to a German company), and an extraordinary 840,000 tons of aluminum (sold to a Hungarian trading company)—trades that had a gross value of about $1 billion.[49]

While the aluminum trades were the most valuable, oil interested Berezovsky more. In 1992 he helped found the International

Business Club of Oil Industrialists. That same year, he created a joint venture that included Logovaz, oil producer Samaraneftegaz (then still known as Kuibyshevneftegaz), the Samara provincial government, and a small Oklahoma-based oil company called GHK Corp. He called it a "financial mechanism" for the export of oil from Samaraneftegaz, with part of the money going to purchase equipment from GHK.[50]

Berezovsky was on home ground. Samara's most important manufacturing enterprise was Avtovaz, the source of the Logovaz cardealership business. The relationship between Berezovsky and Samaraneftegaz was close. He made the oil company one of the shareholders of his Avva investment scheme in late 1993. At the same time, his Swiss financial company, Forus, tried to arrange a $60 million U.S. Eximbank loan for the oil company. The loan fell through after the Americans uncovered evidence that Samaraneftegaz's managers were siphoning off the company's export earnings. Berezovsky, however, continued doing business with the oil producer for at least three years.[51]

One of the local off-takers of Samaraneftegaz crude was the Samara Refinery. Since early 1992 the Samara Refinery had been exporting its products through a trading company called Nefsam, a joint venture between the refinery and a Belgian-based company called Tetraplast, itself owned by a shady entrepreneur linked to Otarik, the famous gangster. Virtually none of the export revenues returned to the refinery. In 1993, by presidential decree, the Samara Refinery began to be integrated into a newly formed holding company called Yukos. The Yukos managers were eager to gain control of the refinery's exports, but the trading company resisted. In October 1993, Vladimir Zenkin, general manager of the refinery and a Yukos vice president, was murdered outside his home. Three months after Zenkin's murder, the head of Tetraplast was critically wounded when his Mercedes was shot up on the streets of Moscow. Shortly after that, another executive of Tetraplast was killed. The takeover of the Samara Refinery proceeded fairly quietly after that.[52]

To avoid paying taxes, most Russian traders resorted to an old

KGB trick—false invoicing of imports and exports. High-quality Russian timber, for instance, was registered as low-quality timber and exported at reduced prices, with the foreign buyers paying the extra amount into the foreign bank accounts of Russian government officials or traders who permitted the transaction. Aluminum, steel, nickel, precious metals, furs, and fish were exported in a similar fashion. Enormous profits were made importing food, clothes, consumer electronics, and industrial equipment at inflated prices, with kickbacks going into the foreign bank accounts of the Russian buyers.

The biggest Russian enterprises became entangled in crooked export deals. Davydov, for example, points to the sales operations of Magnitogorsk, built during Stalin's First Five-Year Plan as the largest steel mill in the world. "Why did Magnitogorsk sell its steel for $20 a ton or $30 a ton, when it used to sell it for $110 a ton? It could have easily sold it for $100 and if [the trader] had been a professional and had known the market, he could have sold it for $111.20. But this is a question of personal interests." Most of Magnitogorsk's steel exports, Davydov explains, were handled by a Canadian company owned by the son of the steel mill's general director.[53]

The new foreign trade companies tended to hide most of their profits abroad. Capital flight from Russia during these years was estimated at $15 billion to $20 billion a year, as crime bosses, corrupt officials, and factory directors set up bank accounts in Switzerland, Luxembourg, Austria, Germany, England, Israel, the United States, and the Caribbean.[54]

"[The government] didn't take any tax payments from them at this time," says liberal parliamentarian Grigory Yavlinsky. "At that time, taxes were an abstract conception."[55]

Even a cursory glance at most Russian foreign trade contracts would have revealed a pattern of fraud, embezzlement, and tax evasion. If a Russian oil producer shipped crude directly to a refinery in Germany, for example, the payment never came back along the same path. Rather than paying the producer directly, the German refinery typically paid an obscure trading company located in some offshore tax haven.

"We have no idea how much money passes through without

paying taxes," said Berezovsky's old friend, Foreign Trade Minister Pyotr Aven. (In December 1992, amid accusations that Russia's export earnings were being shamelessly looted, Aven resigned from the government and joined the Alfa Group, one of the trading companies he had been regulating.)[56]

"At that time [1992–94] there was a problem with tax evasion," admits Yegor Gaidar. Then he adds reassuringly: "But tax arrears were not a serious problem." Of course Russia's traders had no incentive to declare their earnings to the tax authorities. Gaidar acted as if this were a distasteful but entirely predictable phenomenon. "Naturally, as everywhere else in the world, enterprises try to minimize their tax liabilities and use various methods to accomplish this," he explains.[57]

The government went to great lengths to make it easier for the new traders to shelter their windfall profits. For example, Russia had an old tax treaty with Cyprus, allowing Cyprus-registered companies to take profits out of Russia without paying the 20 percent withholding tax (though these companies did have to pay a 4 percent corporate income tax to the Cypriot authorities). The island had been one of the KGB's preferred places to launder money. Within months after the fall of Communism, Cyprus became home to hundreds of trading and financial companies doing business in Russia. Dozens of big Russian oil-trading companies, for instance, reregistered as joint ventures with a Cyprus corporate entity. Despite the Cyprus arrangement's notorious reputation as the primary channel for capital flight out of Russia, neither the Yeltsin government nor the parliament altered the tax treaty.[58]

The biggest trading profits were made in oil, Russia's single biggest export. The state-owned oil producers and Russian tax authorities were lucky if they recouped half of the value of the oil they shipped abroad. The rest was siphoned off by the traders. "And on that basis [the traders] bought up the whole government apparatus," notes Oleg Davydov.[59]

Officials in both the Ministry of Foreign Trade and the Ministry of Fuel and Energy complained that they were facing a wave of bribes and threats to pressure them into issuing export licenses. Russia's organized-crime groups were determined to get in on the action. Their first vehicles were the Special Export firms, trading companies with a license to export such strategic materials as oil and metals. The Special Export firms increased rapidly in number and, by the end of 1993, accounted for half Russia's oil exports. At this point, Oleg Davydov began to reduce the number of Special Exporters, leaving only the big oil producers and the more experienced traders. (Logovaz was among the Special Exporters whose licenses were pulled.)[60]

"We were under tremendous pressure, of course, considering that most of these organizations were representatives of criminal organizations," Davydov recalls.[61]

Davydov's reforms were too late: The oil industry was already one of the main battlegrounds of Russian organized crime. The Special Export firms depended on the collusion of the oil companies' Soviet-era managers, and they killed managers who refused to work with them. The head of the Tuapse refinery (on the Black Sea), for example, was assassinated in 1993. That same year, the head of the Kirishi refinery (in the north) was badly beaten up. A year later, the president of Megionneftegaz, one of Russia's largest oil producers, was killed.[62]

"We Must Tighten Our Belts"

On December 14, 1992, with Gaidar's reforms failing, Boris Yeltsin named a new prime minister: Viktor Chernomyrdin, the former chief of the oil and gas industry. Chernomyrdin was an older man and boasted a career as a successful Soviet industrial manager. He brought with him a new team: older men proudly calling themselves "industrialists" and "government patriots." The change made little difference to government policy.

I went to the former offices of the Communist Party Central Committee on Old Square to meet with one of the key members of the "new" Chernomyrdin team, Yevgeny Yasin. The old economist was one of the "veterans" brought in to correct the mistakes made by the "young reformers." Within months, Yasin was appointed Yeltsin's main economic adviser. I had thought that Yasin, as an older man and a representative of the more conservative part of the Russian political establishment, would be anxious to address the economic mistakes of his predecessors. I was wrong.

"There are no miracles," Yasin began by telling me. "This country must drink the cup to the bitter end." He spoke of using the confiscatory character of inflation to realign the economic balance in society. "Over the near future—at least a year—we will live under conditions of inflation and we must focus on those problems that inflation allows us to solve—to establish a more rational relationship between prices, a different relationship between prices and incomes."

In other words, Yasin was suggesting dramatically lowering the real wages of the average Russian citizen; meanwhile, inflation would destroy the remaining savings of the population as a source of domestic capital. In the absence of significant foreign investment, where was Russia going to get capital to feed the economy?

"There is only one method—to tighten our belts," said Yasin. "We must lower our living standards." The term "tighten our belts" resonated with the popular sacrifices made by the Russian people during World War II. But this time there would be no victory—only impoverishment and early death for those pensioners whose savings would be destroyed by inflation.[63]

Later, Grigory Yavlinsky would remember being struck by how little reformers such as Yasin or Gaidar seemed to care about the Russian people. Fundamentally, Yavlinsky argued, the people who guided Russia under the Yeltsin regime were both heartless and cruel.

"[Gaidar and his colleagues believed] that Russia was populated by *sovki* (rotten Soviets), that everything that exists in Russia should be wiped out and that only then can you build something new," Yavlinsky fumed. "Any methods or means are all right. So let inflation destroy everything—that's no problem at all. Because, in any case, all that stuff is dead, unnecessary. That's the way Gaidar spoke—'The scientific establishment can wait! The northern regions are unnecessary for us! The older generation is guilty...' The paradox of those years is that they were building capitalism using purely Bolshevik methods. A Bolshevik is a man for whom the aim is important but the means are not."[64]

Many members of the Yeltsin government often spoke about their country with such icy detachment that you thought they were describing a foreign land. "The Japanese and the Germans [after World War II] had it easier, because their industry was destroyed, they were living under an occupation regime and already much had been done to clear the ground and allow them to begin anew," Yevgeny Yasin told me. "Unfortunately, Russia doesn't find itself in such a situation."[65]

The Death of a Nation

The result of Gaidar's hasty liberalization of prices meant that more than 100 million people who had achieved some kind of basic material prosperity under the Soviets were plunged into poverty. Schoolteachers, doctors, physicists, lab technicians, engineers, army officers, steelworkers, coal miners, carpenters, accountants, telephone receptionists, farmers—all had been wiped out. The crash liberalization of trade, meanwhile, allowed Russia's natural-resource wealth to be looted by insiders. The Russian state was deprived of its biggest revenue source; consequently it had no money for pensions, workers' salaries, law enforcement, the military, hospitals, education, and culture. Gaidar's shock therapy set in motion a relentless decline—economic, social, and demographic—that would last until the end of the Yeltsin era.

While the rest of the developed world continued to grow, the Russian economy was shrinking. In the Gorbachev era, the Soviet Union had been the world's third-largest economy (after the United States and Japan). Naturally, the Russian economy alone would be significantly smaller than that of the former Soviet Union. But the real decline occurred after the Soviet Union broke up. From the beginning of Gaidar's shock therapy, Russia's gross domestic product shrank by approximately 50 percent in just four years. Eventually, Russia would sink below the level of China, India, Indonesia, Brazil, and Mexico. On a per capita basis, Russia would become poorer than Peru. Decades of technological achievement were lost. Renowned scientific institutions fell apart. The Russian cultural establishment disintegrated. And the country's assets were sold off.[66]

Anyone who traveled to Russia in the early Yeltsin years was treated to the spectacle of ordinary Russian citizens trying to get by. Outside the ramshackle, hollow concrete structures that were the Soviet Union's supermarkets, new private bazaars formed, which included not just brawny *babushkas* selling vegetables, but also little huts offering bad-quality imported goods: CDs of disco music, fake Nikes, Marlboros, cans of Vietnamese pork. These bazaars sprawled out in the mud and the garbage at subway stations, along the big avenues, in populated squares.

On Stoleshnikov Lane, near the legendary Moscow Art Theater, around the corner from the Bolshoi Theater, elderly men and women gathered daily and formed two parallel lines along what de facto had become a pedestrian street. Anyone who ran the gauntlet of these pensioners, neatly dressed in their tattered clothing, was besieged with silent pleas to buy a teakettle, a pair of knitted stockings, three wineglasses, a secondhand sweater, a used pair of leather shoes. Meanwhile, beautiful antique volumes began piling up in the bookstores, selling for ridiculously low prices—Moscow's intellectuals were selling their libraries. In the flea markets outside the city, you could buy the highest Soviet battle decorations, the equivalent of the Victoria Cross or the Congressional Medal of Honor: the old veterans of World War II were selling their medals to buy a few scraps for the dinner table.

With Russia in a slump far worse than the Great Depression, people tapped an old survival instinct. Amid rumors of crop failure and impending food shortages, millions of city dwellers traveled to the countryside to plant cabbages and potatoes in their garden plots. The arable land just outside Moscow was swarming with people digging and planting. It was back to medieval agriculture. Chubais and Gaidar were proud of the fact that mass starvation had been avoided. But it was avoided not because prices had been liberalized, but because the Russian people had returned to the countryside. It was with a shovel and a sack of seed potatoes that Russians escaped starvation in 1992 and 1993.

Any doubts about the first years of the Yeltsin era's being a disaster were dispelled by the demographic statistics. These numbers, even in their most general form, suggested a catastrophe without precedent in modern history—the only parallel was with countries destroyed by war, genocide, or famine.

Between 1990 and 1994, male mortality rates rose 53 percent, female mortality rates 27 percent. Male life expectancy plunged from an already low level of sixty-four years in 1990 to fifty-eight in 1994; men in Egypt, Indonesia, or Paraguay could now expect longer lives than men in Russia. In the same brief period, female life expectancy fell from seventy-four to seventy-one. The world had seldom seen such a decline in peacetime.[67]

Each month thousands of Russians were dying prematurely. Such a drop in life expectancy, labeled "excess deaths," has always been a standard algorithm in demographers' calculations of the death toll of the great disasters—whether Stalin's collectivization in the 1930s, Pol Pot's rule in Cambodia in the 1970s, or the famine in Ethiopia in the 1980s. American demographer Nicholas Eberstadt estimated that the number of "excess deaths" in Russia between 1992 and 1998 was as high as 3 million. By contrast, Eberstadt observed, Russia's losses in World War I were 1.7 million deaths.[68]

Many premature deaths occurred among the elderly—the *babushkas*, church ladies, and old men—people who had seen their life savings disappear in the great inflation of 1992, who had seen their pension checks turn worthless, who did not have families to support them, and who simply could not scrape together enough money for a nutritious diet or medicine. The stress of finding themselves in the ferocious and unknown world that emerged after Communism was also a major (though unquantifiable) factor in killing off the elderly. It was a frightening experience for them—coming in the twilight of their lives, when they were weak and slow—the feeling of seeing the world turn upside down, the streets become unfamiliar, all the comforting supports of life swept away. Many hung on for a while, wandering around town; the men became drunks sprawled in the icy gutter; the women became bone-thin ladies begging at the entrance of churches; then they died. The younger generation had turned its back on its elders and allowed them to perish.

A more visible factor in the rise of mortality rates was the disintegration of Russia's public health system. Hospitals were unsanitary, underfunded, underequipped, bereft of medicine. Suddenly Russia was suffering outbreaks of diseases associated with the most impoverished regions of the Third World: diphtheria, typhus, cholera, and typhoid.

Tuberculosis, the great killer of the Industrial Revolution, was largely wiped out in the twentieth century with the advent of antibiotics and better public hygiene. But in the 1990s, Russia found itself with hundreds of thousands of active TB cases and even more dormant cases. The most worrying aspect of this phenomenon was the appearance of drug-resistant TB—a highly infectious strain of the bacterium resistant to any known antibiotic. The breeding ground of this scourge was the prison system—active TB afflicted up to 10 percent of Russia's huge prison population. Under conditions of overcrowded cells and minimal medical treatment, the disease spread rapidly and was transmitted further into the general population. Each year some 300,000 people (mostly young men) entered the prison system, while a slightly smaller number of convicts were released upon completion of their term. According to two researchers studying Russia's problem, Dr. Alexander Goldfarb of New York's Public Health Research Institute and Mercedes Becerra of the Harvard Medical School, Russia's prisons were releasing 30,000 cases of active TB into society and 300,000 carriers of the dormant bacterium every year. If nothing was done to address the problem, Goldfarb declared, the number of TB cases would continue to double every year, reaching 16 million by 2005 (11 percent of the population).[69]

If the living conditions were appalling for the 1 million young men in Russia's prisons, they were hardly any better for the 1.5 million in the armed forces. Every year, 2,000 to 3,000 young conscripts perished—either by suicide, murder, accident, or hazing incidents. (The precise number of these kinds of deaths was not released by the army.)[70]

The Yeltsin era witnessed an explosion of sexually transmitted diseases. Between 1990 and 1996, new syphilis cases identified every year skyrocketed from 7,900 to 388,200. AIDS was virtually unknown in Russia in the years before Communism fell. Since then, fed by burgeoning intravenous drug use and rampant, unprotected sex, AIDS spread with geometric rapidity through the Russian population. The government had no idea of the precise number of people afflicted, but based on the growth of visible AIDS cases, Dr. Vadim Pokrovsky, the nation's leading epidemiologist, estimated that Russia would have 10 million people infected with HIV by 2005 (almost all between fifteen and twenty-nine).[71]

A significant portion of the increase in mortality rates in Russia was due to lifestyle choices: an unhealthy diet, heavy smoking, and perhaps the highest rate of alcohol consumption in the world. Drug addiction took an increasing toll. Initially, post-Communist Russia had served only as a transshipment point for opium and heroin from Southeast Asia or Central Asia to the West. Soon the drugs began to appear in Russia itself. By 1997, Russia's domestic market had ballooned into one of the largest narcotics markets in the world. According to official estimates, Russia had 2 million to 5 million drug addicts (3 percent of the population). These were mostly young men and women.[72]

For the older generation, the poison of choice was alcohol. It was impossible to tell just how much alcohol was consumed in Russia, since so much of the vodka was produced in bootleg distilleries. One 1993 survey found that more than 80 percent of Russian men were drinkers and that their average consumption was more than half a liter of alcohol a day. In 1996, more than 35,000 Russians died of alcohol poisoning, compared with several hundred such deaths that same year in the United States.[73]

Heavy drinking and crime contributed to a spectacular rise in violent and accidental deaths—the single fastest-growing "cause of death" category. Between 1992 and 1997, 229,000 Russians committed suicide, 159,000 died of poisoning while consuming cheap vodka, 67,000 drowned (usually the result of drunkenness), and 169,000 were murdered.[74]

While Russians were dying in increasing numbers, fewer children were being born. In the late 1990s, there were 3 million state-funded abortions each year—nearly three times the number of live births. (The total number of abortions was even higher.) Abortions had long been used by Soviet women as the primary method of birth control. The average Russian woman had three or four abortions; many women had ten or more. As a result of these multiple abortions, as well as drug addiction, alcoholism, and sexually transmitted diseases, one third of Russian adults were estimated to be infertile by the late 1990s.[75]

Many young Russian women abandoned motherhood not by choice but by coercion. Several million Russian women were forced into prostitution; of these, several hundred thousand became sex slaves abroad.[76]

The rapid decline in births, combined with an even faster growth in mortality rates, produced a relentless decline in Russia's population. In 1992, the Russian population was 148.3 million. By 1999, the population had fallen by 2.7 million people. If it had not been for the immigrants coming into Russia from the even more desperate situations in Ukraine, the Caucasus, and Central Asia, the Russian population would have shrunk by nearly 6 million between 1992 and 1999. These figures did not include the millions of Russians (mostly the healthier, more enterprising members of the younger generation) who had emigrated to Europe and North America unofficially.[77]

The most pitiful victims of Russia's social and economic decline were the children. In 1992, 1.6 million children were born in Russia; that same year, 67,286 children (4 percent of all births) were abandoned by their parents. By 1997, the breakdown of parenting had grown to catastrophic levels. That year 1.3 million children were born, but 113,000 children (equivalent to 9 percent of all newborns) were abandoned. Russia had no real program of adoption or foster care, so most of these children ended up on the street. According to some Western aid agencies, there were more than 1 million abandoned children wandering Russia's cities by the end of the 1990s. The rest ended up in the vast state orphanage network. Here they were often left in dark, overcrowded wards, haunted by malnutrition, insufficient medical care, and routine abuse by the staff and the older orphans. At least 30,000 Russian orphans were confined to psychoneurological *internaty* for "incurable" children; an easily reversed speech defect such as a cleft palate was enough to get a child classified as an "imbecile" and locked up in an institution where he or she would essentially be left to die. It didn't need to be this way—95 percent of Russia's orphans still had a living parent.[78]

When I first went to Togliatti to interview the directors of Avtovaz, I decided to take the train from Moscow. The journey would last twenty-four hours, but I usually liked traveling by train in Russia—rumbling through countryside in those 1930s-era railcars was one of the best ways to meet people.

In the carriage of my Togliatti train was a mother with an ailing seven-year-old child. It was hot. The boy was stripped to his underwear. He was covered with sores—he had a wiry, blistered little body. His mother was evidently taking him home after an unsuccessful attempt to get him treated for some skin disease. The boy was in agony. He kept wanting to scratch himself. He was crying. His mother applied plasters to the worst of the sores. "Mama...Mama... it hurts," he called out.

The boy's suffering continued throughout the night, his cries echoing through the darkened railroad carriage. The next morning the passengers seemed more silent and subdued than usual; there was a palpable sense of people trying to harden themselves against the child's suffering. The boy finally fell asleep in midmorning. I saw his mother sitting in the corridor alone, gazing blankly at the passing Russian landscape.

CHAPTER FOUR

SELLING THE COUNTRY
FOR VOUCHERS

A Friend of the Family

In the winter of 1993–94, most Russians had never heard of Boris Berezovsky. He had already made a substantial fortune selling Avtovaz cars and trading commodities, but he was only one of a score of new business magnates—and he was by no means the wealthiest. The breakthrough occurred when the car dealer became a friend of the Yeltsin family and an intimate member of the presidential circle.

One of the key members of Yeltsin's entourage at that time was the presidential security chief, General Aleksandr Korzhakov. As Yeltsin's drinking buddy and ever-present companion, he helped determine the pecking order among the president's cronies and influenced the appointment of top government officials. I first met Korzhakov in the hallways of the parliament in the summer of 1997. His days of adventure were largely behind him. A large man, puffed up with muscle and fat, he lumbered down the corridor. His attitude was friendly, straightforward. His small eyes sparkled merrily as he recounted his experiences with the Russian president; they narrowed with suspicion when I asked questions that were too searching. One could see why so many people who met Korzhakov could not help but like him. He had the aura of a high-school coach—inspiring, brave, loyal, a pleasure to be with.

Korzhakov was a career KGB man, but he was not a spy; he had spent nearly two decades in the Ninth Chief Directorate, which was responsible for the protection of important people and places (like the Secret Service in the United States). He met Yeltsin in December 1985, when he began serving as the future president's bodyguard. He became one of Yeltsin's closest and most trusted companions—Korzhakov even stuck by his boss when Yelstin was expelled from the Communist Party leadership and went into political exile. After Yeltsin took power, he asked Korzhakov to create the Presidential Security Service (SBP). Yeltsin did not trust the old KGB. He wanted to create (in his own words) "a mini-KGB—all-knowing, powerful, and staffed only by loyal people."[1]

As a national leader, Yeltsin had some unique talents. He was an effective speaker—he stood erect, marshaling his impressive bulk, his deep voice booming over the audience. His delivery was a bit bumbling and slow, but this only made him seem more like a regular guy—a strong man who had put away too much vodka the night before.

But after his courageous stand during the Putsch of 1991, Yeltsin seemed to relax—it was almost as if he considered his main historical task accomplished. In managing affairs of state he was lazy and impatient; he quickly lost touch with events around him. He typically took a nap after lunch. His working day usually ended early. He was asleep by 10 p.m. but suffered from insomnia. Often he woke at about 2 a.m., complaining of aches and pains, and stayed up for hours during the night. But unlike that other insomniac boss of the Kremlin, Joseph Stalin, Yeltsin did not work much in the middle of the night. He just puttered about. Korzhakov soon noticed that the Russian president was prone to long bouts of depression. In the spring of 1993, when the newspapers were full of revelations of corruption among Yeltsin's closest associates, Korzhakov was worried that the president would try to commit suicide.[2]

Yeltsin was a heavy drinker. This problem often led to embarrassing diplomatic incidents, such as his drunkenness at a lunch with Bill Clinton in Washington, D.C., in September 1993. But nothing better illustrated the pitiable state to which he had sunk than a humiliating occurrence on August 31, 1994, during a ceremony marking the departure of the last Russian occupation troops from Germany. The occasion was deeply symbolic. The Russian army, which fifty years earlier had defeated Hitler and conquered half of Europe, was finally going home. But with their homeland in a shambles and dependent on foreign charity, the Russians were leaving not with the aura of victors, but with the stigma of the vanquished.[3]

Yeltsin was in Berlin to mark the occasion with the German chancellor, Helmut Kohl. The Russian president had started drinking early in the morning and by noon his security chief, General Korzhakov, knew he was in trouble. "Everybody was waiting for the ceremony to begin," Korzhakov would later recall. "Kohl immediately understood the [inebriated] condition of Boris Nikolaevich and hugged him in a friendly manner. Then the chancellor threw me a knowing glance. With my eyes I implored him to help our president, at the very least to support him in the literal sense of the word. Kohl understood everything: Gently holding Boris Nikolaevich up by the waist, he made his way with him to the ceremony."[4]

After a drunken lunch, Yeltsin reviewed the assembled troops and crowds. Walking past the Police Orchestra of Berlin, the president impulsively jumped on the rostrum, snatched the conductor's baton, and started waving it around, embarrassingly at odds with the music. After the music stopped, he launched into a drunken version of "Kalinka." The Berlin crowds had never seen anything like it.

Once Korzhakov realized how debilitating alcohol was for Yeltsin, he tried to limit his boss's intake by ordering that no spirits be kept in the presidential kitchen and, if necessary, giving him specially prepared bottles of diluted vodka. But the president resorted to crafty ways of getting his booze. "If Yeltsin really wanted a drink, he would invite one of his trusted friends for a meeting," Korzhakov recalls. "The meetings with [Prime Minister] Chernomyrdin, for instance, always ended with a drink for Yeltsin. But sometimes the president would call in one of the aides in the anteroom (never failing to pick out the weakest-willed one) and order him: 'Go and buy me one.' The aide would immediately run to me and say: 'Aleksandr Vasilievich, what am I to do? Boris Nikolaevich gave me a hundred dollars and requested that I bring him a bottle.' "[5]

Gradually, Yeltsin and his top officials became a single, tight-knit clan—they lived together, worked together, played and drank together. It was a court, in the traditional monarchist sense of the word. The picture of Yeltsin that emerges from General Korzhakov's memoirs (a huge best-seller in Russia) resembles that of the Roman emperor Nero, presiding over a corrupt and deadly imperial palace. Yeltsin and his entourage spent enormous amounts of time squabbling over trivial amenities—apartments, dachas, furniture, phone access, cars, and other status symbols.

They lived in a specially constructed apartment building on Osseneya Street. It was an elite residential unit: large rooms, special communications cables, heated floors, saunas. There was bickering over who would receive an apartment there and who would not. Eventually, the list of residents included Yeltsin, Korzhakov, Yeltsin's daughter Tanya Dyachenko, Prime Minister Viktor Chernomyrdin, Defense Minister Pavel Grachev, secret police chief Mikhail Barsukov, the journalist Valentin Yumashev, Yeltsin's tennis coach Shamil Tarpishchev, and the mayor of Moscow, Yuri Luzhkov. They lived together, met each other in the elevators, walked their dogs together. The residents of this building, like the proverbial *nouveaux riches*, spent a lot of time trying to outdo one another's lifesyle.[6]

"Yeltsin was very quickly compromised by all those things that accompany limitless power: flattery, luxury, absolute irresponsibility," concludes Korzhakov. "All the changes promised to the people essentially turned out to be endless changes in the ranks of the top officials. Meanwhile, after every round of sackings and new appointments, the government was filled with people who were less and less likely to pursue the interests of the state."[7]

The Russian economy was disintegrating and on the streets of Moscow the Great Mob War was raging. Yet the Yeltsin government seemed incapable of doing anything to reverse the slide. The root of the government's impotence was its corruption. Even the highest officials, for all the power they wielded, spent as much time worrying about their personal financial security as they did about their service to the state. Typically, a government minister received excellent medical care, a car and driver, a big apartment, and the right to rent a dacha in a prestigious settlement outside Moscow; but the minister's cash salary rarely amounted to more than $500 a month. Naturally, the temptation to earn money on the side was overwhelming.[8]

Oleg Lobov, deputy prime minister and head of Yeltsin's Security Council, was accused of playing a role in arming Aum Shinrikyo, the doomsday cult that tried to kill thousands of people in the Tokyo subway in 1995 by releasing the deadly nerve gas sarin. During the Japanese government's investigation, cult members said that they had received the blueprints for sarin gas from Lobov in return for a $100,000 payment; Lobov was also accused of having sanctioned military training for Aum Shinrikyo members at a secret army base in central Russia. Russian prosecutors questioned Lobov, who denied the accusations. The investigation never went further, though Lobov was quietly dismissed from his government post.[9]

The minister of security (the head of the former KGB), Viktor Barannikov, was another official who was caught. Barannikov acted as one of the main patrons of commodities trader Boris Birstein and his Toronto- and Zurich-based company Seabeco. Birstein was lavish with his rewards. In early 1993, he invited Barannikov's wife and the wife of the first deputy minister of internal affairs on a three-day shopping trip to Switzerland. The two ladies bought $300,000 worth of furs, perfumes, scarves, watches, and so on (all paid for by Seabeco) and carried their loot back to Russia in suitcases. They were fined $2,000 for their twenty pieces of excess luggage, but this bill, too, was paid by Seabeco. Barannikov and the first deputy minister of internal affairs were fired but never prosecuted.[10]

Prime Minister Viktor Chernomyrdin was widely believed to have profited from the privatization of the natural-gas monopoly Gazprom and from innumerable oil and gas export deals. According to one CIA estimate, by 1996 Chernomyrdin had accumulated a personal fortune of $5 billion. Asked by a congressional committee to comment on this allegation, CIA director John Deutch diplomatically answered: "I would be happy to talk to you about this particular point elsewhere."[11]

The "young reformers" were not immune from scandal either. Finance Minister Boris Fyodorov, one of the leading young reformers and the great hope of the West to keep the market reform alive in 1993, signed the permits for Golden Ada. This was a scheme concocted by a young trader named Aleksandr Kozlyonok and an old Yeltsin crony who was head of the Committee on Precious Metals and Gems to embezzle funds of the Russian government and park them abroad. Golden Ada set up trading operations in San Francisco and Antwerp and, over three years, handled an estimated $178 million in diamonds, gold, and antique jewelry, smuggled out of the Russian treasury. Fyodorov's signature gave Golden Ada the green light to begin operations.[12]

An even more important member of the young reformer camp, privatization chief Anatoly Chubais, also earned a lot of money on the side, not always ethically. On a six-month sabbatical from his government responsibilities in 1996, Chubais headed up a charitable foundation funded with a $3 million interest-free loan from Russia's top businessmen. Chubais used the money to invest in government securities. He earned $300,000 in just a few months from these operations, as well as from "lectures and consultations." A year later, he and five close associates in the Russian government—all avowed "democrats" and free-market advocates—received $90,000 each as an advance on a book contract from a subsidiary of Onexim Bank, one of the main beneficiaries of Russia's privatization program. (The book was eventually published two years later, but it was largely ignored by consumers.)[13]

The whirlpool of corruption affected distinguished foreigners as well. When Chubais was guiding Russia's transition to a market economy, he relied on American legal experts, economists, and public-relations consultants. One of the largest American aid projects—a program to help develop Russia's securities markets—was coordinated by Harvard University and funded by U.S. AID. In May 1997,

U.S. AID canceled the last installment of aid earmarked for the program ($14 million) when it learned that the two American consultants managing the program had allegedly used inside connections to profit from the Russian securities markets. The U.S. Justice Department began an investigation, though no charges were brought against the consultants. At the same time, an audit by the Russian Chamber of Accounts revealed serious misappropriations at the Russian Privatization Center, a private, nonprofit organization linked to Chubais and funded by Western aid money; the Chamber of Accounts discovered that much of this money was distributed directly to Chubais's cronies and to key political bosses in return for their support of market reforms.[14]

Although the venality of Yeltsin's ministers was common knowledge, little action was taken against them. Even when the details of top officials' bribe-taking or embezzlement spilled onto the pages of the newspapers, the officials were almost never prosecuted—usually they were simply dismissed and left to enjoy their ill-gotten wealth in private. By contrast, several overzealous prosecutors and anticorruption crusaders were fired. Among the first was Yuri Boldyrev, whom Yeltsin had appointed state inspector in charge of an anticorruption campaign in early 1992; he was fired less than a year later, after uncovering too much dirt about Yeltsin's inner circle. Vice President Aleksandr Rutskoi was also dismissed from his anticorruption post in early 1993. Andrei Makarov, yet another chief of a committee investigating corruption, was dismissed in November.[15]

General Korzhakov recalls how badly Yeltsin reacted to revelations of corruption. "Reporting to Yeltsin about embezzlement in [top government circles], I noticed that he didn't like hearing about the thievery," he says. "Boris Nikolaevich understood that some people, claiming to be his loyal friends and allies, were simply enriching themselves on the basis of this loyalty."[16]

Few proclaimed their loyalty more eloquently than Boris Berezovsky. Korzhakov was there when the car dealer was introduced to Yeltsin's inner circle. It was during the winter of 1993–94. "We were looking for a publisher for Yeltsin's second book, *Notes of a President*," Korzhakov recalls. The book had been started in early 1992 as a quick description of the August Putsch, eventually becoming a memoir of the first two years of Yeltsin's presidency. Yeltsin employed the same ghostwriter, a journalist named Valentin Yumashev, he had used for his 1989 book. In the eyes of the rest of Yeltsin's entourage at the time, Yumashev was a strange bird. Korzhakov recalls his "sloppy appearance" and calls him a "hippie journalist." The former security chief was surprised at how protective Yeltsin was of his association with the writer.

"I had good relations with Yumashev at that time," Korzhakov says. "Boris Nikolaevich was not delighted by our close relations. He always got nervous whenever I spoke well of Yumashev." Later, Korzhakov would discover that Valentin Yumashev played a key role in the Yeltsin family's finances.[17]

Once *Notes of a President* was finished, the Russian president needed a publisher. "Yumashev declared that a large amount of money was needed to publish it," Korzhakov recalls. "Unfortunately, neither Yeltsin nor I had ever worked in publishing. As it became clear later, several big Russian publishing houses had been ready to publish *Notes of a President* at their own expense and even to pay Yeltsin a significant advance. But Yumashev explained the problem by saying that without the money of some businessman it would be impossible to publish *Notes of a President*."

Yeltsin's London literary agent, Andrew Nurnberg, agreed to represent rights outside Russia. But who was going to publish the book in Russia? General Korzhakov recalls the following conversation: "'Did you notice that all Moscow is covered with Logovaz billboards?' Yumashev asked me. 'Logovaz is Berezovsky.' Yumashev made it seem as if the publication of Yeltsin's book at Berezovsky's expense would be an act of charity and that, of all Russia, only Berezovsky was capable of pulling it off."[18]

Berezovsky's Logovaz advertisements appeared in more places than just on billboards. Throughout 1993, Logovaz placed many ads in the cash-strapped magazine *Ogonyok*, where Yumashev served as deputy editor. Two years later, in 1995, when *Ogonyok* was reorganized, Yumashev became general director and Berezovsky half owner.[19]

In any case, Berezovsky was taken on as the publisher. "Yumashev brought Berezovsky to the Kremlin and introduced him to me," Korzhakov recalls. "Afterward, Berezovsky was introduced to Yeltsin."

Berezovsky arranged a quick and high-quality printing of the book in Finland. He claimed that he had invested $250,000 in the project and that his business partner, Avtovaz boss Vladimir Kadannikov, had committed another $250,000. The book was published under the *Ogonyok* imprint.[20] "Berezovsky portrayed his task as an extraordinarily generous act of charity for the benefit of the president," Korzhakov noted.

Andrew Nurnberg claims that several hundred thousand copies were sold worldwide outside Russia. Even so, Yeltsin's royalties, after Yumashev's and Nurnberg's commissions were paid, were almost certainly under $200,000.[21]

"Yeltsin hoped to earn $1 million and repeatedly complained to me that 'these scoundrels' had robbed him," Korzhakov notes. "They [Yumashev and Berezovsky] understood that they had to fix their mistake. They started filling Yeltsin's personal bank account in London, explaining that this was income from the publication of the book. By the end of 1994, Yeltsin's account already had a balance of about $3 million. Berezovsky repeatedly boasted that he was the person who had arranged the accumulation of Yeltsin's personal funds."[22]

Among the reported contributors to the Yeltsin "book fund" was Golden Ada, the San Francisco–based trading company engaged in embezzling $178 million in gold and diamonds from the national treasury.[23]

Nurnberg says that Yeltsin's bank account at Barclays Bank was closed around 1993–94 and that the overseas revenues from Yeltsin's book were channeled directly into Yeltsin's Russian bank account.[24]

Korzhakov, however, tells a different story. "Over the course of 1994 and 1995, Yumashev visited Yeltsin every month in the Kremlin. No one understood why this badly dressed, sloppy-looking journalist regularly came to visit the president, would speak privately with him and then leave his office after a few minutes. I knew the reason for the visits. Yumashev was bringing Yeltsin the interest payments from the account: about $16,000 in cash every month. Yeltsin stored the cash in his safe."[25]

The book project was important to Yeltsin because he did not have many subsidiary sources of income. "At that time, Yeltsin never accepted any money from anyone," Korzhakov says. "There was never even a question of his accepting a bribe. He even had trouble finding the money to complete the construction of his personal dacha [country cottage]. We had to fight hard to convince him to accept $20,000 that a Japanese newspaper was willing to pay in return for an interview."[26]

He was grateful to Berezovsky for the financial windfall his book represented. When the Russian edition of *Notes of a President* appeared, there was a triumphant book party at the Presidential Club, the luxurious tennis club reserved for the highest government officials. Yeltsin proposed Berezovsky as a member of the club. The tycoon was now part of the Kremlin circle.

"Berezovsky was there often, lobbying for his personal projects," remembers Korzhakov. "He knew, for instance, that every morning I would definitely be there. He followed in my footsteps and even while I was taking a shower would continue telling me about his political and business projects. I must admit that I was glad that the sound of the water muffled his hurried speech."

Berezovsky's fortunes rose immediately upon the publication of Yeltsin's book. "Through his access to Yeltsin, Berezovsky appropriated for himself a unique and leading role among the so-called oligarchs," Korzhakov remembers. "In other words, he became the favorite of the president's family. Berezovsky was constantly demonstrating his special relations with the Yeltsin family; he reminded people of his closeness to the president on any pretext. High government officials feared refusing Berezovsky, who insisted on specially favorable conditions for his own business."

Using his access to the highest officials of the Russian government and his reputation as a close friend of the Yeltsin family, Berezovsky hammered away at the privatization projects that would put key state industries in his grasp.

"[First Deputy Prime Minister Oleg Soskovets] did a great imitation of Berezovsky," Korzhakov recalls. "He took a worn-out leather briefcase, went out of the room, quietly scratched at the door, slinked across the threshold and then, with a hunted look, tiptoed sideways up to the table. This was so reminiscent of Berezovsky's visits that we all rolled on the floor with laughter."[27]

Berezovsky would get the last laugh, but for now he had to put up with the mockery of Yeltsin's government cronies. He needed their help in gaining control of some of Russia's biggest companies.

Blasting Away the Parliament

Before mass privatization could proceed, Yeltsin needed to gain absolute power. In the first years after the fall of Communism, he was not yet the autocratic ruler he would later become. In early 1993, Yeltsin was challenged by a broad alliance composed of Russia's top politicians. His hold on power seemed shaky. Constitutionally, the legitimacy of his government was in question—the dissolution of the Soviet Union, after all, had violated the Soviet constitution and a nationwide referendum. In addition, the policies of Yeltsin's government had proved manifestly destructive to the nation.

"Boris Yeltsin did not come up to our expectations," stated the diamond dealer Georgi Khatsenkov. "He turned out to be completely incapable of managing the country. Instead of stabilizing the situation, he served as the catalyst for Russia's disintegration. We need a Russian Pinochet."[28]

There was one individual who aspired to that role: Vice President Aleksandr Rutskoi. Young and handsome, Rutskoi was a former air force general who had fought in the Afghanistan War. Yeltsin's choice of Rutskoi as vice president was supported by Russian patriots and the military, and Rutskoi's solid support of Yeltsin during the Putsch of 1991 had been an important factor in the Russian president's victory.

But Rutskoi retained some scruples, as well as personal political ambitions. When he saw Russia's wealth being shamelessly looted by Yeltsin's cronies, he rebelled. He promoted himself as Russia's law-and-order candidate. Yeltsin made Rutskoi chairman of a special commission investigating government corruption. But when Rutskoi uncovered massive violations, Yeltsin took him off the job. Rutskoi responded by going public, appearing on evening TV in April 1993 claiming he had eleven suitcases full of documentary proof of criminal dealings by at least four dozen top officials. The Yeltsin clan denied the charges and launched corruption accusations against Rutskoi. A few months later, Yeltsin dismissed him (another violation of the constitution).

A bigger problem for Yeltsin was the Russian parliament. Having been elected in 1990, the parliament was composed of hard-line Communists, but also many Gorbachev-era reformers. It was unsparingly hostile to Yeltsin and his policies, opposing both Gaidar's price reform and Chubais's plans for mass privatization, blocking presidential decrees, and threatening to impeach the president.

On March 20, 1993, Yeltsin appeared on TV and threatened to declare martial law if the political crisis was not resolved. He knew that on March 26 an extraordinary session of the Congress of People's Deputies would initiate impeachment proceedings against him. On March 22, he summoned the chief of the Kremlin guard, Mikhail Barsukov:

"We must be prepared for the worst, Mikhail Ivanovich! Work out a plan of action in case we need to arrest the Congress."

"How much time do I have?"

"Two days maximum."

Barsukov devised (and Yeltsin approved) the following plan. If the Congress voted to impeach Yeltsin, a security officer would read a prepared presidential decree over the public-address system dissolving the Congress. Electricity, water, and heating would be shut off. Failing that, tear-gas canisters would be set off in the balconies of the assembly hall; troops equipped with gas masks would then arrest the incapacitated parliamentarians and take them to specially prepared prison buses outside.[29]

On March 28, the Congress voted. It did not gather enough votes to pass the impeachment measure. Yeltsin's backup plan proved unnecessary. A month later, in a national referendum on April 25, Russian voters voiced their confidence in both Yeltsin and his economic reform program. Considering the fact that the country was falling apart and many people were on the verge of starvation, the vote was testimony to Yeltsin's lingering charisma and the Russian people's reluctance to challenge the Kremlin. In the same referendum, a proposition to dismiss the parliament failed to gain enough votes to pass. The electorate, therefore, was sending a message that it wanted the constitutional conflict between Yeltsin and the parliament resolved without the destruction of either side.

The standoff continued and mutual accusations of corruption escalated. Throughout the summer of 1993, in front of an increasingly bewildered populace, the nation's top politicians accused one another of fraud and embezzlement. The Russian press was full of revelations about rigged export-import contracts with kickbacks going to the secret Swiss bank accounts of top officials. (One garish scandal involved an intercepted telephone conversation in which top government officials were apparently discussing the assassination of Russia's prosecutor-general.) The spurned vice president and anticorruption crusader, Aleksandr Rutskoi, meanwhile, joined forces with the Russian parliament.

Finally, at 8 p.m., Tuesday, September 21, Yeltsin appeared on national TV and announced that he was dissolving parliament. The Russian Supreme Court quickly pointed out that this act was blatantly unconstitutional. Inside the White House, several hundred supporters, many of them armed, had gathered under the leadership of Rutskoi and parliamentary speaker Ruslan Khasbulatov. Yeltsin declared that they had until October 4 to leave. He surrounded the parliament with detachments of paramilitary police from the Ministry of Internal Affairs (MVD). The phones and electricity were cut off; the building was under siege.

On Sunday, October 3, the day before the protesters inside the White House were supposed to leave, more than 10,000 parliamentary supporters gathered by Lenin's statue in October Square. They broke through police cordons and divided into two columns. One group marched to the Ostankino TV tower on the outskirts of Moscow—the building from which most of Russia's national TV networks broadcast. The second column headed toward the besieged parliament building. There, hundreds of militants broke through the MVD blockade and joined forces with the men inside. An attempt to take the Moscow mayor's office, however, was repelled by a line of security troops backed by units of the Most Bank Security Service.[30]

The confrontation turned bloody. Hundreds of anti-Yeltsin militants had surrounded the Ostankino TV tower, some armed with machine guns and others with rocket-propelled grenades. Inside, about eighty men of the elite Vityaz unit of the MVD had all the entrances covered. The militants were beaten back as they repeatedly tried to storm the building throughout the night. Scores of bodies were left on the pavement.

The next morning, Yeltsin ordered tanks to the center of Moscow. They parked on the Novy Arbat Bridge, just under the parliament building, and began blasting away. Smoke and documents burst into the sky after every hit. Soon the top floors of the building were on fire. Militarily, it was a perverse replay of the Putsch of 1991, except that the tanks were on Yeltsin's side this time and the battle was much bloodier. Thousands of Muscovites looked on from their windows and rooftops, watching the grand spectacle of the Russian parliament being dissolved with high explosives.

Meanwhile, using tunnels, special forces units infiltrated the parliament building and began clearing it out, room by room. At sundown Rutskoi and Khasbulatov surrendered to the authorities and, together with their colleagues, were led to the prison buses outside.

The events of October 3–4 left Russians with a bitter aftertaste. Officially, ninety-two people died. Most people put the death toll at several hundred, and perhaps higher. Both sides appeared self-serving and hypocritical. Rutskoi and Khasbulatov had vowed to fight to the death to save Russia, but in the end they meekly surrendered to the authorities. The Yeltsin camp accused the militants of high treason, but released them in a general amnesty a few months after the battle.[31]

In a national referendum on December 12, 1993, two months after the attack on the parliament, Yeltsin's proposed constitution was narrowly approved—by 60 percent of the votes cast, or 32 percent of the eligible voters. There were serious allegations that figures on voter turnout (53 percent) had been falsified in the provinces to reach the requisite number, but for the time being, it seemed that the constitutional crisis had been resolved. Under the new constitution, the president was given enormous powers—something akin to the powers of the kaiser or the czar in pre–World War I Europe. As with those two sovereigns, the Russian president had to deal with an independent parliament, but this institution (now called by its czarist name—the Duma) was virtually incapable of taking action against the president.[32]

On the same day they voted for the constitution, Russians also elected the new parliament. Most of the chosen deputies were fiercely critical of Yeltsin and his government. It seemed to be a protest vote, an expression of anger against Yeltsin's brutal dissolution of the previous parliament. But in Russian politics, hypocrisy was the order of the day. The Russian people might have thought they were electing an anti-Yeltsin parliament, but in fact the assemblage turned out to be an obedient tool of the Yeltsin regime. It was not only the parliament's lack of constitutional powers that made its threatening postures easy to ignore. It was also that the largest parliamentary faction—Vladimir Zhirinovsky's party—was secretly supported by the Yeltsin government; the Zhirinovsky party, in return, supported Yeltsin on all key measures that came up for a vote.

Anatoly Chubais Privatizes Russia

Now that Yeltsin had consolidated his power, the government was free to take another dramatic step toward turning Russia into a market economy. Yegor Gaidar had already freed most prices in 1992, but the vast majority of Russian companies still remained state-owned. Now, with the parliament removed as an obstacle to Yeltsin's will, the way was clear for these enterprises to be privatized.

For most Russians, their country was changing with terrifying speed. It took just four months—between August and December 1991—to abolish Communism and dismantle the Soviet Union. It took only a few weeks in early 1992 for Yegor Gaidar's shock therapy to destroy the old planned economy, as well as the life savings of the majority of Russia's citizens. Now privatization would be carried out with the same ruthless speed.

The program's architect was a thirty-eight-year-old economist from St. Petersburg, Anatoly Chubais, the man in charge of a relatively small agency called the State Property Committee (GKI). This agency was located in a building across from the Rossiya Hotel. The hallways rang with American voices: young advisers on economics, finance, law, and public relations that Chubais had enlisted to help with his privatization crusade.

"Chubais arrived at the headquarters of GKI in 1991," Gaidar, then the acting prime minister, recalls. "At the time there was an absolute administrative and organizational vacuum, since the old mechanisms of Party and government control had been completely dissolved. A significant part of the bureaucracy, on the basis of semilegal procedures, was brazenly distributing property. Very often, at official meetings of the government, dozens of provincial leaders spoke out: 'Give everything to us and we will select the owners.'"

Everyone wanted the right to distribute Russia's natural-resource wealth. The Communists and nationalists wanted most of the property to remain in the state's hands. Regional political bosses wanted to distribute it to their own cronies. Chubais decided to go with a small group of well-connected businessmen. Gaidar believes it was the right thing to do. "Chubais managed, in the heaviest of circumstances, to develop a system and put the privatization process in Russia on a sound footing," Gaidar says. "He is a very precise administrator, excellent at defining the tasks at hand and controlling their execution."[33]

I once visited Chubais when he was first deputy prime minister, and witnessed his phenomenal work capacity. Our interview was scheduled for 9:30 p.m., in Chubais's offices. The huge government administration building (the White House), including the offices of the prime minister, the deputy prime ministers, and their staff, was dark—nothing but empty corridors, silent offices, and locked doors. Everyone had gone home. But Chubais's wing was blazing with light. People were rushing around, negotiating on the telephone, drafting documents, setting up meetings. Chubais himself, when he came into the conference room for our interview, was slightly breathless, visibly a busy man. Yet his speech was clear and concise; here, it seemed, was a superb technocrat.

Under Chubais's supervision, privatization vouchers had been mailed to every Russian citizen (151 million vouchers). The plan was to privatize more than half of Russian industry within two years. The most common privatization scenario was to sell 29 percent of a company for vouchers at a public auction, distribute 51 percent among the managers and workers, and retain the rest for the state, to be sold later for cash or investment promises.

The vouchers were the most important part of the program, giving every Russian citizen a stake in the new economy; everyone could become a shareholder and the best companies would not be bought by the wealthy. The idea was Jeffersonian democracy in action—creating the foundations for a solid middle class and a big domestic market to drive the economy. Russian citizens' savings might have been destroyed by the hyperinflation of 1992, but now they could salvage some personal wealth by owning shares in Russian enterprises.

When he introduced the voucher-privatization concept to the Russian people in the summer of 1992, President Yeltsin declared: "We need millions of owners, rather than a handful of millionaires. Everyone will have equal opportunities in this new undertaking, and the rest will depend on ourselves.... Each citizen of Russia and each family will have freedom of choice. The privatization voucher is a ticket for each of us to a free economy."[34]

The stake represented by a single voucher was so small, and the Russian securities markets so primitive, that voucher owners generally could not buy shares directly at the privatization auctions. Each citizen had a choice either of investing the voucher in his own company, through a company-sponsored share-ownership program, or of investing it in one of the newly created voucher mutual funds.

Because they were desperately poor, unsure even if they would have enough money to buy food, most people sold their vouchers on the street for quick cash. Consequently, the street price of the voucher was ridiculously low. In the threadbare winter months of 1993–94, it was common to see pimply, shabbily dressed young men standing in the metro stations with signs reading I'LL BUY VOUCHERS. The price was 10,000 rubles—about $7, or enough for two bottles of cheap vodka. With each voucher selling on the street for $7, Russia's vast industrial and mineral wealth was being valued at about $5 billion.

Voucher privatization seemed like a good plan, but it turned out to be a complete failure. The first mistake was the timing. Privatization in Russia should have been carried out before Yegor Gaidar's price reform in 1992, as a means of soaking up the ruble overhang and minimizing the inflationary threat. This had been the essence of Grigory Yavlinsky's 500-Day Plan.[35] "I considered privatization to be precisely the instrument we needed to deal with the monetary overhang—not inflation, which reached 2,500 percent in 1992, but privatization," Yavlinsky recalls. "I concluded that the correct course was to use the accumulated savings for privatization. I proposed starting with the smallest things—privatizing trucks, small plots of land, apartments, small shops—and to move gradually to larger-scale privatization."

Yavlinsky insisted that the privatization had to be done for money. "Privatization addresses the issue of ownership, the replacement of the [old Soviet] management," he said. "Of course the money people earned from the Soviet times was quite small, but spending even their small savings, they became true property owners." You should never get anything for free, he argued. You won't respect it properly. All valuable things are gotten by the sweat of the brow. "You always have to sell [a property], because only that way do you create a real owner. Even if you buy a piece of a company for a tiny sum, it is still your money—you would value it, you would look for a manager, you would try to build up the business. But when people got their vouchers in the mail, they didn't care either way."

Moscow mayor Yuri Luzhkov agreed, saying that privatization didn't make sense when the prices were so low. "We say privatization is necessary to create a new owner, so that new owner can use the property better and expand production," Luzhkov argued. "Take the privatization of a factory—the new owner is supposed to manage the factory better. But all this is possible only when the factory is sold for real money. I will give you a concrete example. I bought Zil, a giant automobile company, for $4 million. I don't need to use it as an automobile factory. It covers 240 hectares of land and I can get $4 million back by turning it into warehouses or anything else, for that matter.

"My [chemistry] institute sold for $200,000. This institute, first of all, had a cohort of specialists, every one of whom should have been earning $200,000 a year. Second, it had an experimental production base on which it was possible to develop the newest technology. And this enterprise was sold for $200,000! The price of a single spectrophotometer! The new owner decided he didn't need to keep the institute working. He said: I'm firing everybody, I now have free real estate, which I will use to supplement my earnings—I am renting out this space and getting $500,000 a year. That's business for you!"[36]

Yavlinsky and Luzhkov were ignored. Gaidar argued that he and Chubais simply did not have enough time to privatize first; they had to free prices to put food on the shelves, even if this meant that people would have to use their remaining savings to buy several weeks' worth of produce. Still, even a disorderly fire sale of trucks, apartments, shops, and garden plots, and, later, a sell-off of small enterprises such as brick factories, lumber mills, and textile workshops, would have been far less destructive than Gaidar and Chubais's policy.

Chubais's second mistake was the speed with which he privatized Russian industry. If privatization had begun earlier, it could have proceeded much more gradually. Chubais put everything on the block at once—the biggest oil companies, metals and mining companies, huge timber companies, automakers, machine-tool makers, tractor factories, big engineering companies, huge shipping fleets, the country's main ports—and ended up by flooding the market. As any investment banker knows, getting a good price for a company's initial public offering of shares depends on limiting the supply. First you sell a small portion of the company, then you offer more shares when you see that the demand is there. The sale of Internet companies in the United States in 1995–2000—companies with little or no present earnings, but a lot of long-term potential (the same sales pitch that was given regarding Russian companies)—was handled this way and produced high prices as a result.

Like Yavlinsky, Oleg Davydov, foreign trade minister in 1994–97, believes that the first target for privatization should have been shops, restaurants, and small workshops. Next, the government could have sold light industrial enterprises such as textile mills and food-processing outfits. "The market always begins with the service of people," Davydov notes. "Small enterprise."[37]

Instead, Chubais privatized Russia's biggest and most profitable exporters. "All the criminality appeared there, in that sector, which had such a huge export potential," Davydov laments. He argues that Russia's industrial giants in such strategic sectors as oil and gas, metals, aluminum, and aerospace, should have been converted into state-owned corporations. They would learn to function as autonomous enterprises and then they could be privatized gradually, as the market developed. There are many such examples in Europe and Japan.

The final mistake of the Chubais-Gaidar policy was to set the face value of the privatization vouchers at a low figure (10,000 rubles). According to early 1992 prices, the face value of the vouchers gave all of Russian industry an implicit value of $100 billion, which was clearly at odds both with the size of the Russian economy and with the tendency of equity markets to exceed the value of a nation's gross domestic product. The $100 billion price tag Yeltsin's economists attached to Russian industry in 1992 paled in comparison to the value attached to such markets as Mexico and Hong Kong, whose equity market capitalization at the time was $150 billion and $300 billion respectively. By the end of 1993, when Russians actually were able to use their vouchers, inflation and devaluation of the ruble had destroyed 95 percent of the voucher's face value, and Russia's industrial and natural-resource wealth was valued at a mere $5 billion.[38]

By assigning such a ridiculously low face value to the vouchers, the government was telling investors that it considered the securities nearly worthless. The Russian state was the loser. It sold its best assets for an insignificant sum. Russia's 151 million citizens lost out as well, since very few of them received equity stakes in the privatized enterprises. Well-connected investors, on the other hand, got an incredible bargain. Perfectly good companies could be bought for one times earnings or less. For investors playing the Russian equity market that first year, annual returns of 300 percent were not unusual. Significant fortunes could be made in a matter of weeks.

One such insider was Berezovsky. Eventually he would use the privatization process to accumulate Russia's largest private fortune. But his participation in its first stage—voucher privatization—was minimal. He focused on other projects in 1993–94 and left the voucher market to other operators.

The Ultimate Emerging Market

In the late 1980s, Mikhail Kharshan, thirty-nine years old, had started his first business: an adventure-travel agency for foreigners. In 1992, after the announcement of Anatoly Chubais's privatization program, Kharshan established the First Voucher Investment Fund, a mutual fund dedicated to taking vouchers from the population and exchanging them for equity stakes in companies. Most of Russia's 600 voucher mutual funds attracted investors with loud advertising campaigns, promising returns of several hundred percent a year. Kharshan's First Voucher Fund did not make such extravagant claims, but it managed to become the largest mutual fund in Russia, receiving 4.5 million vouchers (3 percent of the total); the fund ultimately invested these vouchers in eighty different companies—hotels, fertilizer plants, metals companies. Kharshan had a significant advantage over his rivals: He was able to sell the shares of his mutual fund through thousands of post offices—an instant retail network of impressive proportions. He achieved this feat by paying the Ministry of Posts and Communications $1 million a month, as well as by bribing a few officials and using old KGB connections.[39]

Kharshan's methods were often unorthodox. One of his favorite stories was how he acquired his stake in the Solikamsk Paper Co., in the northern region of Perm. With a capacity of 460,000 metric tons, Solikamsk was a very large paper mill and virtually the only producer of newsprint in Russia. Together with the company's manager, Kharshan "convinced" the local privatization committee to hold the voucher auction only on the grounds of the plant itself. Since the paper company was situated in an area of defense plants, it was officially off-limits to outsiders. But Kharshan arranged for a high government official in Moscow to call the former Communist Party boss running the local administration. Kharshan's people were let through by security guards, and he was able to buy 10 percent of the company. "We didn't shoot anyone and we didn't violate any laws," explained Kharshan, grinning. "These are the normal business practices in Russia."[40]

Another big player in voucher privatization was Alfa, the well-connected trading company. Its owners had made a fortune in 1992 and 1993, exporting oil and metals, importing sugar, and trading other agricultural products. In December 1992, Foreign Trade Minister

Pyotr Aven, the government official overseeing trading companies like Alfa, resigned his post and became president of Alfa Bank, sharing power with two other well-connected young men. That same month, Alfa set up its own voucher mutual fund, called Alfa Capital. Having collected 2.5 million vouchers, Alfa Capital became one of the largest mutual funds in the country. It bought stakes in forty-six companies, including cement plants and other construction-materials producers, food-processing plants, and chemicals companies. Mikhail Aleksandrov, Alfa Capital's twenty-seven-year-old investment strategist, gushed to me in the midst of this buying spree: "There are some deals where you can't believe the company is in Russia—the machinery is all so new and clean and the assets are so good."[41]

Aleksandrov, who had been trained in finance at one of the top Soviet institutes and had worked briefly for KPMG Peat Marwick and CS First Boston, bragged about how Alfa had beaten the American conglomerate RJR Nabisco for control of the Pskov cracker factory. Pskov made something similar to a Ritz cracker and sold 56 million packs a year. Alfa Capital bought a 14 percent stake in the voucher auction in April 1993. But RJR Nabisco had its eye on the company and bought a 10 percent share in the fall. Control still rested with Pskov's management and workers, who retained 51 percent of the company. Alfa Capital arrived first, bought out the management and workers, and obtained 65 percent of the company—all for under $4 million.[42]

Aleksandrov noted that Western investors would never be able to get assets at such cheap prices. "When an American investor comes to Moscow, everybody will try to cheat him," he chuckled. A foreign investor needed a Russian stockbroker's help to purchase shares. But once he knew that a foreign buyer was interested in a particular company's shares, the Russian broker would often buy the shares himself and then resell them at three times the price. In Russia, front running was not illegal, and when it came to foreigners, it was that much easier. "No one would try to cheat us," Aleksandrov observed. "We would catch them and bash their face. We could always hire some Ivan to shake the money out of them."[43]

Stockbrokers and mutual fund managers often paid in blood for their activities. One of the victims was Andrei Orekhov, founder of the Grant brokerage house. Orekhov tried to run a clean operation and play by Western market rules; in late 1994, he even master-minded the successful defense of the Red October Chocolate Factory in Russia's first hostile-takeover attempt. But that winter, the Russian stock market crashed, disappointing many investors, who evidently did not understand that markets can go down as well as up. On April 18, 1995, Orekhov was leaving his home in Moscow, together with his six-year-old daughter, Maya, whom he was taking to school, when gunmen opened fire on his car. Orekhov and his driver were seriously wounded; the little girl was killed.[44]

So large was the number of Russian companies being privatized that many enterprises could be bought at one times earnings or less. The problem was identifying the real jewels. "The audited balance sheets are nonsense; only 5 percent of the privatized companies are really profitable," mutual fund manager Mikhail Kharshan noted. The GKI (state property committee) and the Russian Federal Property Fund had more detailed information on companies' balance sheets and properties, but they were prohibited from releasing it. Kharshan boasted that he managed to get the data anyway—smuggled out on computer disks from GKI. As a result he purchased some companies at one times earnings or 10 percent of sales.[45]

Foreign entities such as CS First Boston, Brunswick-Warburg, and stand-alone outfits like the Leukadia Group managed to buy minority stakes in some good companies. But that didn't mean they received the same rights as shareholders had in the United States or Western Europe. Most Russian managers felt that outside shareholders were not putting anything into the company and were simply intruding in other people's business. They regarded Chubais's privatization program as a hostile takeover. Some companies would not even disclose their corporate charter, let alone open their shareholder register. Claudia Morgenstern, a senior investment officer at the World Bank's International Finance Corp. who was advising the Russian government on its privatization drive, said: "Like everything else in Russia, it's not what the law says, it's people's attitude and how they choose to act."[46]

Many of the best Russian companies were bought by their Soviet-era managers. Typically, these were men who had secretly diverted their companies' cash flow and, using that money, either bought vouchers to tender in the primary auction or tried to buy shares from their own workers on the secondary market. Since workers often received their meager wages months late, they were eager to sell their shares to a management-controlled financial company for quick cash.

The most notable success in these discreet management takeovers was Gazprom, the Russian natural-gas monopoly. Sitting on a third of the world's natural-gas reserves, the sole gas supplier to most of the former Soviet Union, and the dominant supplier to Western Europe, Gazprom was the richest company in Russia. It may have been the most valuable private company in the world. If it had been a Western company, Gazprom would have been worth between $300 billion and $700 billion in market capitalization on its gas reserves alone. Instead, it sold in the voucher auctions for an implicit price of just $250 million.[47]

The auction had been rigged. First, most of the tenders occurred in remote Siberian towns, where Gazprom ran the show. Second, foreigners were forbidden to buy shares. Third, management received the right of first refusal on any sale of Gazprom securities on the secondary market, as well as the right to refuse to register shareholders it didn't approve of.[48]

In the autumn of 1993, Gazprom managers asked Mikhail Kharshan's First Voucher Investment Fund to buy 530,000 vouchers so they could tender them in the upcoming auction of their company. (This amount represented one eighth of the vouchers that were eventually tendered in the Gazprom auction.) Kharshan boasted how he ran up the price of the vouchers almost twofold on the stock exchange in order to charge the Gazprom managers at least twice what they would have paid had they tried to accumulate the vouchers slowly. "They're bureaucrats, and what's more, they're provincial fellows," Kharshan said, chuckling. "They're not very quick."[49]

In the end, however, the Gazprom managers got the better end of the deal. With the Gazprom-voucher auction rigged in favor of company managers and employees, the shares sold for a terrifically low price. In the obscure Republic of Mari El, for instance, the price of the shares implied a value of just $80 million for the entire company, while in the northern industrial city of Perm, the implied valuation was $79 million. The average price at which Gazprom shares sold on the voucher auction valued the entire company at $250 million. Flush with the vouchers they had accumulated through brokers like Kharshan, Gazprom managers ended up with about 15 percent of the company, as well as voting control of the government's 35 percent stake.[50]

The result of this manipulation of the market and of the careless way in which privatization vouchers were sold to the population was that Russia's industrial wealth was auctioned off at extremely low prices. Below is a table of the largest companies subjected to voucher privatization.

Voucher Value of Russian Companies Compared with Market Value (in Million U.S. Dollars)[51]

Company	At Voucher Auction Prices (1993–94)	At Russian Stock Market Prices (August 1997)
Gazprom (natural gas)	250	40,483
Unified Energy Services (electricity)	957	17,977
Lukoil (oil)	704	15,839
Rostelecom (telecommunications)	464	4,172
Yuganskneftegaz (oil)	80	1,656
Surgutneftegaz (oil)	79	6,607

These six industrial giants, the crown jewels of Russian industry, were valued by the Russian secondary market at least twenty times higher than the price at which they were sold in the voucher auctions.

The People's Car

The irony of the intricate maneuvers carried out by Russian managers at the voucher auctions was that they did not need to capture a majority stake in a Russian company to control it. "Under Russian conditions, there is no difference between control and ownership. If you have control over an enterprise, you control the cash flow," noted the First Voucher Fund's Mikhail Kharshan.[52]

Berezovsky understood this perfectly. He was more interested in solidifying his control over the cash flow of Avtovaz than he was in acquiring stakes in other companies through the voucher auctions. He knew that privatization was not simply a matter of buying shares of a company on the market. Soon he would perfect the art of "virtual privatization"—gaining control of companies in which he had a minimal equity stake.

But Berezovsky would not let an opportunity like voucher privatization pass by without exploiting it in some way. Characteristically, he chose an unusual way to play the market. Berezovsky realized that the voucher program was the foundation for a securities market in Russia, and he was one of the first individuals to understand how to raise capital in this market. His plan was to start an investment fund taking cash from rank-and-file investors.

The first Russian stock exchanges were primitive affairs. They did not trade shares of Russian enterprises—typically, they had only one liquid security available: the privatization voucher itself. In early 1994, I accompanied Sergei Andreyev, a trader for the Russian stockbroker Grant Financial Group, to the largest exchange in Moscow, the Russian Commodities and Raw Materials Exchange. Andreyev's mission was to purchase 10,000 vouchers for CS First Boston. The exchange, which traded commodities in the morning and vouchers in the afternoon, resembled a bus station. There was even an ice-cream stand in the corner. Crowds of shabbily dressed people milled about with suitcases and bags. These were traders who had come from distant cities with suitcases of vouchers; they were waiting to strike a deal, count the money, and leave with suitcases of cash.

The afternoon I accompanied him, Andreyev made note of the voucher transaction prices on the board. He said he would use the board price as a benchmark and then buy from one of the traders in the corner to avoid paying the exchange a commission. The voucher price at that moment was hovering just above 20,000 rubles. "Too high," Andreyev said. "Most of these guys are provincials and they have to sell. They didn't surrender yesterday and they're not surrendering today, but they'll have to give in before they leave at the end of the week."

Accompanying us on this visit to the exchange was a tough guy named Aleksandr Fomenko, a former KGB bodyguard who was working for the Grant Financial Group. While we were waiting for a transaction to take place, Fomenko told me of a case in late 1993, when several executives from a Siberian trading company came to the exchange with suitcases of vouchers, which they sold for $150,000 in foreign currency—cash. They were robbed as soon as they left the exchange.

Most investors were robbed as well, though not at gunpoint. If they didn't sell their privatization vouchers for $7 on the street corner, most Russian citizens usually had no choice other than to invest them in one of the voucher mutual funds. Some funds, like Kharshan's First Voucher Investment Fund or the Moscow Real Estate Fund, were legitimate operations, but they typically produced paltry dividends for their investors—not even close to compensating them for the effects of inflation.

Investment fund managers routinely committed gross violations of their fiduciary responsibility. When an investment fund made a superb purchase in one of the voucher auctions, shareholders in the fund were lucky if they received the full capital appreciation of the investment. In a typical scenario, an investment fund would purchase a block of shares in a company at one of the voucher auctions. Several months later, the fund would be approached by a buyer willing to pay twenty times the purchase price. The fund then sold the shares at two times the price to a company owned by "friends" of the fund's managers; this company, in turn, sold the block of shares to the ultimate buyer at ten times the price it had bought them for. "It is a typical *paganaka* [a poisonous mushroom]—investment funds do this kind of thing all the time," observed Andrei Orekhov, general director of the Grant stock brokerage.

Most mutual funds were pyramid schemes, robbing Russians not only of their vouchers but of their cash savings as well. By 1994, those Russians who still had any cash savings were desperate to put their remaining funds in a safe place. With inflation running at 215 percent or more, keeping money in the bank was a bad idea, since banks typically paid less than 50 percent on savings accounts. The only alternative was to invest in one of the new cash mutual funds, most of which were primitive Ponzi schemes. One of the first and most obvious rip-offs was a mutual fund called Tekhnichesky Progress, which, in the summer of 1993, promised a 500 percent annual return and began paying out. It attracted some 300,000 small investors and then disappeared. Other mutual funds promised equally outlandish returns, collected investments from millions of rank-and-file investors, and then vanished. No one knows exactly how much money was lost in the pyramid schemes, but it amounted to billions of dollars. Those Russian citizens who managed to avoid total ruin in the inflation of 1992 were conned out of their money the second time around.

The largest pyramid scheme was MMM, originally a trading company (a cooperative) established in 1989 by the three Mavrodi brothers. They had cut their teeth on the lucrative commodities trade. By 1992, MMM comprised twenty different companies, including the original trading house, a bank, and a brokerage firm.

I wanted to meet the boss of the firm—thirty-seven-year-old Sergei Mavrodi. But I managed an interview only with his thirty-year-old brother, Vyacheslav. The MMM headquarters was on the outskirts of Moscow, an hour's drive from the center of the city. We had to maneuver through a moonscape of shabby high-rise apartment blocks, across some railroad tracks, past crumbling warehouses, before we finally arrived at the headquarters: a nondescript four-story building that might have been an old school.

I found Vyacheslav Mavrodi in an agitated state. Two months

earlier (on January 30, 1992) the company had been raided by tax inspectors, accompanied by twenty policemen armed with machine guns. "They presented no warrants and no charges," Mavrodi told me, twitching with indignation. "They took all our documents. The tax authorities now say they are conducting a normal investigation on whether we underpaid our taxes."

Evidently, the company was still under great pressure. In the middle of our interview, Mavrodi took a phone call. It was his brother, Sergei. "Yeah, the usual honor guard accompanied me this morning," Vyacheslav Mavrodi complained to his brother. "From the house all the way here."[53]

After intimidating the company, the Russian government mysteriously left MMM alone, and it was allowed to organize the country's largest mutual fund, attracting the savings of millions of ordinary Russians by promising returns of up to 3,000 percent in rubles. It turned out to be a classic Ponzi scheme in which latecoming money was used to pay off earlier investors. The Russian government had no watchdog to investigate the scheme, let alone close it down. In the summer of 1994, MMM was finally revealed as a pyramid scheme and collapsed.[54]

I went to see some of the MMM victims six months later. It was one of those bitterly cold Russian days—if you stood outside for too long, the cold from the ground would penetrate the soles of your boots, first freezing your feet and then the rest of your body. I took a gypsy cab to an address in a shabby suburb of Moscow. There I saw a large, lifeless building, in which the local MMM office occupied the corner space. The office was shuttered. About fifty middle-aged and elderly people were gathered outside—a small demonstration. Were they finally rising up against the fraud and corruption? No. The MMM protesters, I discovered, were just milling about without clear intent.

"Mavrodi is a crook and the government should make him give us our money back," one elderly gentleman told me.

"It's not Mavrodi's fault," a *babushka* broke in. "It's the government's doing—they forced him to go bust."

The boss of the MMM operation, Sergei Mavrodi, was arrested

and jailed by Russian authorities for tax fraud and released two months later, but in 1994 he won a by-election to the parliament and received immunity from prosecution.[55]

So it was in Russia in those days—it was unclear who was more at fault: the government or the financial wheeler-dealers. It was clear who the losers were, however: ordinary Russian citizens.

Apart from MMM, the most famous cash investment fund being peddled to Russian citizens in 1993–94 was something called Avva (the All-Russian Automobile Alliance). This was Berezovsky's brainchild. The Avva plan was to create a joint venture between Avtovaz and General Motors to build an assembly line in Togliatti to produce a new Russian car. Production would begin in 1996 (the year of the forthcoming presidential election) and would eventually reach 300,000 vehicles a year. The cost of the project was optimistically estimated at $800 million, of which $300 million would be procured from the securities markets. Avva, claimed its boosters, would provide Russia with a new "people's car." Like the other famous "people's car"— the Volkswagen Beetle, which spearheaded the German economic miracle after World War II—Avva would signal the economic rebirth of Russia. Not only would Avva be democratic Russia's first big privately financed industrial project, but it would also revitalize the automobile industry, which was central to Russia's economic welfare.[56]

Even by Russian standards, the rise and fall of Avva was spectacular. The company's launch in October 1993 was attended by enormous publicity. Several weeks after having sent in the tanks to blast his political enemies out of the parliament building and around the time he accepted Berezovsky's offer to publish his memoirs, President Yeltsin signed a decree awarding Avva lucrative tax breaks and customs duty exemptions. Banque Nationale de Paris offered a $150 million line of credit. Avtovaz, GM, and Avva signed a memorandum of understanding on the construction of the new assembly line in Togliatti.[57]

The Avva project allowed Berezovsky to stride onto the national stage as a great statesman and industrialist. Representing a large investment in the crucial automaking sector, a joint venture with GM, and a new method of financing industrial growth (through private investment and the capital markets), Avva symbolized a bright new age. As the general director of the Avva project, Berezovsky was able to present credentials as a man working for Russia's future. "If Russia is not to become a Third World country, it must have a well-developed industrial base," he declared grandly. "If Russia wants to secure its economic and political independence, it must develop its industry."[58]

In December, Avva began selling securities to the Russian public. The first securities were certificates of deposit; priced at $7 (10,000 rubles) each, accessible to even the most impoverished Russian. The certificates were sold by traders with suitcases in the metro and bus stations, alongside the traders offering to buy privatization vouchers (whose face value coincidentally was also 10,000 rubles). The Avva certificates had no term limits—they could be redeemed for cash or equity once the new auto plant was built. The holders of the certificates could participate in a lottery, where the prizes were thousands of new Avtovaz cars (worth about $7,000 each), discounted 30 percent to 100 percent.[59]

Convinced by the advertising campaign that the plant would be built, and eager to participate in the automobile lotteries, more than 100,000 Russians bought a total of $50 million worth of Avva certificates. By early 1994, the certificates were trading briskly on Russia's main exchanges; along with privatization vouchers, the Avva certificates were the country's most liquid security.[60]

Several months later, GM backed out of the Avva project, alarmed by gangsterism and corruption at Avtovaz. Apart from Banque Nationale de Paris and the $50 million invested by Russian citizens, no other financing for the project came through. There was not enough money even to begin construction. Only 650 Avtovaz cars were offered at the first Avva lottery. Investors began to suspect that Avva was a sham. By the autumn of 1994, the securities had become worthless.[61]

In the spring of 1995, Avtovaz chairman Vladimir Kadannikov admitted to the press that Avva had not collected enough money to build the new assembly line. He suggested that the Avva funds be used to fund other, more modest Avtovaz investment projects. The people's car was dead.[62]

When I asked Rene Kuppers, general director of the Swiss financial company Forus and one of Avva's founding shareholders, about Avva in early 1999, he reacted with contempt. "Bah...I don't know what the company is doing today," he snorted. "It was an investment which didn't work out, so we wrote off our participation to zero."[63]

There was one footnote to the Avva story. In late 1996, Avtovaz announced that it had found a way to use the Avva funds to build a new car after all. The plan was to create a joint venture among Avtovaz, Avva, GM, and Valmet, a Finnish manufacturing company, to assemble 30,000 Opel Astras in Finland (one tenth of the volume envisioned in the original Avva project). The initial investment was slightly more than $100 million, with the Russians putting up the cash and the Finns forgiving a part of Russia's sovereign debt. In early 1997, Foreign Trade Minister Oleg Davydov and Prime Minister Viktor Chernomyrdin inspected the first cars rolling off the Avva-Valmet assembly line. "I was with Chernomyrdin," Davydov remembers. "They showed him this car, he sat in it, drove around, and they gave it to him as a gift. But [the Avva project in Finland] is over. They stopped assembling the cars and everything closed down."[64]

Periodically, Berezovsky would reassure the Russian public that Avva was still alive and well. In an interview with the newspaper *Sevodnya* in June 1996, he declared that the money deposited in Avva was still intact; the assets had even increased in value, he claimed, to nearly $140 million. None of this, however, made any difference to Avva investors (they never got their money back) or to Avtovaz (it never got any significant investment from the scheme).[65]

For Berezovsky, Avva was a great success. If nothing else, it created yet another hollow investment vehicle to reinforce his control over his cash cow, Avtovaz. At least part of the Avva money was used to buy a 34 percent stake in Avtovaz (the Berezovsky-linked financial company, AFK, owned another 19 percent, while the pyramid scheme MMM owned another 10 percent). The automaker was now solidly in the hands of its managers—Kadannikov, Glushkov, and the rest of the Logovaz cabal. All the companies underlying Berezovsky's car-trading business—Avtovaz, Avtovaz Bank, AFK, Avva, and Logovaz—were now bound together in a network of interlocking shareownership arrangements. In the center was Berezovsky.[66]

The Avva investment scheme gave Berezovsky at least $50 million interest-free to pay salaries, rent, and security-service fees. When Avva began falling apart a year after its inception, no one in the Kremlin paused to remember the promises Berezovsky had made amid the hoopla of the project's launch. He was already onto bigger and better things. And the contacts he had made with President Yeltsin had now become an iron bond.

As the Yeltsin family's favorite businessman, Berezovsky was now receiving the protection of General Korzhakov, the SBP chief. The car dealer made sure of Korzhakov's goodwill by doing him favors as well, including offering the SBP staff Avtovaz cars at half price ($3,000 to $4,000). Throughout 1994–95, during the tail end of the Great Mob War and the feeding frenzy of Russia's privatization, General Korzhakov was there for Berezovsky, loyally protecting the tycoon from his rivals and providing him with political support as he accumulated state-owned assets. He helped derail the police investigation after Berezovsky came under suspicion for murdering the famous TV personality Vlad Listyev. Korzhakov also sent his men to attack the security service of Berezovsky's archrival, Vladimir Gusinsky, in late 1994. This incident—popularly known as "Faces in the Snow!"—shocked Moscow. Later, General Korzhakov would assert that the attack came at the tail end of repeated requests by Berezovsky to assassinate his business rival.[67]

CHAPTER FIVE

THE LISTYEV MURDER

"The Clearest Illustration of Corruption in Russia"

As far back as the late 1980s, it had become apparent that the nascent Russian market economy operated according to one simple principle: Commercial success depended on political influence. If you had good political connections, you could become fabulously wealthy. If you did not, you would almost certainly fail. The further Russia's market evolved, the more prevalent this rule became. Soon, virtually every major businessman had to have a "roof." For small-scale businessmen the roof often meant the protection of a local gangster boss; the bigger businessmen used top government officials as their roof.

When he published Yeltsin's memoirs and entered the president's inner circle, Berezovsky took a major step forward in his business career. But this act alone was unlikely to provide him with sufficient political protection in the long term. His "act of charity" would soon be forgotten. He needed to find an ongoing means of earning the gratitude of the Kremlin; he needed to find a way of gaining the upper hand over his political patrons. So Berezovsky decided to take control of the primary vehicle molding public opinion: television.

His target was the state-owned Channel 1. As Russia's only nationwide TV network, Channel 1 was an extraordinarily powerful institution—its political clout was equal to that of ABC, CBS, NBC, and Fox combined. It reached 180 million viewers, both in Russia and beyond. For many people, Channel 1 was their only source of news and analysis. But the network's power went beyond its geographic reach. For decades Russians had received the official Party line from Channel 1—the network essentially told the population what to think.

In the first years of the Yeltsin regime, Channel 1 was losing hundreds of millions of dollars a year. The managers of the network were frequently bribed to provide airtime to one production company and deny it to another. Businessmen and politicians could bribe a TV producer to show a pseudodocumentary program either flattering them or crucifying their enemies. The network was also the scene of mass embezzlement of government funds. The Russian government picked up the tab for running Channel 1; the operating costs—the transmission costs, the salaries, much of the programming—amounted to about $250 million annually. The network did produce some advertising revenues—about $80 million a year—but few of these funds went to the network itself. They were taken either by the TV production companies broadcasting on the network or by media sales companies (advertising wholesalers). Thus, Russia's hard-pressed taxpayers were paying for the accumulation of private fortunes.[1]

"Essentially everything that was going on at Channel 1 was the clearest illustration of corruption in Russia," Berezovsky would later recall. "There were lots of little companies that bought chunks of airtime. So, on the one hand, there is government funding [$250 million], which pays for the signal and the programming. On the other hand, there are private companies which...on the basis of government funding, produce the programming, broadcast their programming, and receive revenues from advertising."[2]

The advertising business of Channel 1 was dominated by Sergei Lisovsky, a thirty-six-year-old entrepreneur who had made his first big fortune by starting a chain of Moscow discotheques. Lisovsky's flagship disco, U Lis`sa, was partly owned by Otarik, the gangster-businessman who would be assassinated during the Great Mob War.

It was here, in April 1993, that mob boss Globus was killed by a sniper's bullet, unleashing that war. Lisovsky had greater ambitions than just running discos, so he started a publishing business, a video retail business, a film company, a TV company, and an advertising company. The advertising company, called Premier S.V., was Lisovsky's cash cow. By the end of 1994, Premier controlled more than half the TV advertising business in Russia, serving as the media sales agent (wholesale broker of advertising airtime) for Channel 1, Channel 5, and Channel 6.[3]

According to the Moscow organized-crime squad, one of the founding shareholders of Premier S.V. had earlier been arrested on racketeering charges. Two colleagues in the graphics and design part of the business had been murdered. The chief financial officer of Premier S.V. was Aleksandr Averin. Known in the underworld as "Avera Junior," he was the younger brother of Viktor Averin, the right-hand man of Mikhas, the boss of the Solntsevo Brotherhood.[4]

This was the crowd Berezovsky would have to deal with to break into the advertising business at Channel 1. By late 1993, he had established two media subsidiaries, Logovaz-Advertizing and Logovaz-Press, but his first attempts to carve out a niche in the mass media market were unsuccessful. In the summer of 1994, with his political influence growing inside the Kremlin, he got his chance. His point man was his trusty Logovaz partner, the man who, according to police sources, served as his go-between with Russian organized-crime groups: Badri Patarkatsishvili.[5]

Sometime in June or July, at the height of the mob war, with Berezovsky recuperating abroad from the failed car bombing and Badri managing affairs in Russia, Logovaz gained a foothold in the Channel 1 advertising market. Despite its bloody battles with Solntsevo-connected gangsters like Sylvester and Igor Ovchinnikov during the mob war, Logovaz signed a deal with Sergei Lisovsky, the Solntsevo-connected advertising mogul. Lisovsky had gathered the major Channel 1 advertising outfits together to form Advertising-Holding to monopolize the sale of advertising on the network. Berezovsky's fledgling Logovaz-Advertizing was included as one of the founding shareholders of Advertising-Holding. (Lisovsky held a 49 percent stake in the new company and served as its general director.)[6] The new monopoly's task was to push the independent production companies out of the advertising market (previously Channel 1 had allowed production companies to put their content on the network for free and to claim a good part of the advertising revenue). The plan was for Channel 1 to buy the programming itself and get paid back with advertising revenues. The airtime would be sold to advertisers (on a commission basis) by Advertising-Holding. The arrangement involved numerous conflicts of interest and opportunities for self-dealing. Berezovsky's Logovaz, for instance, had both an advertising company that bought airtime from Advertising-Holding and a share in Advertising-Holding itself. Lisovsky's position was even more dubious: He controlled Advertising-Holding, bought airtime from it, and was one of the biggest content providers for the network. Such conflicts of interest offered the possibility to transfer revenues out of the network and into the intermediaries.[7]

Berezovsky was not satisfied with just a piece of the advertising business at Channel 1. He wanted to control the network itself. "I did not intend to make money [with Channel 1]," he told me in 1996. "I regarded it as an important source of political influence. All the subsequent events—especially the presidential elections—have confirmed that."[8]

In the late summer of 1994, having largely sat out the voucher-privatization process, Berezovsky began lobbying General Korzhakov and President Yeltsin to privatize Channel 1. He argued that, with the 1996 presidential elections less than two years away, Channel 1 would be a key weapon in the campaign. He promised that it would be the "president's channel."[9]

"Berezovsky's plan, which he repeatedly explained to me," Korzhakov would later recall, "was to leave the majority stake in the TV channel in the hands of the government, while a group of politically loyal investors would take 49 percent. He argued that only in this way would the channel become financially self-sufficient, while simultaneously allowing the president to control it."[10]

On November 30, 1994, President Yeltsin signed the decree privatizing Channel 1. The new company would be called ORT (Russian Public Television).

The Enemy

That Berezovsky recognized the power of television in 1994 was testimony to his ability to be several moves ahead of his peers. There was, however, one man who had understood the importance of TV before him: Vladimir Gusinsky. A prominent business magnate, Gusinsky was the owner of Most Bank—one of Moscow's largest retail institutions—and NTV, Russia's first independent TV network. Berezovsky regarded Gusinsky as his archrival. The origins of the epic feud are unclear. In one interview, Gusinsky placed its beginnings in the winter of 1993–94—the time of the mob war. Another version is that the two men competed over whether the national airline Aeroflot would be serviced by Gusinsky's Most Bank or Berezovsky's Avtovaz Bank. Others place the feud in the auto-dealership business—a rivalry between Logovaz and dealers allied with the Moscow city government. Still others locate its origins in television. "When NTV was founded, Berezovsky spent a lot of energy trying to get the channel shut down," General Korzhakov would later remember.[11]

In many ways the two men were mirror images. Berezovsky in particular seemed to chart much of his business career by following Gusinsky. In 1989, when Gusinsky established a joint venture with an American firm, Berezovsky formed his own joint venture with the Italians. In 1991, when Gusinsky established a bank, Berezovsky formed his own bank. When Gusinsky's bank started handling Aeroflot's account, Berezovsky took the account for himself. After Gusinsky started one of Russia's leading newspapers, Berezovsky, too, bought a newspaper. When Gusinsky established a TV network, Berezovsky decided he had to have a TV network as well.

In contrast to Berezovsky, who liked to take over existing enterprises, Vladimir Gusinsky created entirely new companies. He added value to the Russian economy. Irrespective of how he had accumulated his capital, he established a retail banking network, constructed buildings, started newspapers and magazines, created a TV network. Unlike Berezovsky, Gusinsky could legitimately claim to have played a constructive role in the Russian economy.

Gusinsky's past is murky. His official biography reports that he was born into a Jewish family in Moscow in 1952. In the 1970s, he made money as a small-time black-market trader and gypsy-cab driver. In the 1980s, he became part of the Communist establishment, organizing festivals and theatrical events for the Communist Youth League. In 1986, the Youth League chose him to be the director of the cultural program of the Goodwill Games (an imitation of the Olympics, organized by the American TV magnate Ted Turner) in Moscow. Gusinsky made valuable acquaintances in the American business community, and in 1989 convinced the business subsidiary of the American white-shoe law firm Arnold and Porter to form a joint venture called Most ("bridge" in Russian). The joint venture, which offered consulting services to American companies doing business in the Soviet Union, did not last long. Gusinsky soon dropped his American partners and reincorporated Most as a bank and holding company.[12]

Gusinsky's patron was Moscow mayor Yuri Luzhkov, a tough political boss from the Soviet era. Gusinsky had met Luzhkov in the late 1980s, when he established several businesses selling clothing and jewelry, "importing computers," and "producing construction materials."[13]

"[Luzhkov] at that time was responsible for the cooperatives [private businesses] of Moscow," Gusinsky recalled. "All the cooperative owners of Moscow knew that if you needed something, you had to go to Luzhkov, and it wasn't necessary to call beforehand."[14]

Luzhkov was an excellent man to have on your side. He was a highly effective city boss. He liked to refer to himself as a *khozyaistvennik* (business manager). Unlike most Russian managers, he was comfortable discussing business plans, cash flows, and economic statistics; he neither drank nor smoked—another rarity in Russia. In contrast to Yeltsin's young reformers, who gave away the bulk of the state's industrial holdings, Luzhkov preferred to hang on to city-owned property and sell it only for a decent price. In real estate, for example, the city of Moscow received more than $1 billion annually from renting or selling its properties—ten times more than the Russian government received from its holdings throughout the country. The city owned and operated two big auto plants, an oil company, the local phone and electric utilities, a TV network, one of Moscow's airports, a fast-food chain, part of the local McDonald's chain, dozens of hotels, and hundreds of shops and restaurants. While Yeltsin's government presided over one catastrophe after another, Luzhkov's city government was manifestly an institution that worked. The city was able to meet its social welfare obligations. Utility, transportation, and public-housing costs were kept low; pensions were paid promptly. The roads were relatively well maintained. The subway trains rolled on time. Every year, hundreds of new housing blocks, schools, and hospitals were built.[15]

In a country where everything was collapsing, Moscow was an oasis of prosperity and success. Luzhkov's strategy was not to fight organized crime (he did not have the power), but to tax it. He presided over a city plagued by casinos, corruption, and gangland shoot-outs, but he also convinced even the shadiest businesses to contribute to his ubiquitous civic projects. An estimated $350 million went to build the opulent underground shopping mall at Manezh Square, another $100 million to rebuild the Luzhniki soccer stadium, more than $100 million to expand the Ring Road highway, and an estimated $1 billion to reconstruct the mammoth Cathedral of Christ Savior. One of Luzhkov's greatest achievements was the restoration of the historic city center. The prerevolutionary architecture of Moscow had deteriorated badly under the Soviets; under Luzhkov the buildings were quickly restored. Churches, which had served as warehouses or offices during the Soviet era, were reconsecrated and restored. Moscow was in the grip of a building boom. New office buildings were going up; construction cranes were everywhere.[16]

Luzhkov's support allowed Gusinsky to make the Most Group one of the largest business conglomerates in the country. Most Bank was one of the pioneers of retail banking in Russia. The key to its success was its access to the funds of the Moscow municipal government. Apart from the bank, the Most Group included media companies, an insurance company, a security agency, a trading company, construction materials producers, and a real-estate development company.[17]

Western journalists marveled at Gusinsky's wealth—they reported on his BMW, his armored Mercedes, his corporate jet, his Chelsea apartment in London, his villas in Portugal and Spain, his children in Swiss boarding schools, his lavish funding of Russian political candidates.[18]

Perhaps most extraordinary was not Gusinsky's wealth, but his security service. All big Russian businesses had sophisticated security services, but Most's numbered around 1,000 men, and perhaps more. According to some estimates, it was one of the largest armed formations in the Russian capital (after the army and the police). More important, the Most security service had superb intelligence capability, with top surveillance and disinformation professionals hired from the KGB.[19]

One of the chiefs of this private army was General Philip Bobkov, the former first deputy chairman of the KGB. In April 1995, when the distinguished Russian-American political scientist Dimitri Simes confronted Gusinsky in Washington about the presence of Bobkov and other top KGB men employed by the Most Group, Gusinsky replied: "We'd be ready to hire the devil himself if he could give us security."[20]

Later, Gusinsky told an American reporter about the former KGB officers on his payroll: "These people are like fine cars. They'll go wherever you drive them. They're professionals, not politicians... Bobkov is a good car. With a good car, whoever is at the wheel is in control of where the car is going."[21]

Gusinsky was intent on becoming Russia's Rupert Murdoch. By late 1994, he had established a respectable newspaper called *Sevodnya*, concluded a joint venture with *Newsweek* to publish a weekly news magazine called *Itogi*, created Russia's version of *TV Guide*, and established *Ekho Moskvy* as the most influential radio news station in the country.

The centerpiece of his media empire was NTV, the independent television network established in the autumn of 1993 by President Yeltsin's special decree. Gusinsky raided the state-owned Channel 1 for the best onscreen talent, such as TV reporter Yevgeny Kiselev, who quickly became the most influential anchorman in the country; the Channel 1 producers were encouraged to defect by the prospect of higher salaries and equity in the new network. The result was impressive. NTV immediately stood out among Russian TV stations. Even though the network presented a lot of pornography, horror, and gore (even by American standards), few could deny its excellent production values.

The creation of NTV was opposed by Russia's established TV production companies, whose executives were jealous of the special favors granted to the newcomer. These producers, led by a Berezovsky ally named Irina Lesnevskaya, succeeded in getting Yeltsin to freeze implementation of the decree forming NTV in late 1993. There was an immediate response from the Gusinsky camp. According to Lesnevskaya, she received a call from Igor Malashenko, the director of NTV, the very same day the decree was tabled.

"Did you do this?" demanded Malashenko.

"I don't know if it was me or not, but I did have a part in it," Lesnevskaya claims she replied.

"Gusinsky wants to see you," said the TV manager.

"I am not acquainted with him," Lesnevskaya replied. "If he wants to, he can come to REN-TV."

Gusinsky then picked up the phone, Lesnevskaya recalls, and "very politely, almost gently" said that he was not able to come to her and asked if she and the heads of all the other independent TV production companies would visit him. Lesnevskaya, her son, Dimitri (also a TV producer), and three others went to see Gusinsky.

"Over the course of forty minutes [Gusinsky] ran around the room, shouting and screaming," Lesnevskaya asserts. He was outraged that "he hadn't taken into account one stupid wench [Lesnevskaya?]...that he had been pushing through this decree [to create NTV] for half a year and now the decree has been stopped.

"He tried to frighten us. He insulted us. He told me [Lesnevskaya]: 'You have one son. What could be more important for a mother than that her son should be alive and well? But on the 101st-kilometer mark of the Moscow highway, there are always some kind of incidents. He could be accidentally hit by a car, his car could turn over, burn up...'"

Lesnevskaya says she and the other TV producers then dropped their opposition to the creation of NTV; the presidential decree went into force soon thereafter.[22]

Once Gusinsky told a journalist what kept driving him forward at such breakneck speed. "There is a traveling Russian circus and one of the little booths features a squirrel on a wire-mesh wheel," Gusinsky explained. "I feel like that squirrel, busy running on that wheel. The squirrel thinks it is turning the wheel, but the wheel is turning the squirrel. And if the squirrel stops running, its paws get caught between the wires, and the squirrel breaks its paws."[23]

"Faces in the Snow!"

In the late summer of 1994, when Berezovsky returned from his convalescence in Switzerland, he went to see General Korzhakov in the Kremlin to discuss security matters. Although Berezovsky denies it, Korzhakov says that at this meeting and several subsequent ones the tycoon tried to convince him to organize the assassination of Gusinsky. "Berezovsky insisted that Gusinsky was an enemy of the president and represented a mortal danger to Yeltsin," Korzhakov later recalled. "He would usually say: 'Aleksandr Vasilievich, whatever we do, however much we work, as long as these people exist, all our efforts will be in vain.'"[24]

Korzhakov says that, as proof of Gusinsky's bloodthirsty nature, Berezovsky cited the car bomb that had nearly killed him. It was Gusinsky's idea, the work of the "scum" that ruled Moscow, Berezovsky apparently declared.[25]

Answering this accusation in public, Gusinsky told a Moscow newspaper: "It was I and the Most Security Service who evacuated

Berezovsky and his family after the assassination attempt, and some-time later I visited them in Switzerland."[26]

Still, according to Korzhakov, Berezovsky was adamant that Gusinsky must be killed. I asked the security chief what language Berezovsky used in making this request. "In this situation, Bere-zovsky would use special terminology," Korzhakov says. "Instead of the word 'kill,' he preferred to say 'terminate.' It was a term out of the gangster lexicon."[27]

If Berezovsky did ask General Korzhakov to kill Gusinsky, why did the security chief not arrest him on a charge of attempted mur-der? It may be that at this early stage in the Yeltsin era this kind of dirty laundry (one businessman asking a government official to as-sassinate another businessman) was not supposed to be aired in pub-lic. Another factor may have been that such was the contempt for the rule of law in Russia at this time that the only question was not whether Berezovsky had made the assassination request or not, but whether the presidential security chief would carry it out. In any case, Korzhakov says he adopted a contemptuous attitude toward the affair. "Berezovsky thought that the SBP had been created pre-cisely to 'eliminate' people whom Berezovsky didn't like," he says. "From this moment on, I adopted the firm conviction that Bere-zovsky was psychologically deranged and already began to keep an eye on him."

Korzhakov claims he felt sorry for Berezovsky. The big security chief clearly considered the fidgety financier a weakling who had to be protected. "This man had never served in the army, never en-gaged in any sports. I think that explosion [the car bomb] really shook him up. After all, he had seen his driver decapitated in front of him. I think Berezovsky never fully recovered from the shock."[28]

Korzhakov goes on to say that once Berezovsky saw that he was not going to act on his request, he went over his head—to the Rus-sian president. Berezovsky began to convince Yeltsin's favorite daughter and most trusted adviser, Tanya Dyachenko, that Gusinsky was a murderous mafioso and a ruthless enemy of the government.[29]

Neither Gusinsky nor his patron, Yuri Luzhkov, had made any

moves against the Yeltsin regime. Aware that his success in running Moscow was a slap in the face of Yeltsin's incompetent ministers, Luzhkov was careful never to challenge Yeltsin directly. The mayor had gotten his start in politics under Yeltsin's patronage and the two men remained personal friends. In October 1993, when he could have swung the balance of power against Yeltsin, Luzhkov stood firmly by the Russian president's side. Still, the city of Moscow was the only political entity that was at least partly independent of the Kremlin (just as Gusinsky's NTV was independent of state media policy). For Yeltsin, that was threatening enough.[30]

General Korzhakov says that in the late autumn of 1994, he noticed that Berezovsky's lobbying efforts with Yeltsin's daughter had paid off. The anti-Gusinsky message had gotten through to the president.

"Once at dinner," Korzhakov recalls, "the president raised his voice and turned to Barsukov and me: 'Why can't you deal with this guy Gusinsky! What is he doing here! Why is he wheeling around all over the place! Everyone is complaining about him, even my family. How many times has it happened that Tanya or Naina is driving along and the road gets cleared to let this Gusinsky go by. His NTV has gotten too big for its boots. It is acting insolently. I order you to do something about him.'"[31]

Korzhakov understood. "This tirade showed that Berezovsky had found the right path to Yeltsin's ear," he would later recall. The next day, Korzhakov dutifully executed the order. He knew that Gusinsky left his suburban residence every morning and took the Uspenskoe Shosse into central Moscow to work. Every day Gusinsky's convoy—the tycoon's armored Mercedes and several other cars carrying heavily armed bodyguards—could be seen speeding down the avenue at 90 mph, swift as eagles, ignoring stoplights and traffic. Because Yeltsin and his family often used the same route into the city, Korzhakov claimed that Gusinsky's motorcade presented a security threat to the president. The security chief decided to put the motorcade under what he called "demonstrative observation."[32]

On Friday, December 2 (two days after the ORT decree was

signed), Vladimir Gusinsky left his dacha and headed toward the center of Moscow. His motorcade consisted of the Mercedes, a Jeep, and a Ford, containing, altogether, a dozen bodyguards and drivers. Almost immediately, Gusinsky's bodyguards noticed that they were being followed.

A high-speed chase ensued. Korzhakov's cars tried to wedge their way between Gusinsky's car and those of his bodyguards. The Gusinsky convoy swerved away and increased its speed. Gusinsky's security men thought they were under attack by gangsters. There were no distinguishing marks on the vehicles identifying them as cars belonging to the government.[33]

The motorcade finally arrived at the Most Group headquarters. Gusinsky ran into the building, while his bodyguards remained in the parking lot. The SBP men surrounded the building and closed off the exit ramps. From his office above, Gusinsky called the Moscow organized-crime squad, RUOP. The tough men from RUOP soon arrived, confronted the SBP men, noted their license plates, discovered that the cars belonged to the Russian armed forces, and slunk away.[34]

At five o'clock, the SBP detachment was reinforced. Two dozen masked men arrived in buses. They presented no documents, sported no identifying insignia. They were dressed in full combat gear, with face masks, and were armed with handguns and submachine guns. Gusinsky's drivers and guards had locked themselves in their cars. After numerous threats, they were dragged out and pushed to the ground. "The driver of Gusinsky's armored Mercedes locked himself inside the car," Korzhakov would later recall. "He categorically rejected all the demands that he get out. Then a grenade was put on the roof of the car. He jumped out as if he had been stung. But the grenade was harmless—it hadn't even been primed."[35]

Gusinsky's security men were overwhelmed. Moscow TV cameras captured the image of a masked SBP man kicking a prone Most security guard in the groin. This part of the operation became known in popular parlance as "Faces in the Snow!" The Most men spent more than an hour facedown in the parking lot, while the SBP men trained guns on them.[36]

Desperate, Gusinsky called his friend Yevgeny Savostyanov, head of the Moscow FSB–KGB. Korzhakov's men were listening in on the phone call. "Zhenya [a nickname for Yevgeny], save me," Gusinsky screamed into the phone. "Some kind of bandits have arrived. I called in the cops and they arrived, but couldn't do anything. You're my only hope."[37]

The FSB–KGB men soon arrived and confronted the attackers. These guys were more self-confident than the Moscow RUOP men. Their guns appeared. One SBP vehicle had its tires shot out and an SBP man was clobbered over the head with a pistol grip. But, like their predecessors, the FSB men got nowhere; intimidated, they eventually turned around and left the SBP to do whatever they wanted.[38]

The siege lasted until 10 p.m., when Korzhakov's men finally left. About a dozen Most men were taken to the local police station and booked for illegal possession of firearms; three Most men ended up in the hospital.[39]

Korzhakov later claimed that a search of Gusinsky's cars revealed an unregistered Makarov pistol, three pump-action shotguns, a fake identification card of a Moscow police officer, an unregistered radio tuned to the police frequency, and a radio scanner capable of intercepting phone calls. According to a later legal investigation, the weapons were accompanied by the proper documentation. Philip Bobkov and the other KGB specialists who ran Gusinsky's security operation were not amateurs.[40]

Moscow was shocked by the episode. Several days later, Berezovsky's newspaper, *Nezavisimaya Gazeta*, offered an explanation. It said nothing about Berezovsky's purported role in inciting the raid. Citing top Kremlin sources, the newspaper reported that the raid had been ordered by Yeltsin after he had read an intelligence brief about Gusinsky's political ambitions. The brief claimed that Gusinsky and his political patrons [meaning Luzhkov] were scheming to have their ally General Boris Gromov appointed minister of defense, so that the respected veteran of the Afghanistan War would become a national political figure and run for president in 1996.[41]

The Gusinsky-Luzhkov group was destroyed in days. General

Gromov was dismissed from his post as deputy minister of defense. Yevgeni Savostyanov, head of the Moscow FSB–KGB, who had come to Gusinsky's aid during the siege, was fired as well. Gusinsky sent his wife and children to the West. Soon afterward, Gusinsky also fled Russia. He would remain in the West for at least three months.[42]

Meanwhile, Korzhakov swaggered before the Moscow press. "Hunting geese is an old hobby of mine," he quipped. (The Russian word for goose is *gus*—the root of the name Gusinsky.)[43]

In January 1995, there were rumors of a warrant out for Gusinsky's arrest. Again, it was Berezovsky's *Nezavisimaya Gazeta* that reported the situation in the most lurid colors. Again citing a "well-informed source in the Kremlin," the newspaper reported a coordinated campaign being prepared against Gusinsky (and his patron, Luzhkov), beginning with compromising materials in the press and ending with arrests and criminal trials.[44]

Gusinsky responded by giving an interview to *Euromoney*, in which he claimed that the December 2 attack on Most was part of an assault on Moscow mayor Yuri Luzhkov. Gusinsky's deputy, Sergei Zverev, added: "I think that one of the reasons [for the assault] was that our competitors wanted to damage us."[45]

Gusinsky's prime competitor was, of course, Berezovsky. The auto dealer had succeeded in driving his rival out of the country. Luzhkov withdrew into the shadows and loosened his business ties with Gusinsky. On Russian Orthodox Christmas, in early January 1995, Luzhkov and Yeltsin's prime minister, Viktor Chernomyrdin, made peace when they both participated in laying the cornerstone for the Cathedral of Christ the Savior. Several months later, Luzhkov established the Bank of Moscow and transferred many of the municipal accounts from Gusinsky's Most Bank to the new institution.[46]

By this time, it was no longer Vladimir Gusinsky who was in danger of being arrested. It was Boris Berezovsky—in connection with Russia's most famous assassination: the murder of TV producer Vladislav Listyev.

Assassination

The original idea to privatize Channel 1 had come not from Boris Berezovsky but from Vlad Listyev—Russia's most popular talk-show host and the country's most successful TV producer. An opinion survey had identified him as one of the most trusted people in the country, next to the head of the Russian Orthodox Church and the straight-talking paratroop general Aleksandr Lebed. Vlad Listyev had risen to prominence through his talk show *Look*, which began in 1988 and quickly turned into a refreshing forum for honest discussion and criticism of the Communist Party. *Look* remained one of Russia's top-rated shows for six years, rivaled by *Field of Miracles*—another Listyev vehicle—a game show modeled on *Wheel of Fortune*. It was Listyev who approached Berezovsky with the idea of helping a group of independent TV producers privatize Channel 1; as the network's leading producer and the originator of the privatization idea, Listyev was a natural candidate to lead the new company.[47]

As the time for the privatization approached, however, Listyev grew concerned about Berezovsky's efforts to dominate the network. There were indications that Berezovsky wanted someone else to be Channel 1's general director. Some Logovaz staff members were pushing a close Berezovsky ally, the producer Irina Lesnevskaya, to assume the post. But Vlad Listyev was named the general director and Berezovsky was appointed deputy chairman.[48]

"The privatization of Channel 1 took place at the beginning of 1995," General Korzhakov would later recall. "There was no auction—either public or private—for the ownership of the 49 percent stake—Berezovsky himself decided who would get shares and how many they would get."[49]

In some cases, the choice of shareholders was rudimentary. Some of the private banks, for example, were summarily informed by Berezovsky that they were shareholders of ORT. The new owners of the network were chosen in secret, on the basis of a personal agreement among those involved. Since Russian law stated that all privatizations had to be carried out through a public auction, the privatization of ORT was technically illegal.[50]

Among the private shareholders were such well-connected institutions as Menatep Bank, Stolichny Bank, Alfa Bank, Natsionalny Kredit Bank, Gazprom, and the National Sports Fund. Clearly the choice of co-owners was dictated not by a company's financial resources but by its relationship with Berezovsky himself, since the ORT shareholder register omitted such Russian powerhouses as Lukoil, Onexim Bank, and Inkom Bank.[51]

The total share capital of ORT was $2 million. Berezovsky's companies bought 16 percent of the shares. Berezovsky had voting control over a further 20 percent. Thus, by investing a mere $320,000, he gained control of Russia's most important TV network. Would he be expected to fund ORT's operating deficit out of his own pocket? Not necessarily. The state, with 51 percent of the shares, could be expected to continue its massive subsidies to the network.[52]

Immediately after ORT was privatized, the general director, Vlad Listyev, decided to focus on the operation that was costing the network millions of dollars in lost revenues: advertising sales. He began negotiations with the head of Advertising-Holding, Sergei Lisovsky. The advertising magnate apparently offered to pay for the ORT franchise and retain monopoly control. But the negotiations dragged on.

"Around New Year I already knew that Vlad would be killed," one of Listyev's close friends and business associates would tell me. "He had gotten involved with people who had built up their careers with completely criminal methods."[53]

On February 20, 1995, Listyev announced that he was breaking Lisovsky and Berezovsky's advertising monopoly and instituting a temporary moratorium on all advertising until ORT could work out new "ethical standards."[54]

"The end of advertising [at ORT] meant millions of lost earnings for Lisovsky and Berezovsky personally," noted General Korzhakov.[55]

Listyev knew he was playing a dangerous game. According to one field report compiled by detectives of the Moscow organized-crime squad, RUOP, Listyev knew he was being stalked and would probably not survive the spring. According to this report, in late February Listyev explained to his closest friends why he would be killed. When he embarked on his project to end the advertising monopoly, Listyev was approached by the advertising mogul Lisovsky, who demanded $100 million in damages and threatened to kill him. Listyev said he had found a European company willing to pay even more for the ORT advertising franchise: $200 million. Listyev turned to the primary backer of ORT, Boris Berezovsky, to handle the transaction and pay the $100 million to the disgruntled Lisovsky. The money was transferred to one of Berezovsky's companies. When Listyev asked Berezovsky to release the funds, the car dealer refused. Berezovsky made a vague promise to provide funds in three months.[56]

There were other versions of what was happening at ORT at this time. According to one of Moscow's larger private investigative agencies, Listyev's prohibition of advertising on ORT was merely a tough bargaining position. He was simply soliciting bids for the ORT advertising franchise; Lisovsky offered ORT $100 million, but Listyev was holding out for $170 million.[57]

According to his own later explanation, Berezovsky and his aides were engaged in some unusual negotiations with various organized-crime groups at this time. Sometime in early 1995, Moscow police allegedly interrogated a gangster boss sitting in jail. The gangster declared that he had been approached by Berezovsky's aide, Badri, with a contract for Listyev's assassination. Before the gangster managed to fulfill the contract, he was arrested in a Moscow-wide anti-crime sweep and thrown in jail. Berezovsky himself was observed by police negotiating with another known gangster. On February 28, the day before Listyev's murder, Berezovsky met with a thief-professing-the-code (mob boss) identified by police only as "Nikolai" and gave him $100,000 in cash.[58]

On the night of March 1, Listyev returned home from work to find his assassin waiting for him. In the grubby lobby of his apartment house, he was accosted by this person and shot dead.

"Moscow Is Ruled by Scum"

Berezovsky had left the day before the murder, tagging along with Prime Minister Chernomyrdin on a state visit to Britain. When he was informed of the murder, he immediately chartered a private plane and flew back to Moscow. There he attended the memorial service for Listyev at Ostankino.[59]

"On Friday, while I was attending the memorial service for Vlad at Ostankino, I was called by my assistants and told that Logovaz was about to be searched and visited by the OMON [paramilitary police]," Berezovsky recalls. "I was terribly surprised."

Yeltsin at this time was out of Moscow, so Berezovsky decided to use other political patrons. "I turned to [First Deputy Prime Minister] Oleg Soskovets, who happened to be right there, asked him to help me contact the Ministry of Internal Affairs [which had overall responsibility for the police]." Soskovets contacted Minister of Internal Affairs Viktor Yerin, who assured him that no one would touch Logovaz or Berezovsky.[60]

At 3 p.m., when Berezovsky returned to Logovaz House from the memorial service, the courtyard was swarming with detectives from the organized-crime squad and Kalashnikov-toting members of the OMON paramilitary unit. They presented a search warrant and a warrant empowering them to interrogate Berezovsky as a witness in the Listyev case.

Berezovsky refused to let them search his building. He demanded to know their intentions. The standoff continued until midnight. Finally, the detectives asked Berezovsky and his assistant, Badri, to come down to the police station for questioning. Berezovsky knew that if he went he faced the real possibility of being thrown in jail—then his ability to block the investigation would be severely limited. So he picked up the phone and called Russia's acting prosecutor-general, Alexei Ilyushenko. Russia's top law-enforcement official told his deputy to order the police detectives to take Berezovsky's and Badri's statements at Logovaz House, rather than at the police station. They did so and left.[61]

Berezovsky's troubles were not over. He knew that he was one step away from being arrested. He knew that with the police evidence mounting against him, his survival depended largely on his ability to convince President Yeltsin that the affair had been part of a massive conspiracy against him. Yeltsin was still away, so Berezovsky went to Korzhakov's office in the Kremlin and asked to make a direct video-taped appeal to the president. He asked Irina Lesnevskaya, one of the main TV producers on Channel 1, to accompany him; Lesnevskaya was both an indirect shareholder in ORT and a good friend of President Yeltsin's wife.[62]

The appeal is fascinating (see Appendix I). Berezovsky and Lesnevskaya sit side by side. Both speak into the camera, addressing the Russian president directly, intimately, as "Boris Nikolaevich," using his patronymic. Berezovsky, in his usual dark business suit, speaks quickly, nervously. Irina Lesnevskaya is emotional, almost tearful.[63]

"I know who killed Vlad [Listyev]," Irina Lesnevskaya begins. She explains that Vlad Listyev had been killed by rival media magnate Vladimir Gusinsky, by the mayor of Moscow, and by the old KGB.

"Boris Nikolaevich, now everything must be done so that the case would be handled personally by Korzhakov and the FSB, above all not by the police," she pleads. "Because currently a theory is being prepared and witnesses are already on hand to the effect that Vlad was killed by Berezovsky.... When we were called into the prosecutor's office and asked questions, we understood that they have only one theory: that the murderer is either Berezovsky or his chief deputy, Badri Shavlovich [Patarkatsishvili], that all the evidence points only to Berezovsky."

Lesnevskaya puts the Listyev murder in a purely political context, appealing to Yeltsin's paranoia. "Now we have come to Korzhakov, because we understand that what has happened is coup d'état," Lesnevskaya continues. "This is worse than the White House [the battle over the parliament in 1993]. This is worse than the GKChP [the Putsch of 1991]. This is inside the city. An enormous organization has been created. It controls everything—all the Mafiya organizations, all the criminals—decides who should live and who should die...."

Berezovsky, sitting nervously at Lesnevskaya's side, nodding in agreement, breaks in. "Boris Nikolaevich," he addresses Yeltsin. "I would like to draw your attention to the fact that when you are absent from Moscow, when there is no one in Moscow who is close to you, Moscow sinks into absolute lawlessness.... Boris Nikolaevich, you no longer rule in Moscow. Moscow is ruled by scum!"

This was an appeal to Yeltsin's political self-interest (the president always rose to the challenge of destroying his political enemies). Berezovsky left Lesnevskaya to make most of the specific allegations.

"I have no doubts that this...jesuitical plan to murder Vlad was thought up and executed by the Most Group, by Mr. Gusinsky, by Mr. Luzhkov, and by that organization that is supporting them— the enormous pyramid with all its branches—the former KGB," Lesnevskaya declares in the videotape.

The police investigation is all part of the same conspiracy. Why is Berezovsky being set up? To get at President Yeltsin of course! "The country is outraged [over the Listyev murder]: 'Get rid of all the power ministers!' [people are saying]. 'Get rid of Yeltsin!'" Lesnevskaya tells the president. "And you no longer have a leg to stand on. They have identified the murderer—Berezovsky and Badri...."

Though other individuals were under suspicion in the Listyev murder—on the same day that they tried to search Logovaz House, police tried to search the offices of advertising magnate Sergei Lisovsky, as well—there was strong evidence pointing to Berezovsky. How could he explain handing a known gangster boss $100,000 two days before the Listyev killing? Berezovsky admitted to Yeltsin that he had paid the gangster boss, but claimed that he had handed over the cash to discover who was behind the car bomb outside Logovaz the previous summer. Moreover, he had met the gangster in the presence of two detectives from the Moscow police department and had asked two of his own security agents to record the meeting on video "to prove that I was being blackmailed."[64]

But this alibi is problematic. In their videotaped appeal, Berezovsky and Lesnevskaya spend most of the time trying to prove that the four witnesses to Berezovsky's transaction with the gangster boss—and hence the foundation of his alibi—were in fact agents of the political opposition. The two police officers are not reliable witnesses since, according to both Berezovsky and Lesnevskaya, the Moscow police force is controlled by Mayor Luzhkov, who is carrying out some kind of "jesuitical plan" to overthrow Yeltsin. Even more curious is the role of the other two witnesses—Berezovsky's security men. Lesnevskaya, with Berezovsky's silent acquiescence, claims these two men are Gusinsky's agents. (If this is correct, the two security men would likely be witnesses for the prosecution against Berezovsky, rather than witnesses for his defense.) Moreover, the two security guards, according to Lesnevskaya, are actually the perpetrators of the murder of Vlad Listyev—though not on Berezovsky's orders, but on Gusinsky's. (Berezovsky does not accuse anybody specific of the killing, but supports Lesnevskaya's general assertion that a dark conspiracy is under way to overthrow Yeltsin.)[65]

Berezovsky made the backbone of his defense political. "Boris Nikolaevich, I do not doubt your intentions, your endless dedication to the tasks which you perform," he told Yeltsin. "But your people are being set up. One by one, your people are being eliminated from under you."

Apart from appealing to Yeltsin's constant fear of political enemies (such as Luzhkov), Berezovsky cleverly gathered the support of people the Russian president knew and trusted, such as Valentin Yumashev, the presidential biographer, who was present in Logovaz House when the police were trying to search the premises. Berezovsky even appealed to Yeltsin's feverish concern about his image in America, casually mentioning that during his showdown with the police he was entertaining "representatives of Radio Liberty," who "witnessed this whole shameful affair." But the most powerful argument was still Berezovsky's contention that he had been set up for the Listyev murder because he was Yeltsin's loyal friend.

The appeal worked. The men in charge of the Listyev investigation, Moscow prosecutor-general Gennady Ponomarev and his deputy, were summarily fired. The police were ordered to lay off Logovaz and Berezovsky. "He openly used his political contacts to escape the inquiry mandated by law," Korzhakov would later observe.[66]

Berezovsky may have been correct that he was framed for the Listyev murder, but was rival TV magnate Vladimir Gusinsky really behind the incident? Quite apart from the fact that Gusinsky was never questioned by any law-enforcement agencies about the killing, the accusations made against the NTV boss in the Berezovsky-Lesnevskaya videotaped appeal hardly seem credible. Why would Gusinsky want Listyev dead? Even though Listyev was Russia's most popular TV personality, his murder could hardly be expected to set off a revolution, as Lesnevskaya implied in her appeal to Yeltsin. Did Gusinsky merely want to stop the creation of ORT as a rival TV network? Killing Listyev was clearly not the way to do it. Was the purpose of the murder solely to frame Berezovsky? Unlikely, since Gusinsky had been on the run for several months, hiding in Western Europe since Korzhakov's SBP had raided Most Group headquarters in December 1994. Finally, if Gusinsky really had organized the murder and was framing Berezovsky as part of a conspiracy to get rid of Yeltsin, the security services personally loyal to the president—the SBP and the FSB–KGB—would have had every reason to clamp down on the TV magnate. Instead, they did nothing and Gusinsky returned peacefully to Moscow several months later.

When I asked him about Listyev's murder, Berezovsky categorically denied having anything to do with it. He blamed unnamed advertising and production companies hurt by Listyev's reorganization of the network. In other words, he set the Listyev murder entirely in a commercial context. It was an explanation quite different from the one he gave President Yeltsin in his videotaped appeal.[67]

Berezovsky Becomes a Media Mogul

The public outcry over Listyev's killing was immense. Tens of thousands of mourners attended the funeral—the biggest turnout since Andrei Sakharov's funeral in 1989. People threw flowers and wept openly on the streets.

Yet, for all the importance the Russian nation attached to Listyev's assassination, the subsequent investigation was a farce. The first prosecutors to look into the case were fired. Five months later, the federal prosecutor's office announced that it had identified the people who ordered the killing. The next day the prosecutor's office recanted, saying that the investigation was continuing. Two months later, Alexei Ilyushenko, the acting prosecutor-general and the man who had helped Berezovsky avoid arrest, was dismissed; he was later arrested on charges of corruption, fraud, and embezzlement connected with one of the subsidiaries of Berezovsky's future oil company, Sibneft.[68]

In the summer of 1997, Moscow newspapers reported that a man named Igor Dashamirov had been arrested in Tbilisi, Georgia, and had been extradited to Moscow for questioning in connection with the Listyev murder. Dashamirov was reportedly a member of the Solntsevo Brotherhood. But again there was no follow-up.[69]

One evening in early 1999, I visited Pyotr Triboi, the investigator from the prosecutor's office charged with solving the four-year-old Listyev murder case. Triboi's office had the musty silence of a tomb. A pale bureaucrat in a Soviet-era gray suit, Triboi admitted that the investigation was at a standstill. None of the leads had panned out; everything had been checked and nothing had produced results. Triboi seemed resigned to his failure to solve the case. Even more worrisome, he didn't seem to understand some of the key details of the case, specifically the nuances of the advertising business of ORT—the business that almost everyone thought was at the root of Listyev's murder.[70]

Today, the Listyev killing remains a mystery. Whatever the nature of the Berezovsky-Listyev-Lisovsky negotiations and whoever was responsible for the murder of Vlad Listyev, Boris Berezovsky and advertising mogul Sergei Lisovsky both emerged unrivaled in their control of ORT. A few months after Listyev's assassination, the network announced the end of the advertising moratorium. A new company, called ORT-Advertising, would take over the advertising franchise, with the monopoly right to sell advertising time on a commission basis. The boss of ORT-Advertising was none other than Sergei Lisovsky.

Later, when I asked Berezovsky about the theory that he and Lisovsky were jointly responsible for Listyev's death, Berezovsky quickly drew a line between himself and the advertising mogul. "At that time, when we were considering the reorganization [of advertising], we were working against Lisovsky, because we were dissolving Advertising-Holding," Berezovsky said. "It was only later [after Listyev's assassination] that I decided to make Lisovsky head of advertising at ORT."[71]

In October 1995, Berezovsky gave an interview to the newspaper *Kommersant*, in which he described his main achievement at ORT as attaining "an independent position on the advertising market." The next step, he said, was "liberation from the dictatorship of the producers."[72]

That the leading TV producer at ORT, Vlad Listyev, was now dead made that task easier. A week before Berezovsky's interview with *Kommersant*, another leading producer, Dimitri Lesnevsky, gave an interview to Radio Liberty. Lesnevsky's mother and business partner, Irina, had appeared with Berezovsky in the frantic videotaped appeal to Yeltsin after the Listyev murder. Now, evidently, her son had grown disillusioned with the man his mother had so fervently defended to the Russian president. Lesnevsky spoke to the correspondent about the conflict between the independent TV producers and Berezovsky.

"The conflict arose from the very beginning, since the idea of privatizing Ostankino [Channel 1] originated with the independent TV companies," he said. "We approached Mr. Berezovsky with this idea two years ago and essentially gave him the channel as a present. We were confident of his financial capabilities and we were confident that he was a proper person. But over the last two years doubts have arisen, the doubts have grown stronger, and today we understand what kind of a person our partner is."[73]

Berezovsky, meanwhile, solidified his control over the network. With voting control of 36 percent of ORT's shares, he effectively had veto power over any decisions. His power was much greater of course. In the absence of a strong shareholder representing the state, he effectively ran ORT.[74]

It was an extraordinary coup. At the same time that Berezovsky was taking control of Channel 1, he acquired several other valuable media properties: Channel 6 (which had been privatized with the participation of Ted Turner, though the American TV magnate later sold out), the magazine *Ogonyok*, and the newspaper *Nezavisimaya Gazeta*. Suddenly, this former car dealer became one of the arbitrators of Russian politics. He sat at the head of the nation's prime TV network and interpreted daily events for the Russian people. He became an architect of Russian national policy. ORT's news service carried out Berezovsky's political dictates at key moments, singing the praises of Yeltsin during the election of 1996, touting Lebed after he struck a secret deal with the Kremlin during that election, debunking Lebed once he quarreled with Berezovsky several months later, attacking rival businessman Vladimir Potanin at a key moment in the privatization process, attacking Gusinsky occasionally and Mayor Luzhkov more or less consistently, and presenting Berezovsky in statesmanlike roles. Berezovsky's media empire meant that he no longer had to survive in the squalid, bloodstained struggles that characterized the auto-dealership business. He had emerged as the supreme oligarch, first among equals in the Russian business world.[75]

CHAPTER SIX

PRIVATIZING THE PROFITS
OF AEROFLOT

"Our Interest Here Is Clear"

"Privatization in Russia goes through three stages," Boris Berezovsky told me in 1996. "The first stage is the privatization of profits. The second is the privatization of property. The third is the privatization of debts."

In other words, it was not necessary to buy an enterprise to control it. The company could remain in state hands. All one had to do was co-opt the management and then funnel the company's revenues through your own middlemen, thus "privatizing the profits" without spending time and money privatizing the enterprise itself. Berezovsky explained that this first stage, the privatization of profits, "led to the disintegration of the enterprises" and "the primary accumulation of capital" by the middlemen. "Once enough capital has been accumulated, those people who have accumulated it begin to think of how to spend it," he continued. "Some buy property abroad, others go to play in Monte Carlo, and others use their money to acquire the disintegrating enterprises."[1]

It was a strikingly astute explanation of what was happening in Russia. Almost all the big business empires of the early 1990s had been created in this way. Berezovsky himself had honed his model to perfection. First, in 1989, he began privatizing the profits of the

170

automaker Avtovaz, buying cars from the factory at a price that guaranteed him a profit and the automaker a loss. Next, in 1992, he entered into the commodities trade, exporting oil, timber, and aluminum; like all the other big traders of the time, he privatized the profits of the producers by paying them a nominal price while selling the commodities abroad at a huge markup. In 1993, with the beginning of the voucher auctions, Berezovsky proceeded to the second stage of his model—privatization of property. Together with Avtovaz boss Vladimir Kadannikov, he used the Avva investment scheme and a number of other financial vehicles to buy a controlling stake in Avtovaz. With ORT, he began by privatizing its profits through the TV network's advertising monopoly in 1994 and went on to privatize its equity in 1995.

Though he did apply the second stage of his model (privatization of property) by buying equity stakes in Avtovaz and ORT, Berezovsky continued to pay more attention to the privatization of profits. He knew that control of a company's cash flow was more important than the ownership of its equity. With the collusion of company management, he established intermediaries to handle the target company's finances, sales, and marketing on the most disadvantageous terms. By taking control of company management Berezovsky could accomplish a "virtual privatization" without bothering to invest money in acquiring an equity stake. Nothing illustrated his mastery of this type of procedure better than the takeover of Aeroflot.

"Aeroflot is in a transitional phase between the privatization of profits and the privatization of property," he declared in an interview with the business newspaper *Kommersant* in November 1995. "We want to participate in both processes."[2]

The company was one of the crown jewels of Russian industry, boasting the best assets of the former Soviet airline monopoly and dominating Russia's international travel market. The general director of Aeroflot, an industry veteran named Vladimir Tikhonov, had begun to modernize the airline, upgrading the fleet, signing lease agreements with Boeing, McDonnell Douglas, and Airbus for new planes. He had arranged a $1.5 billion U.S. Eximbank loan to purchase Pratt and Whitney engines for twenty new Russian-made Il-96M airliners. Fuel costs, meanwhile, were dropping precipitously. The airline's future looked good.[3]

"Aeroflot is the largest airline in Russia and it is natural that business should be interested in it as a source of big profits," Berezovsky told *Kommersant.* "Our interest here is clear."[4]

He began moving in on Aeroflot early in 1995, demanding that Tikhonov place all of Aeroflot's Russian accounts with Avtovaz Bank–Moscow, in which Berezovsky had a significant share. Tikhonov reluctantly agreed. Still, Berezovsky wanted complete control of Aeroflot's finances. The national airline was being partly privatized, with 51 percent of the shares remaining with the government and 49 percent going to management and workers, but Berezovsky did not want equity—he wanted management control, so he could "privatize" Aeroflot's profits.[5]

In the autumn of 1995, he approached several top government officials, including First Deputy Prime Minister Oleg Soskovets and Minister of Transportation Vitaly Yefimov, requesting a management change.[6]

By late October, Tikhonov had been fired. The new general director of Aeroflot was Yevgeny Shaposhnikov, a marshal of the Soviet air force and a former defense minister. Shaposhnikov had played a key role in saving Yeltsin during the Putsch of 1991: as air force chief, he had refused to carry out orders to bomb Yeltsin's White House and had ordered his airborne units to avoid political conflicts.[7]

In contrast to Vladimir Tikhonov (a respected veteran of the airline industry), Shaposhnikov knew nothing about running a business. Soon the key positions in Aeroflot were occupied by Logovaz officials. Nikolai Glushkov, one of the founding members of Logovaz, became first deputy general director (effectively the boss of the airline), while other Logovaz men took charge of finance, commerce, and sales. All these men were as ignorant as Shaposhnikov about running an airline, but they did have, as Berezovsky put it, "the ability to count money and make profits."[8]

Andava

With the management of Aeroflot now under his control, Berezovsky could begin to privatize its profits. His primary vehicle was the Swiss financial company Andava S.A., which served as Aeroflot's foreign treasury center. I first learned of Andava in the winter of 1996–97, while exploring rumors of mass embezzlement at Aeroflot. The airline had traditionally maintained hundreds of foreign bank accounts, managed by a loosely coordinated network of Aeroflot executives, to pay its bills abroad. Andava was charged with centralizing these operations in a single overseas financial center.

A treasury center abroad sounded like a sensible idea. A properly run treasury center could add millions to a Russian company's bottom line. It was strange, therefore, that Aeroflot made no mention of Andava in its annual reports at the time. Among the "main events" mentioned in the Russian version of the 1997 annual report were new charter flights to Toronto and a representative office in the town of Khabarovsk (in Eastern Siberia), but there was nothing about the rationalization of the company's financial operations or about a foreign treasury center or about Andava.[9]

There were reasons to keep the Andava relationship secret. Although Andava was run by a European manager with a respectable record—William Ferrero, the former head of Volvo Group Finance–Europe—its ownership structure was unorthodox. Unlike other treasury centers servicing multinational corporations, Aeroflot's treasury center—Andava—was not owned by the airline; it was owned by Boris Berezovsky and his Logovaz partner, Nikolai Glushkov.[10]

I visited Andava in Lausanne. The office was located in a small gray building, not far from the center of town. I had to go around the back and ring the buzzer. No response. It was lunchtime. I returned two hours later. Ferrero's gray Porsche 911 was parked outside. Someone was coming down the stairs, so I did not have to ring the outside bell this time. I entered and climbed two flights of stairs. There was a door with ANDAVA S.A. marked on it. I rang the bell. A bearded little man in a cardigan opened the door. I introduced myself. "Mr. Ferrero is on the phone," the man said. "Please wait here." He led me to a waiting room. I could hear Ferrero's voice from down the hall. Otherwise the office was silent. I leafed through several Aeroflot brochures and a pamphlet describing Berezovsky's charitable organization, Triumph-Logovaz. An Aeroflot poster hung on the wall. The office appeared to consist of about six rooms—all clean and newly furnished. Ferrero and the bearded little man were the only people there.

Finally Ferrero appeared. He was a tall gentleman, with a blond crew cut and a clipped mustache—if he had worn a monocle, he could have played the role of a Prussian officer. That impression faded when I noticed that Ferrero had a nervous, gawky way of moving.

He told me that his outfit handled Aeroflot's money in the most conservative way possible. "We invest [Aeroflot's hard-currency profits] with a triple-A bank, because we can't allow the money to be used in our other operations," Ferrero declared. "There was one [newspaper] article that said that I had doubled Aeroflot's money in one year, or something like that. But if I were capable of doubling Aeroflot's money in a year, I wouldn't be running this business—I would be on an island somewhere, under the palm trees."[11]

In my preceding telephone conversations with him, Ferrero had seemed a man of leisure, not a key financial manager for a billion-dollar multinational corporation. We would talk for hours. Sometimes, when I called him, he answered his phone himself. He did not seem to have any secretaries, just a lot of time for long, leisurely discussions.

Andava had been established on February 3, 1994, in Lausanne, Switzerland, as a company jointly owned by André & Cie. and an outfit called Avva International, registered in Switzerland, with Berezovsky as its director. (Eventually, André would sell out its stake and Berezovsky and Glushkov would take majority control personally.) In keeping with the corporate model Berezovsky and André had already employed in creating Forus Services, Andava was based in Lausanne, affiliated with a holding company in Luxembourg and other such enterprises in offshore tax havens.[12]

Berezovsky already had a financial company in Lausanne: Forus Services. Why did he need a second one? The fact is that in 1993–94 Forus was not meeting expectations. Two big investment projects it was working on at the time—a $100 million credit from the Italian Import-Export Bank for new equipment at Avtovaz and a $60 million credit for oil company Samaraneftegaz—collapsed. Forus's directors—Berezovsky and Glushkov from the Russian side, together with Alain Mayor from André—decided to try another tack. Rather than search for hard-currency credits abroad, why not simply take control of Russian companies' foreign revenues?[13]

"It was a new opportunity," Alain Mayor told me, "to run operations of the FINCO type—finance, commerce, personal relationship–type of operations."[14]

The initial business plan for Andava was to service the Avva investment scheme that Berezovsky had cooked up in Russia. "We wanted to create a company that could later become the centralized treasury for Avva," recalls Christian Maret, the head of André's office in Moscow. "If the Avva project had succeeded, that [treasury] would have been Andava. That's why it's called Andava [André and Avva]."[15]

When the Avva project collapsed, Andava decided to become the foreign treasury center for Avva's parent, the automaker Avtovaz. Throughout the next year, Andava handled about $100 million of Avtovaz's foreign revenues. The change in business strategy was explained with characteristic candor by Alain Mayor, who reiterated his belief that personal relationships were the key in Russia—wherever your partner goes, you follow.[16]

"In these deals, we start because of favorable circumstances, because the partners are there, because we sense an opportunity to do interesting work with these partners," he told me. "We create the tools to be able to work together, but we don't necessarily have a 100 percent determined goal at the beginning."[17]

Andava's new business relationship did not last long. In the summer of 1995, Avtovaz declared that it would no longer supply Logovaz with subsidized cars for reexport. After six years, the automaker had finally decided to liberate itself from Berezovsky's grip. The tycoon had promised that his financial companies—Forus, Andava, AFK, and Avva—were going to fix Avtovaz's finances and bring in massive Western investment. But the automaker continued to lose cash and pile up debts; it was being strangled by its intermediaries, principally Berezovsky-owned dealerships and financial companies. No amount of financial engineering could fix the fact that Avtovaz was selling its cars to Berezovsky's Logovaz at a price significantly below the cost of production. Moreover, many of the sales to Logovaz were not for cash, but for some kind of promissory notes; by 1995, Logovaz owed the automaker at least $40 million. Once Avtovaz declared its intention to terminate the Logovaz dealership contract, Berezovsky swung into action: He got Mikhail Khodorkovsky's Menatep Bank to bail him out on the Avtovaz debts. But it was too late. The contract that had given Berezovsky his first fortune was torn up.[18]

An event that could have been disastrous to Berezovsky's growing business empire ultimately proved largely inconsequential—by the time his relationship with Avtovaz came to an end, the tycoon had already found a much more lucrative cash cow: Aeroflot. The intermediaries he had used to handle the finances and sales of Avtovaz were now redirected to service the airline. It helped that Nikolai Glushkov had jumped ship to Aeroflot. Glushkov had been Avtovaz's finance chief and the point man in its relationship with Logovaz (of which he was a founding partner). In the autumn of 1995, Glushkov left his job at Avtovaz and became first deputy general director of Aeroflot (effectively the chief manager). Once again, he was in a position to send business Berezovsky's way. The conflict of interest Glushkov had had at Avtovaz was thus replaced with another conflict of interest at Aeroflot.[19]

In May 1996, Marshal Shaposhnikov, Aeroflot's inexperienced general director, sent a letter to the airline's 152 foreign offices, ordering them to remit up to 80 percent of their foreign currency revenues to Andava in Lausanne. (Much later, Marshal Shaposhnikov

Boris Yeltsin meeting with the "oligarchs" in September 1997. From left: Mikhail Khodorkovsky (Menatep), Vladimir Gusinsky (Most), Alexander Smolensky (SBS-Agro), Vladimir Potanin (Unexim), Vladimir Vinogradov (Inkombank), and Mikhail Fridman (Alpha). To Yeltsin's right is his ghostwriter and chief of staff, Valentin Yumashev. Berezovsky was absent from the ranks of the oligarchs, since he was a government official at the time. *Courtesy of Associated Press.*

The architects of Russia's market reforms, Yegor Gaidar and Anatoly Chubais, conferring in 1992. *Courtesy of Alexander Makarov/RPG/SYGMA.*

During the 1991 Putsch, Yeltsin came to the parliament and made his historic speech on top of a tank, with General Korzhakov standing by his side. *Courtesy of Associated Press.*

Yeltsin with his daughters, Yelena Okulova (left) and Tanya Dyachenko. *Courtesy of CORBIS SYGMA.*

Tanya Dyachenko and Berezovsky attending the ORT annual meeting. *Courtesy of SOVFOTO/ EASTFOTO.*

In October 1993, two years after defending the parliament against Communist hard-liners, Yeltsin ordered the building razed, sending in tanks to expel recalcitrant lawmakers. *Courtesy of SOVFOTO/EASTFOTO.*

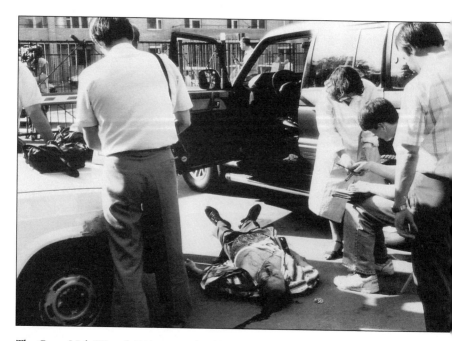

The Great Mob War of 1993–94 involved hundreds of assassinations and shoot-outs in the center of Moscow. *Courtesy of CORBIS SYGMA.*

Sergei Mikhailov (alias: Mikhas), boss of the Solntsevo Brotherhood (right), in police custody in Switzerland. *Courtesy of Associated Press.*

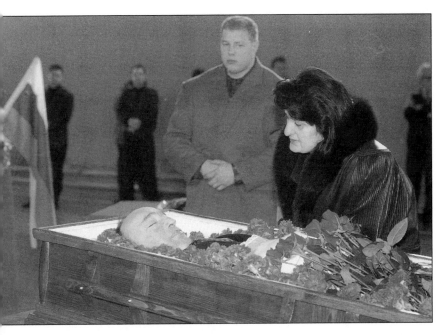

The gangster-philanthropist Otari Kvantrishvili in his coffin, mourned by his wife and friends. *Courtesy of Associated Press.*

A car bombing in Moscow in 1994. The vehicle, believed to be Berezovsky's armored Mercedes, was hit by a bomb planted next to the house at the top of the photo. Berezovsky escaped through the back door; a body—apparently that of Berezovsky's decapitated driver—is barely visible in the front seat. *Courtesy of Georges de Keerle/CORBIS SYGMA.*

The funeral of television producer Vlad Listyev in March 1995. *Courtesy of GAMMA.*

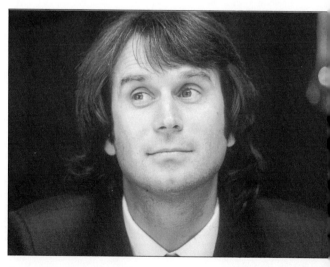

Sergei Lisovsky, Berezovsky's partner in the advertising business at ORT. *Courtesy of SOVFOTO/EASTFOTO.*

Berezovsky and his partner in the oil business, Roman Abramovich. *Courtesy of Associated Press.*

General Lebed (left) and liberal parliamentarian Grigory Yavlinsky. *Courtesy of Georges de Keerle/SYGMA.*

Chechen terrorist commander Shamil Basayev during the 1995 hostage-taking raid in the town of Budyonnovsk. *Courtesy of Laski Diffusion/GAMMA.*

Berezovsky conferring with Chechen hard-line leader Movladi Udugov in Moscow, November 1996. *Courtesy of AFP.*

would tell a Russian newspaper that he had been told to sign this letter by Berezovsky's man, Glushkov.)[20]

"When the foreign representative offices of Aeroflot questioned why the airline was not receiving the money it was collecting and why it was going into private accounts, they were told that the money was being collected for Yeltsin's election campaign," General Korzhakov asserts. "In fact, Berezovsky kept the money for himself."[21]

In the autumn of 1998, I asked Aeroflot's general director, Valery Okulov, why the airline worked with Andava. Okulov replied: "Andava's operations help us procure credits from abroad."[22]

But the only significant credits Aeroflot managed to procure were either guarantees by the U.S. Eximbank (for the purchase of Boeings) or were arranged by André & Cie. Andava played no part in any of these operations.[23]

The treasury operations that Andava performed were routine foreign exchange operations, expediting payments to leasing companies and fuel suppliers, and occasionally obtaining small-scale bank guarantees. The money from Aeroflot offices around the world was kept in Andava's account at the Lausanne branch of the Union Bank of Switzerland. Andava managed the money on a trust basis.[24]

In the past, these operations had been carried out by Aeroflot's own Center for International Payments. If it wanted to execute its international transactions more efficiently, Aeroflot could participate in the International Air Transport Association's clearinghouse system, used by airlines around the world to handle their international fuel, maintenance, and landing fees, and the necessary foreign exchange operations.[25]

But Aeroflot chose to use Andava instead. This cost the airline dearly. The fees were unusually high. Most big international banks would jump at the opportunity to perform the kind of treasury management that Andava did for Aeroflot; the fees are typically a fraction of a percent of the assets managed. But, according to the May 1997 contract between Andava and Aeroflot, Andava received a commission of 3.125 percent on the funds it handled for Aeroflot. On an annual turnover of about $400 million, that was $13 million a year. Not bad for an operation with only a couple of employees.[26] Next, Andava lent the money it earned from Aeroflot back to the airline—in the course of 1997, for instance, Andava lent the airline $25 million in short-term loans—a minuscule amount compared to what the airline needed.[27]

Andava, in other words, performed no task that Aeroflot could not perform itself. The one delicate operation—the mobilization of foreign bank loans—was performed adequately by André & Cie. So why was Andava necessary? And why were up to 80 percent of the export earnings of Aeroflot kept in Andava? "I don't know," says Maret. "I can only guess."[28]

With the huge problem of Russian capital flight (the root cause of the destruction of the ruble), the Russian Central Bank tried to enforce stringent rules on the repatriation of capital by Russia's main exporters. Russian law stated unambiguously that export revenues of Russian companies had to be repatriated and that half of this sum had to be converted into rubles. If a company needed more foreign currency, it could always sell its rubles for dollars on the currency markets in Moscow.

"We are not subject to that regulation," Andava's general director, William Ferrero, told me. "We are exempted by a Central Bank license."[29]

Aeroflot applied for that license on July 5, 1996, not long after Berezovsky gained control. But officials at the Central Bank, evidently puzzled by the Aeroflot-Andava relationship, demurred.[30]

By early 1997, information about this murky operation began emerging, first in *Forbes*, then in the Russian newspapers, and finally in the Swiss press. Aeroflot was operating a "slush fund" in Lausanne, one Russian newspaper declared. The prosecutor's office began an investigation.[31]

Yeltsin's Son-in-Law

Yevgeny Shaposhnikov, the old air force marshal who was head of Aeroflot, began to question the Andava operation. He was fired on March 12, 1997. His replacement as general director was Valery

Okulov. An athletic, handsome man, Okulov had worked for twenty-two years in Soviet civil aviation, rising to the position of navigator on Aeroflot flights (number three in the plane behind the pilot and copilot). His professional career was unremarkable except for one fact—he was married to Yelena Yeltsin, the eldest daughter of the Russian president.[32]

Within two months of Okulov's appointment as general director of Aeroflot, the Russian Central Bank gave the airline its exemption from Russia's currency-repatriation rules.

"Using his political connections, Berezovsky was able to get a Central Bank license," Korzhakov would recall.[33]

Korzhakov referred me to a transcript of phone conversations in March 1997 between Berezovsky and the Central Bank chairman, Sergei Dubinin. These conversations had been intercepted by an undisclosed security agency; the tapes were sent to the Russian prosecutor-general's office, where their authenticity was confirmed. In the conversations, Berezovsky, though serving as deputy secretary of the Security Council at the time (and hence prohibited by law from engaging in commercial activities), blatantly lobbies the Central Bank chairman to accept the proposal of Nikolai Glushkov—a man of "impeccable reputation," says Berezovsky—to license the Andava-Aeroflot relationship.

"We are losing time," Berezovsky declares to the Central Bank chief. "Time does not stand still. And the affairs [at Aeroflot] are not proceeding as they should, though they are going somewhat better, of course, thanks to the initiative of Glushkov. But for things to go normally for the government, and in the eyes of the government, we need certain decisions on your part—decisions which we have not been able to receive for already a year."[34]

Dubinin promises that he will order his underlings to move on the license application. After receiving this promise, Berezovsky immediately calls Nikolai Glushkov. The following conversation takes place:[35]

BEREZOVSKY: Allo! Kolya! I have just spoken with Dubinin. He is the worst sort of cunt, of course. But that's OK. He says that his people have two basic problems. One problem is the transparency of the operation. That's precisely what your people don't want. Second, he wants guarantees, because he says [Andava] is a small outfit. What kind of guarantees can it give? Therefore, it needs a guarantee either from an insurance company or from a consortium of banks. Do you understand?

GLUSHKOV: Guarantees for what, Borya?

BEREZOVSKY: This is what I don't understand, Kolya.

GLUSHKOV: Because, at present, we are holding our accounts in some kind of third-rate banks, who can give no guarantee of any kind....

BEREZOVSKY: Kolyenka, let's not discuss this issue now. There's a woman [at the Central Bank] named Artamonova. Call Dubinin's office and they will transfer you. I want you to do it right now.

GLUSHKOV: Okay, I'll do it now.

BEREZOVSKY: Talk directly with her and tell her clearly that there has been a conversation with Dubinin. Dubinin has ordered that this issue be resolved in the quickest possible manner. Therefore, set up an appointment and let's meet.

GLUSHKOV: Okay.

Andava received its license. But the fact remains that the Andava-Aeroflot relationship flourished for more than a year before it was formalized with a Central Bank license. On what legal basis did the activity proceed?

"I don't know. I assume it was fully legal," says Andava's Ferrero.[36]

Unlikely, says Oleg Davydov, who was Russian foreign trade minister at the time. "You need permission to leave your money abroad," he says. "Gazprom and Lukoil keep money in such accounts, but in order to do so, you need to get permission."[37]

The clients of Andava—first Avtovaz, then Aeroflot—were big, prestigious companies. They should have had no trouble convincing the Central Bank to grant them the same rights as the oil and gas companies Gazprom and Lukoil.

"I think that in fact Aeroflot made its application when it started working with us and then it took a long time [to get the actual license]," says Ferrero. "These things here [in the West] can take twenty-four hours. Over there [in Russia] it can take months or worse. There was a period when the license was in fact there. There was a whole period in which the application had been submitted, everything had been announced, but the license [officially] wasn't there. There has been some confusion about this 'finesse,' if one can call it that."[38]

Yet there is no provision under Russian law stipulating that a company can operate during this "waiting period" in the same way as it can when it actually gets the Central Bank license.

"We were planning to stabilize the ruble in 1997," says former Foreign Trade Minister Oleg Davydov. "For this reason, those who were concealing hard currency abroad were harming the state greatly."[39]

The Dublin Connection

By the mid-1990s, close observers of Russia already knew that many big companies were being looted by their managers and intermediaries. The capital-flight statistics (an estimated $15 billion a year) indicated that a huge portion of the profits of Russian enterprises were evading both the tax collectors and the shareholders. The looting ruined many of Russia's best enterprises, starving them of investment capital, while the capital flight eventually doomed Russia's efforts at monetary stabilization. Few outsiders understood the precise mechanisms by which a particular company was looted and the capital funneled abroad. For this reason the Aeroflot case remains invaluable—it is the most detailed example to date of how to loot a company you do not own.

As Aeroflot's foreign treasury center, Andava was able to dispose of the huge foreign currency profit the airline earned every year. In 1997, for instance, Aeroflot received $897 million in foreign currency revenues, but spent only $646 million in foreign expenses. What happened to the difference? Andava transmitted no foreign currency back to Russia, Ferrero said. In fact, much of Aeroflot's foreign currency profit was going to pay the usurious interest payments arranged by Berezovsky's financial companies.[40]

The oldest of these was the Lausanne-based financial company Forus Services S.A. (the outfit American entrepreneur Page Thompson visited to obtain his letter of credit). I visited Forus's headquarters in a modern office complex next to the Lausanne Airport. The office consisted of several rooms. It seemed as lifeless as the office of its sister company, Andava, across town. The door was answered by a secretary. There were no signs of other employees or of any work being done. I asked to see the general director, Rene Kuppers. "Sorry," the secretary told me. "Monsieur Kuppers is traveling in Russia today."

Forus acted as a financial consultant and intermediary between Aeroflot and foreign money markets. Together with André & Cie., Forus did arrange four syndicated loans from Western banks to Aeroflot between May 1997 and the end of 1998. They were short-term loans (180 days), ranging in size from $40 million to $59 million. Forus received a commission for these transactions, though, with André & Cie. handling most of the work, it was unclear why Forus's presence was necessary.[41]

Forus was also involved in the murkiest part of Aeroflot's operations: the overflight or frequency-right payments, which allowed foreign airliners to service the Russian market or fly over Russia en route to somewhere else. This was a large sum of foreign currency: $284 million in 1997 alone. Although these fees accounted for one fifth of the airline's revenues, they were mentioned in just one sentence in the company's annual report.[42]

What did Aeroflot do with the money? The Russian prosecutor-general's office told me that between 1996 and 1998 Forus received a total of $380 million in Aeroflot overflight payments. Half of that sum went to repay the syndicated loans André & Cie. was organizing. The prosecutor-general's office could not find out what had happened to the rest of the money, and neither Aeroflot nor Andava could give me an explanation. Forus's general director, Rene Kuppers, acknowledged that his company did get overflight payments in 1996 (shortly after Berezovsky took control of Aeroflot, but before the syndicated loans) and claimed that the money was used to repay small loans ($5 million to $15 million) that Forus had made to Aeroflot.[43]

Andava, meanwhile, was effectively a loan-sharking operation, setting up subsidiaries clearly designed to milk Aeroflot's cash flow. In April 1996 Berezovsky's loyal agent in Switzerland and Andava board member, Hans-Peter Jenni, flew to Moscow to sign the registration documents creating a company called FOK (United Finance Company); Andava S.A. was the sole founder of this outfit. FOK's mission was similar to Andava's—to gather Aeroflot's funds to pay the airline's foreign bills. FOK was also remarkably similar in its mission to AFK, the financial company Berezovsky had helped set up to handle Avtovaz's payables and receivables in 1994. (FOK and AFK shared the same general director, a man named Roman Sheinin.) Besides Aeroflot, two other companies used FOK's services: AFK and Transaero (a private company, partly owned by Berezovsky, that was Russia's second-largest international airline).[44]

The bulk of FOK's business was with Aeroflot. Nikolai Glushkov was the sole signatory on behalf of Aeroflot in the loan contract with FOK. He was able to sign the contract despite being a major shareholder of Andava, the sole founder of FOK. (Berezovsky was an even bigger shareholder, of course.)[45]

In the last half of 1996, FOK handled $139 million worth of Aeroflot's payables; in the first half of 1997, it handled a further $82 million. For using this service Aeroflot had to pay "penalties" to FOK of $29 million in 1997.[46]

Why was the implicit interest rate so high? The trick was having more than one intermediary handle a relatively simple bill-paying operation. When it received a foreign invoice—say, a fuel bill from BP or Mobil—Aeroflot would not pay the bill directly. Instead, it passed it on to FOK for payment. FOK, in turn, passed it on to a second intermediary, an obscure outfit called Grangeland Holdings Ltd., which paid the bill.[47]

Registered in Dublin, Ireland, Grangeland had virtually no records. Its registration documents described the company in terms of every commercial activity imaginable. The last year that Grangeland filed its required annual financial statements with the Irish Companies Registration Office was 1995; it declared $10 in assets. Less than a year later, this outfit would be handling hundreds of millions of dollars of Aeroflot's money.[48]

Grangeland was simply a post-office box. Its shareholders included two unknown companies registered in Panama. Grangeland listed only two directors responsible for the company's activities: both employees of the small Swiss accounting firm that serviced Andava and Anros S.A. (the anonymous vehicle that owned half of Logovaz). Grangeland's banker was the Union Bank of Switzerland in Lausanne, the same bank employed by Andava. Andava denies any knowledge of Grangeland, although the Russian prosecutor-general's office says it is in possession of documents received from Swiss prosecutors proving that Andava had in fact bought Grangeland in 1996.[49]

The whole Grangeland-FOK-Aeroflot operation was a chain of short-term loans. Grangeland's payment of an Aeroflot bill was considered a loan to FOK, which, in turn, was treated as a loan from FOK to Aeroflot. Between FOK and Grangeland, Aeroflot paid horrendous interest costs. Grangeland charged an annual percentage rate of 30 percent in hard currency. FOK charged an annual percentage rate of 65 percent in ruble terms for the bills it handled for the airline—since the ruble declined about 10 percent against the dollar during the FOK-Aeroflot operation, this was equivalent to an annual interest rate of about 50 percent in dollar terms. The contract between FOK and Aeroflot—barely more than a page—was vague, speaking of Aeroflot's making its final payment to FOK "taking into account the 30 percent refinancing rate of FOK." In other words, this left open the possibility that FOK's 50 percent dollar interest rate would be added to Grangeland's 30 percent—forcing Aeroflot to pay a final interest rate of 95 percent in hard-currency terms.[50]

The result of many financial companies' (Andava, AFK, FOK, and Grangeland) performing essentially the same task for the same group of companies (chiefly Aeroflot and Avtovaz) was significant interest earnings accumulated abroad. That these intermediaries often shared the same managers, the same banks, the same accountants, and the same owners made the job easier.

If Aeroflot was gouged by FOK, Russian tax authorities were not getting anything from the Moscow-registered financial company either. In its 1996 annual report, FOK stated that it lost 97 percent of its pretax profits as a result of foreign exchange losses in its dealings with Grangeland. (The ruble was stable against most hard currencies at this time.)[51]

Besides FOK, Grangeland, and Andava, there were other strange middlemen sucking the cash out of Aeroflot. In 1996–97, for instance, Aeroflot paid $17 million to an obscure advertising company called NFQ to produce its in-flight magazine, even though the airline had its own advertising and publishing department. "This contract was far from the best as far as the interests of Aeroflot were concerned," Aeroflot's general director, Valery Okulov, would later admit.[52]

There were also irregularities in the sale of Aeroflot tickets abroad. After the Berezovsky team took control of Aeroflot, the old practice of concluding agreements directly with travel agents was gradually replaced by a new system, involving an intermediary called Aeroflot Tour Group, managed by a former Logovaz executive named Leonid Itskov. In 1999, Aeroflot general director Valery Okulov went public with his unhappiness about Aeroflot ticket agents' losing their business to "unsanctioned" consolidators and middlemen, especially those operating in the United States and Canada. He did not mention how much money the airline lost from the diversion of this business.[53]

Meanwhile, Berezovsky's managers at Aeroflot courted American investors. They hired Salomon Brothers to help issue shares in the United States in the form of American Depository Receipts. The ADRs never materialized. That did not prevent foreigners from buying Aeroflot's shares in Russia. But Aeroflot's shareholders, including investors in the U.S., had no idea what was really happening with Aeroflot's advertising contracts or its ticket sales operations. They were not told that the bulk of Aeroflot's foreign currency revenues were being handled by two unknown Swiss companies: Andava and Forus. Neither did Aeroflot inform its shareholders that one of the owners of the Swiss companies was none other than Nikolai Glushkov, Aeroflot's chief financial officer and the man with the unilateral right to sign financial contracts for the airline.[54]

Aeroflot's financial results should have tipped off investors that something was amiss. Although murky, the results indicated that the airline was not flourishing under the tutelage of Berezovsky's managers. Despite solid foreign currency operating profits and massive government subsidies in the form of overflight payments, Aeroflot lost $82 million in 1996 and $93 million in 1997. Aeroflot had a liquidity problem. Over the course of 1997, its cash and cash equivalents decreased by $165 million; it developed a net current-asset deficiency of $159 million; the debt-to-equity ratio increased from 169 percent to 207 percent. In the spring of 1998, Aeroflot begged its old partner André & Cie. to increase a syndicated loan by $17 million, to cover a cash-flow shortage. A few months later, Canadian authorities impounded an Aeroflot plane because the airline had failed to pay a $6 million bill to a Canadian hotel operator.[55]

Only in 1999 did the roof over Berezovsky and his managers at Aeroflot finally collapse. The Russian prosecutor-general's office raided the offices of Aeroflot, NFQ, and FOK, among others, and opened a criminal case against Berezovsky and Glushkov. In Switzerland, the federal prosecutor's office began its own investigation and froze the accounts of Andava and Forus. Yeltsin's son-in-law Valery Okulov also turned against the Logovaz team: Nikolai Glushkov left his post in late 1997, shortly after a particularly damning article about Andava appeared in *Moskovsky Komsomolets*. In February 1999, Okulov fired the two remaining Logovaz kingpins, commercial director Aleksandr Krasnenker and sales director Leonid Itskov. He also announced that he was suspending the treasury service contract of Andava.[56]

News reporters asked Okulov whether the two men had been engaged in criminal activity. His answer was: "It is rather difficult to differentiate between incompetence and embezzlement."[57]

But all that lay ahead. In the autumn of 1995, Boris Berezovsky's takeover of Aeroflot, coming on the heels of his rupture with Avtovaz, seemed a stroke of genius. The policy he would follow at the Russian national airline remains a textbook example of how to privatize the profits of a company you do not own. Within two months after he had established control of Aeroflot, Berezovsky pulled off an even greater coup: He privatized one of Russia's biggest oil companies. True to form, he paid almost no money for it.

CHAPTER SEVEN

THE RACE FOR OIL

"The Meaning of Life Is Expansion"

For Russia, 1995 was a year of bloodshed and decline. On the TV screens every evening, Russians were presented with the carnage of the Chechen War. The business community, only just recovering from the ravages of the Great Mob War, was rocked by the assassination of several top executives in the oil and aluminum industries. The economy continued its precipitous decline; the nation's population continued to shrink. Abroad, Russia's influence sank to a new low. NATO was expanding eastward, absorbing Poland, the Czech Republic, and Hungary. When NATO warplanes bombed Serbian positions in Bosnia that summer, Russia protested on behalf of its traditional ally, but could do nothing.

Berezovsky, however, was prospering. In September 1995, he told a British journalist how he had once visited the famous dissident Andrei Sakharov and asked him the meaning of life. "The meaning of life is expansion," Berezovsky remembered the great man's replying. "From the moment you start reproducing—I'm talking about the sexual instinct—it's expansion, in a sense. You're replicating yourself. The same with the desire to spread religious or philosophical ideas: that's expansion too. I think this is my driving motive."[1]

Berezovsky was expanding fast. At the beginning of that year, he had been merely a well-connected car dealer and commodities trader, under suspicion for the Listyev murder. His company, Logovaz, was in turmoil. The general director, a longtime Berezovsky partner named Samat Zhaboev, resigned in March (the same month as the Listyev killing); according to press reports, Zhaboev was suffering from ill health. Four months later, on July 22, Zhaboev's old deputy at Logovaz, Mikhail Gaft, fell to his death from the balcony of his apartment. (Police judged the death to be an accident.) But Berezovsky remained relentless in his drive for power. Among the companies that fell into his hands in 1995 were some of the nation's preeminent enterprises. Such institutions as Channel 1, traditionally the spearhead of the state's propaganda machine, and Aeroflot, the national airline, were among the most visible symbols of Russian state power. Now they were under Berezovsky's control. Financially, however, the tycoon achieved his greatest triumph that autumn, when he turned his attention to the lifeblood of the Russian economy—oil.[2]

The Russian Rockefeller

For decades, Russia had produced more oil than any other country, far outstripping its closest rivals, Saudi Arabia and the United States. This wealth in oil fields, pipelines, and refineries was controlled by the Soviet government. With the fall of Communism in 1991, it was up for grabs. The big multinational oil companies rushed in to claim a share of Russia's vast oil fields. "This is the last frontier for an oilman to find major new reserves," T. Don Stacy, president of Amoco's Eurasian subsidiary, said in 1995.[3]

"It's the new prize for world oil," declared oil historian Daniel Yergin. "The biggest risk is the risk of not being there."[4]

By 1995 such Western majors as Royal Dutch/Shell, Exxon, British Petroleum, Amoco, Chevron, and Texaco had concluded major production-sharing contracts. But virtually without exception, they were floundering. From the Arctic wastes of Timan Pechora to the shores of the Caspian Sea, from Sakhalin Island in the Pacific to the steppes of Kazakhstan and the forests of Western Siberia, foreign oil companies were finding the former Soviet Union a very difficult place to do business. Permits were being held up, contracts renegotiated, promised pipeline access withdrawn. The foreigners were puzzled by the delays, but knowledgeable Russians knew that native businessmen had gotten to the oil fields first and had no intention of sharing profits with foreigners.

No man loomed larger over the industry than Vagit Alekperov. In just five years, Alekperov had created Lukoil, Russia's first integrated oil company, accumulating more oil and gas reserves than Exxon, and turning the company into a multinational giant. Not only did Alekperov expand Lukoil aggressively, he also emerged as the power broker in almost every big oil deal in the former Soviet Union. His career revealed how the Russian oil business worked and how a private empire could be built on Russia's premier natural resource.

Alekperov grew up in Baku, the fabled oil town overlooking the Caspian Sea. After studying engineering in the local petrochemical institute, he made his mark in Western Siberia. In 1983 he was appointed director of oil production in the remote town of Kogalym. In seven years Kogalym became one of the most productive oil fields in the Soviet Union. Alekperov became a legend in his oil town, famed for his steely composure in the face of danger. One story involved a crucial pipeline leak. Hot oil was gushing from the pipe. Repair workers with welding equipment gathered, but were afraid to work on the damaged pipe lest a spark ignite an explosion. Alekperov arrived, assessed the situation, and lay down next to the pipe. "Now weld it," he ordered.[5]

In 1990 Alekperov, forty years old, was appointed deputy minister of fuel and energy in Moscow. The Soviet oil industry was falling apart, crippled by declining government investment and the disintegration of the planned economy. Alekperov decided to go to the West to see how such big multinational oil companies as BP, Agip, and Chevron did business.[6]

"All he wanted to talk about was the company that would later become Lukoil," recalled Thomas Hamilton, an American oil executive representing BP in the negotiations with Alekperov. "He came with a long list of questions about putting an integrated oil company together. 'Why won't this work?' 'Why can't I do this?' 'How do I implement this kind of plan?'"[7]

Alekperov asked BP to help him create Lukoil and offered a strategic partnership. BP refused. "We thought it was a pipe dream," remembers David Reardon, one BP executive at the discussions. "We couldn't imagine the privatization of the Soviet Union's prize oil resources."[8]

Alekperov returned to Moscow to discover that he was regarded as somewhat of a saboteur by Communist hard-liners. He was about to be fired when his boss at the ministry became embroiled in the 1991 Putsch against Mikhail Gorbachev. With the failure of that coup and the dismissal of his boss, Alekperov was left in charge of the Oil Ministry in the final months of the Soviet Union.[9]

He moved quickly. First he handed the country's oil assets to the state-owned holding company Rosneft and placed his friend Aleksandr Putilov in charge. Two of Alekperov's subordinates at the ministry headed up the national oil companies of Turkmenistan and Kazakhstan. Alekperov then gathered some of the best Russian oil properties and some of the country's brightest engineers and incorporated them as Lukoil. When the company was privatized, first in the 1994 voucher auctions, then in a series of smaller cash auctions, Alekperov and his top managers bought a quarter of Lukoil's equity. Through his own shares and his right to vote on behalf of the company pension fund and labor union, Alekperov had undisputed control of the company.[10]

By 1993, he was ready to expand beyond the Siberian oil fields. He chose Baku, where a consortium of Western oil companies (including BP, Amoco, Exxon, Pennzoil, and Unocal) had been negotiating for two years with the Azerbaijan government for the rights to 4 billion barrels of offshore oil. But the $8 billion deal was stalled. The Azerbaijani negotiators seemed paranoid about being taken advantage of, but this paranoia did not stop them from demanding large bribes. The Russians were complicating matters by demanding a piece of the action.[11]

In the spring of 1993, a former KGB general and Soviet Politburo member named Geidar Aliyev took power in Azerbaijan after a bloodless coup supported by the Russians. Two months later, Russian oil minister Yuri Shafranik arrived in Baku, accompanied by Alekperov. "Shafranik appeared before the top officials of the Azerbaijan government and declared that Russia had a right to participate in the oil project," recalls Rasul Gouliev, who was chairman of the Azerbaijan parliament at the time. "Then he introduced Alekperov and warmly recommended Lukoil as a business partner. We all understood what he was saying."[12]

The Azerbaijani leader, Geidar Aliyev, agreed to sell Lukoil a 10 percent stake in the Baku oil consortium for a nominal $15 million. The Western oil companies were informed that they had a new partner. Amoco's T. Don Stacy recalls the bewilderment he and others felt: "Our first feeling was that Lukoil was on the other side of the table. But we found that they were businessmen, like ourselves, wanting to make a profitable venture, rather than to take advantage of either side."[13]

With Lukoil in the Baku consortium, the pieces quickly fell into place. The deal was signed, Russia provided pipeline access to the Black Sea, and the first Baku oil began flowing to the West two years later. "We want to make Lukoil into the biggest oil company in the world—in both production and profits," the dapper forty-five-year-old Alekperov said in late 1995. "The seven sisters now have a brother."[14]

Russian industrialists typically went through three stages in their relations with foreign businessmen. The first stage was characterized by ignorance and bluster ("We don't need your help—we're the best already"). The second stage was paranoia—the Russians had become aware that they did not understand how global business worked and were afraid to strike any deal at all, since they assumed they would be skinned by the unscrupulous foreigners. The third stage was panic—the Russians understood that they could no longer survive without striking a deal, so they agreed to do anything, sign any deal, no matter how onerous the terms.

Alekperov, by contrast, was both astute and decisive. "Other Russian oilmen lack Alekperov's finesse and understanding of the commercial game," noted Thomas Hamilton. "They don't know how to make the trade-offs he's willing to make."[15]

Among those trade-offs were gifts of very cheap Lukoil stock to influential Russians. One key shareholder was Yuri Shafranik, the Russian oil minister who cut Lukoil into the Baku oil deal. Another big shareholder was Aleksandr Putilov, head of the powerful state-owned oil company Rosneft.[16]

Alekperov, together with oil minister Shafranik, soon began to put the "Baku squeeze" on Chevron and the Republic of Kazakhstan in their giant Tenghiz project. The reserves in this field (9 billion barrels) were equivalent to about three quarters of the reserves of Prudhoe Bay, Alaska, when that field was discovered in 1967. By 1995, Chevron had sunk $700 million into Tenghiz and had committed itself to investing $10 billion more over the next two decades. The initial results were disappointing.[17]

"You can't do such projects alone," Alekperov declared. "You have to share."

Alekperov and Shafranik saw to it that Chevron received only a third of the Russian pipeline capacity it had been promised for its Tenghiz exports; Chevron was forced to ship its exports via the Russian railroads. "If we are included in the Tenghiz consortium, we are obligated to solve the pipeline issue," Alekperov said in late 1995.[18]

Within months, Chevron let Lukoil into the Tenghiz deal. It was a wise decision. It was impossible to succeed in Russia without a strong native partner. The largest American oil company, Exxon, tried to persevere in Russia on its own, with the giant Timan Pechora project, but got nowhere until it admitted Lukoil as a partner.

Even as he successfully outmaneuvered the world's best oil companies, Alekperov ran his business like an old-fashioned Turkish pasha. "The one thing that always strikes you is that he has exceedingly good-looking translators," Thomas Hamilton observed.[19]

The flight attendants on Alekperov's company jet in which I flew in 1995 were demure beauties, the epitome of Eurasian loveliness. "Presidential quality for the presidential plane," whispered the chief of Lukoil's pension fund, eyeing a particularly striking girl.

Alekperov also knew how to give gifts. Lukoil, for example, financed the aircraft-design bureau, Yakovlev, in building the country's first executive jet, based on the 100-seat Yak-42. Lukoil had no interest in aircraft manufacturing, but these luxuriously upholstered executive jets, valued at $19 million each, made wonderful presents. He gave one of the jets to Moscow mayor Yuri Luzhkov. He gave another to the president of the natural-gas monopoly Gazprom. In 1995, as Lukoil's negotiations over the Tenghiz oil project were heating up, Alekperov gave a jet to the president of Kazakhstan, Nursultan Nazarbaev.[20]

For free?

"Nothing is free," Alekperov said, chuckling, "but the forms of payment can differ."

Dead Man in the River

Vagit Alekperov had shown what a smart and energetic manager could do with a Russian oil company. Whatever the methods he had used in accumulating his empire, there was no denying that Alekperov was a capable industrialist. He was Russia's first modern oil magnate.

Berezovsky, too, wanted to be an oil magnate, but he had difficulty finding a place at the table. In 1995, the only big oil properties that had not been privatized were held by the state-owned oil company, Rosneft. Boasting majority stakes in some of the biggest oil fields and best refineries in Russia, Rosneft was a formidable company—the second largest in the nation. It was proof that President Yeltsin had not given away everything to private businessmen.

Berezovsky was determined to get a piece of it. But he had a problem. Though Rosneft was a state company, it was within the

Lukoil sphere of influence. Vagit Alekperov's old partner, Aleksandr Putilov, was Rosneft's general director; the Ministry of Fuel and Energy, meanwhile, was headed up by another Lukoil partner. Lukoil itself owned a stake in one of the best Rosneft properties—the Omsk Oil Refinery.

Early in 1995, Berezovsky was approached by Roman Abramovich, a twenty-nine-year-old oil trader, with a plan. Abramovich was a successful commodities trader, supervising a network of companies operating out of Switzerland; he was also a business partner of Boris Yeltsin's son-in-law Leonid Dyachenko. Abramovich proposed that he and Berezovsky separate out the two best Rosneft assets—oil producer Noyabrskneftegaz and the Omsk Refinery—and create a new company, called Sibneft.[21]

Berezovsky had no ambitions to become a professional oilman, but he knew that oil was the key to major funding in Russia. "Sibneft interests me because it gives me greater maneuverability," Berezovsky would tell me later. "Oil is not only a business in itself, insofar as you can sell it or refine it, but the oil business is also a serious financial business, insofar as it is a good guarantee for loans. Among other things, Sibneft gives me the opportunity to operate other businesses, insofar as it gives me the opportunity to attract investment."[22]

With his quarry in sight, Berezovsky approached General Korzhakov and First Deputy Prime Minister Oleg Soskovets (responsible for industrial policy overall) with the idea of creating Sibneft. "I have good relations with Korzhakov and Soskovets," he told the Russian newspaper *Kommersant* in the fall of 1995. "I deeply respect these people. In general, I think it is normal for business and government to have good relations. All over the world, government helps business. More than that, capital controls the government, though here in Russia that is still ahead."[23]

He also lobbied Pavel Borodin, head of property management within the presidential administration, and, more important, Prime Minister Viktor Chernomyrdin. In return for Chernomyrdin's support, Berezovsky reportedly volunteered to coordinate the funding of his "Our Home Is Russia" political party during the 1995 parliamentary elections; he also provided Chernomyrdin with favorable TV coverage on ORT. His argument to the prime minister was that ORT needed huge subsidies to function and that a privatized Sibneft would allow Berezovsky to use the oil profits to fund the TV network.[24]

Sibneft was incorporated in the summer of 1995. The one government institution that would have resisted the creation of Sibneft, the Ministry of Fuel and Energy, was left out of the backroom deal.

"The presidential decree creating Sibneft was a surprise even to specialists," noted General Korzhakov. "The plan of the development of the oil industry of the country had already been approved by the government and the appearance of such a powerful company was in no way foreseen."[25]

The Omsk Refinery, one of the two key components of the new company, was left out of the negotiations. Boasting modern equipment, huge throughput capacity, and a great geographic location, Omsk was easily the best refinery in Russia, servicing many of the country's biggest oil producers. Though Rosneft retained a majority stake (soon to be transferred to Sibneft), the refinery also had important outside investors: Lukoil owned 10 percent of the stock, and foreign investors such as CS First Boston also owned small stakes. Omsk, in other words, was well positioned to survive on its own.[26]

The general director of the Omsk Refinery was Ivan Litskevich, a respected industry veteran. He had dealt with Berezovsky's partner, Roman Abramovich, before, when the Omsk Refinery had sold Abramovich oil products, which the trader then sold abroad. But Litskevich protested when he heard that Abramovich and Berezovsky planned to take over his refinery and make it part of Sibneft. On August 19, 1995, he was found drowned in the Irtysh River. Police found no evidence of foul play.[27]

Deprived of the leadership of its longtime general director, the Omsk Refinery quickly fell into line. Berezovsky and Abramovich now faced one final obstacle: a company called Balkar Trading that owned most of the rights to export crude from Sibneft's chief oil producer, Noyabrskneftegaz. Balkar Trading was named for a lawless region—Kabardino-Balkaria—near Chechnya. The company was run by

Pyotr Yanchev, a rival of Berezovsky's in the Avtovaz car-dealership market.[28]

At the time that Sibneft was being incorporated, Balkar was one of the largest oil trading companies in the country. One of its biggest contracts was to provide American oil giant Mobil with 25 million tons of Noyabrskneftegaz crude oil (worth roughly $3 billion) over five years. Naturally, Yanchev did not want to surrender such a lucrative arrangement to a newly formed holding company called Sibneft. In his fight to remain in the game, Pyotr Yanchev could count on some powerful patrons: Yeltsin's tennis coach, Shamil Tarpishchev; Russia's acting prosecutor-general, Alexei Ilyushenko, whose wife served as a paid consultant to Yanchev; and Ilyushenko's deputy, who was Yanchev's father-in-law.[29]

Ironically, it was Ilyushenko who had had overall responsibility for solving the Listyev murder; immediately after the murder, Ilyushenko had done Berezovsky a favor by calling off the investigators during their attempt to search Logovaz headquarters. But by the autumn of 1995, Berezovsky had no more use for Ilyushenko. At the end of September, one month after Sibneft had been incorporated, Yanchev was arrested on charges of bribery and embezzlement and thrown in jail. Ilyushenko was dismissed as acting prosecutor-general and, several months later, arrested for taking bribes. Most of Balkar's big contracts selling Noyabrskneftegaz oil were annulled. The Mobil executives went home.[30]

Berezovsky and Abramovich could now take over Sibneft without much protest. There remained only the technicality of how to acquire the equity of the new oil giant. The means were soon found: that infamous mechanism of pseudoprivatization called the "loans-for-shares" auctions.

Loans for Shares

The idea for the loans-for-shares auctions was first mentioned to me in 1994 by Boris Jordan, the American investment banker, then head of CS First Boston's Russian equity operations; Jordan argued that the Russian government should organize a gigantic debt-for-equity deal. In early 1995, he left CS First Boston and established a Russian-based company called Renaissance Capital; his partner was one of Russia's leading bankers, Vladimir Potanin. Together they developed the debt-for-equity idea. The plan called for Russia's leading banks to lend the cash-strapped government $2 billion, collateralized by big stakes in some of the country's best industrial companies. The banks would hang on to these equity stakes until the companies could be privatized in a second round of auctions in 1997.[31]

Russia's leading private banks were ready to assume the role of industrial empire-builders. "In Russia, you can't be comfortable earning big profits on trade," Mikhail Khodorkovsky, head of Menatep Bank, told me. "No one understands it. You won't be given any peace. You'll be invaded by racketeers and assaulted by all the government authorities. Even your own parents will stop respecting you."[32]

The real reason was more prosaic. Having made a fortune handling the accounts of Russia's major exporters—most of them still majority state-owned—banks like Menatep needed to ensure that these companies would continue pouring money into their accounts. The banks wanted to use the money earned from handling the accounts of the major exporters to buy the exporters' equity. But Russia's new banking titans could not afford to pay very much for their stakes: They had made billions of dollars in just a few years, but a lot of that money had been frittered away on marble lobbies, fleets of Mercedeses, mansions in Moscow, and villas abroad.

No one was more adept at taking over big companies at virtually no cost than Berezovsky. The tycoon had invested only a few million dollars to obtain a controlling stake in Avtovaz; he had paid just $320,000 for a controlling stake in ORT; and he had paid nothing at all to take control of Aeroflot. He would have to put up real money to purchase a controlling stake in Sibneft, but he was determined to pay just a fraction of the market price.

Out of these considerations—the desire to take over Russia's big exporters and the determination to pay almost nothing for the equity—the concept of the loans-for-shares auctions was born. It was left up to Vladimir Potanin to make the proposal to the government.

Potanin, thirty-six, was one of the golden boys of the Soviet establishment. He was smart, energetic, sensible, and efficient. He was also well connected. His father was an official in the Ministry of Foreign Economic Relations. Potanin joined the Communist Youth League, studied at a prestigious institute, joined the Communist Party at twenty-six, and took a job at his father's old ministry. After the Berlin Wall fell, Potanin and some colleagues from the ministry established an international trading company called Interros to compete with the old state-owned trade monopolies. Helped by the support of high officials in the Ministry of Foreign Trade and the Communist Party Central Committee, his company quickly obtained key foreign trade licenses, as well as prestigious exporters as clients.[33]

After Yeltsin took power, Potanin moved into banking. He established two banks, MFK and Onexim Bank, on the ruins of the Soviet Union's primary foreign trade banks. The best clients of these bankrupt institutions—more than forty top Russian exporters—transferred their accounts to Potanin's new banks, while the foreign trade debts were left with the Russian government. Only a few years after their creation, Onexim Bank and MFK ranked as Russia's third- and fourth-largest banks respectively, with subsidiaries in Switzerland, Cyprus, and the Caribbean.[34]

Even a sophisticated financial institution like Onexim Bank retained traces of the primitive origins of the Russian banking system. I noticed this when I visited Potanin at his headquarters at Onexim Bank—a huge, ugly, modern building. First I had to show the heavily armed guards my letter of invitation; then I had to pass my briefcase through an X-ray machine and walk through a metal detector. After getting through a final barrier—Italian-made security doors of thick, grenade-proof glass—I finally arrived in the lobby. Why all the security? "Gangsters," explained Potanin when we met upstairs. "The government is too weak to protect us, so we have to protect ourselves. The most talented government employees, including those in the security services, have gone private."[35]

That Onexim Bank's clients included the producers of Russia's most sensitive exports—oil, diamonds, precious metals, and armaments—confirmed its privileged status. Potanin never became entangled in political intrigues, never publicly criticized anyone in the government. In contrast to Berezovsky, Potanin was discreet about his ambitions. "I never considered that banks have a decisive influence on the government or that government should be at the service of the banks," he told me.[36]

But the loans-for-shares concept was precisely about the government's being at the service of a few chosen banks. The other top Russian bankers eagerly signed on to Potanin's idea. As the most articulate and presentable of the bankers, Potanin presented the loans-for-shares idea to the Council of Ministers in March 1995; he was accompanied by Menatep's Mikhail Khodorkovsky and Stolichny Bank's Aleksandr Smolensky. Potanin did all the talking. Led by Anatoly Chubais and Prime Minister Chernomyrdin, the Council of Ministers agreed to the plan.[37]

Immediately, Russia's leading bankers began to fight over the right to lend the government money in return for gaining control of prize equity stakes. Colonel Valery Streletsky, whose Department "P" of the Presidential Security Service was responsible for tracking government corruption, discovered how the victors of the loans-for-shares auctions were selected.

"They were chosen as a result of a multitude of meetings, in the course of several months during the summer and autumn of 1995, between half a dozen leading businessmen (Berezovsky, Smolensky, Khodorkovsky, Potanin, and so on) and government officials," he says. "They did not always meet all together. Sometimes they met in separate groups, talked things over, came to an agreement. If they failed to come to an agreement, they went their separate paths through friendly officials in the government apparatus. Then they would all meet together again. In this way the spheres of influence were divided." As a result of the horse-trading that preceded the auctions, several of Russia's leading commercial bankers were left out of the charmed circle: Vladimir Gusinsky (Most Bank), Vladimir Vinogradov (Inkom Bank), Pyotr Aven (Alfa Bank), Vitaly Malkin (Rossiisky Kredit Bank).[38]

"A key factor at this stage in the privatization process was the attitude of Tatyana Dyachenko to this or that banker/oligarch," Streletsky recalls. "She would go to the president [Yeltsin] and say: This man is a good man and that man is a bad man; this man should be supported and that one should not be supported." Mostly, Colonel Streletsky says, Tanya advanced the interests of Berezovsky. She was good friends with Valentin Yumashev, the ghostwriter for Yeltsin's memoirs, *Notes of a President.* "They saw each other every day, constantly called each other on the phone and had a common group of friends," General Korzhakov recalls. It was Valentin Yumashev who introduced Berezovsky to Tanya Dyachenko.[39]

The friendship between the car dealer and the president's daughter soon blossomed. Berezovsky, according to General Korzhakov, lavished Dyachenko with presents of jewelry and automobiles. "The first major present that Tatyana Dyachenko received from Berezovsky in 1994 was a Niva [a Russian version of a Jeep]," Korzhakov says. "The vehicle was customized to include a special stereo system, air-conditioning system, alarm system, and luxury interior. [This car would have had a market value of over $10,000 in the Russian market.] When the Niva broke down, Berezovsky immediately gave Tatyana Dyachenko a Chevrolet Blazer [a sport-utility vehicle selling in Russia for about $50,000]."[40]

At that time, Tanya Dyachenko's husband, Leonid, was an oil trader who was working with the Omsk Refinery; he was particularly closely associated with Berezovsky's partner, Roman Abramovich.[41]

Berezovsky had chosen his friends well. Gradually, as Yeltsin's old comrades disappeared—old warhorses who had been with the president in difficult times but now fell victim to corruption scandals and political intrigues—the president began relying more on his family. Tanya would soon become his most trusted political adviser. Indeed, she seemed to be his favorite daughter. The only memorable part of the *Notes of a President* was a strange story of Yeltsin's offering baby Tanya his own breast to prevent her from crying on a train journey the two were taking together.[42]

How could Yeltsin's daughter, who had no knowledge of business or affairs of state, be permitted to have a say in the distribution of Russia's prize industrial assets? "Tatyana Dyachenko is the only person to whom the president listens, whom he boundlessly loves and whom he boundlessly trusts," Streletsky continued. "Perhaps the president personally did not arrange who won the loans-for-shares auctions and who lost. That's where [Prime Minister] Chernomyrdin came in—he carried out all the instructions of the president, guessed his wishes and was ready to do anything. This was all one chain [the businessmen—Tanya—Yeltsin—Chernomyrdin], and Tatyana Dyachenko was the connecting link."[43]

One of the first victors in the loans-for-shares auctions was an old business partner of Berezovsky's, Mikhail Khodorkovsky. Back in 1992–93, when Berezovsky made his first foray into commodities trading—with several big export deals involving oil, timber, and aluminum—his banker on those deals reportedly was Khodorkovsky's Menatep Bank. In the summer of 1995, as the scheme for the loans-for-shares auctions was being worked out, Menatep helped Logovaz pay off its debts to the automaker Avtovaz. During the auctions themselves, Menatep and Berezovsky's financial companies would provide each other with mutual guarantees to ensure each other's victory.[44]

Mikhail Khodorkovsky's office was a Victorian castle in the middle of Moscow. The building was surrounded by a tall wrought-iron fence, with sharp spikes. The grounds were swarming with security guards, some in well-tailored suits, others in black uniforms and boots. Passing cars were carefully scrutinized.

"I personally do not own a single share in my company. I just have my salary and my car," Khodorkovsky said the first time we met. He was the boss of one of the largest business empires in the country and was one of the richest men in Russia. His holdings included Menatep Bank, a dozen other banks, large quantities of

Moscow real estate, a steel mill, Russia's largest producers of titanium and magnesium, as well as a slew of food-processing, fertilizer, textile and chemical companies.[45]

Still, as he sat drinking coffee by a tiled stove in his office, a Ussuri tiger skin at his feet, the thirty-one-year-old Khodorkovsky insisted on a humble demeanor—he had the manner of an earnest graduate student. He had the classic career path of a Yeltsin-era business magnate. In 1987, as a top leader of Moscow's Communist Youth League, Khodorkovsky established a trading cooperative financed with Communist Party money; the next year, he established a bank. In 1990–93, Khodorkovsky entered the Russian government, serving first as economic adviser to the Russian prime minister and then as deputy minister of fuel and energy. The Menatep Group, meanwhile, continued to grow. His trading company made big profits in oil, grain, sugar, and metals. Menatep Bank grew rich off the accounts it handled for the city of Moscow and various federal government ministries.[46]

Khodorkovsky was careful to obtain the right connections abroad, as well. Menatep's first deputy chairman was Konstantin Kagalovsky, who was married to Natasha Gurfinkel-Kagalovsky, the head of Russian operations for the Bank of New York. (Gurfinkel would resign her post in 1999, during the U.S. government's investigation into money-laundering at the Bank of New York.) Khodorkovsky tried to make a name for Menatep in the West. In 1994 he spent $1 million on full-page ads in *The Wall Street Journal* and *The New York Times*. He hired Arthur Andersen to audit the books and issued ADRs through the Bank of New York.[47]

Some of Khodorkovsky's largest foreign trade deals, however, could hardly endear him to the United States. The Russian tycoon had worked closely with the fugitive American commodities trader Marc Rich. In the course of 1994–96, Menatep sold hundreds of millions of dollars' worth of oil to Cuba, in return for Cuban sugar. Khodorkovsky also helped establish an institution called the European Union Bank on the Caribbean island of Antigua, a notorious money-laundering haven.[48]

In the loans-for-shares auctions, Khodorkovsky was interested in Yukos, a recently formed holding company that was Russia's second-largest oil company. Among Yukos's holdings was the producer Samaraneftegaz, whose exports were being handled by Berezovsky, and the Samara Refinery, which had been rocked by a series of assassinations two years before. As one of the best-endowed oil companies in the world, Yukos was the single largest prize in the loans-for-shares auctions. A 45 percent stake was being auctioned. In early November 1995, Menatep brazenly warned outside bidders to stay away. "There should be no two opinions about this," Konstantin Kagalovsky, first deputy chairman of Menatep, told the press, "We will get Yukos."[49]

A month later, on December 8, a consortium of Inkom Bank, Alfa Bank, and Rossiisky Kredit Bank offered $350 million for the Yukos stake—by far the highest bid. But the institution in charge of registering applications for the Yukos auction was none other than Menatep Bank. It rejected the consortium's bid, saying that part of the consortium's security deposit was in the form of T-bills, not cash. A Menatep front company won the Yukos stake by paying just $9 million over the starting price of $150 million.[50]

Meanwhile, the originator of the loans-for-shares idea, Vladimir Potanin, had his eye on another prize: the metals giant Norilsk Nickel. This company was one of Russia's premier exporters and a major shareholder of Onexim Bank. Located just under the North Pole, Norilsk had been established in 1935 as a concentration camp—one of the worst Soviet gulags. The first directors of the Norilsk conglomerate were generals of the NKVD, Stalin's infamous secret police. In the twenty years it was in operation, the Norilsk concentration camp cost the lives of more than 100,000 prisoners—victims of starvation, sickness, exposure, and execution squads.

Underneath this patch of Arctic tundra lay 35 percent of the world's known nickel reserves, 10 percent of its copper, 14 percent of its cobalt, 55 percent of its palladium, 20 percent of its platinum, as well as significant quantities of coal and silver. Norilsk Nickel mined what was probably the richest ore base in the world, with an exceptionally high metal content.[51]

After 1991, Norilsk managers took control, set up their own foreign trading companies, and began looting the company's foreign cash flow. Minority shareholders, like Onexim Bank and CS First Boston, were powerless to stop the theft. With the loans-for-shares auctions, Potanin got his chance: The government was selling its equity stake (38 percent of common shares; 51 percent of the voting shares).[52]

In the Norilsk auction, on November 17, the highest bidder was a company called Kont, fronting for Rossiisky Kredit Bank and offering $355 million for the Norilsk stake. Onexim Bank, in charge of registering applications for the auctions, disqualified the Rossiisky Kredit bid because of "insufficient financial guarantees." The victor was a subsidiary of Onexim Bank, which paid $170.1 million, just $100,000 over the starting price.[53]

Several weeks later, Onexim Bank was also placed in charge of registering applications for the auction of a 51 percent stake in the oil giant Sidanco. Again, Rossiisky Kredit Bank made a bid, but was rejected by Onexim Bank, allegedly because Rossiisky Kredit had not paid the necessary deposit; Rossiisky Kredit claimed that its representatives were not even let in the Onexim Bank building on the day of the auction. Onexim Bank's sister company, MFK, won the auction by paying just $5 million over the $125 million starting price.[54]

By now, it was clear to the general public that the loans-for-shares auctions were not normal free-market auctions, but rigged deals favoring a select group of businessmen.

The Sibneft Auction

Boris Berezovsky's loans-for-shares auction—for 51 percent of Sibneft—was the last of the series. Sibneft was a glittering prize. It was one of the largest private oil companies in the world, with hydrocarbon reserves equivalent to those of Amoco or Mobil, though its actual production was a third of these American giants. Having pushed through the incorporation of Sibneft in the summer of 1995, Berezovsky was not about to let anyone else buy its shares.[55]

His vehicle for purchasing the Sibneft stake was a specially formed company called Neftyanaya Finansovaya Kompaniya (NFK). Berezovsky said he was merely a "consultant" to the NFK bid. In my interview with him less than a year later, he dropped this pretense and frankly admitted his ownership of NFK.[56]

On December 28, 1995, the auction for Sibneft took place. The starting price was $100 million. There were two competitors: Inkom Bank, whose metals subsidiary, Sameko, offered $175 million for the stake, and Berezovsky's NFK, which offered $100.3 million.

It had been decided months earlier that Berezovsky would win the auction, but how was the rival bid to be neutralized? State property chairman Alfred Kokh would later tell me what happened.[57]

"The auction begins. Suddenly there is the sound of footsteps. The door opens. Some gentleman walks in and puts a fax down on the table of the committee. 'I, Ivan Ivanovich Ivanov (I don't remember his name), the director of the Sameko factory, am withdrawing my bid.' That was it. There was no competition. There remained only one real bidder and they won."

What was so strange about that?

"If I were of sound mind and memory," Kokh replied, "I wouldn't think of submitting an auction bid only to withdraw it tomorrow—all the more so if we are talking of $100 million or $200 million. The director of Sameko—Ovodenko, I think his name was—was the one who submitted the bid [and he was the one who withdrew it]. Something must have happened over the course of those several days so that, not giving a damn about his owner and boss [Inkom Bank], not giving a damn about the fact that he would look ridiculous and would be a laughingstock to the end of his days, he—"

"You think he did this against the wishes of Inkom Bank?" I asked.

"I am 100 percent sure of that. Absolutely."

"Why did he do it?"

"Probably he considered his life more important than his boss," Kokh answered, chuckling. "He was made an offer he couldn't refuse."

Still, the auction had to have at least two bids to be valid. Luckily, a company called Tonus—a front for Menatep Bank, which was simultaneously providing the financial guarantees for Berezovsky's NFK—bid $100.1 million. NFK bid $100.3 million (just $300,000 over the starting price) and won the auction.[58]

Boris Berezovsky thus managed to buy control of Sibneft at an implied market capitalization of less than $200 million. Two years later Sibneft was selling on the Russian stock exchange at a market capitalization of $5 billion. What had happened in those two years to justify the 2,400 percent appreciation? Sibneft remained the same operation it was before the loans-for-shares: a ramshackle, sloppy oil giant, producing the same amount of oil for the same customers. But once the company's shares began to trade on the free market, they traded at a price closer to their inherent value.

In the following months, Inkom Bank's boss, Vladimir Vinogradov, protested to the press that the loans-for-shares auctions had been "fixed." Inkom Bank, though one of the largest commercial banks in Russia, had clearly been left out of the game. In the winter of 1994–95, it was among the list of shareholders of Berezovsky's ORT (the ultimate insider's club) but was dropped at the last minute. It had been involved in the preparations for the loans-for-shares auctions but was subsequently eliminated from the list of designated winners. It tried to bid for Yukos but lost. It tried to bid for Sibneft but lost.

As soon as it publicly protested the results of the Sibneft and Yukos auctions, Inkom Bank was subjected to official harassment. On January 15, the Central Bank began investigating the bank's solvency. Even though Inkom Bank was current on its payments, there was widespread press speculation that it was insolvent; it was even briefly taken under Central Bank administration. By the autumn of 1996, the bank finally agreed to stop fighting the loans-for-shares results; privately, Vladimir Vinogradov continued to claim that Berezovsky and Smolensky were behind the campaign to discredit his bank.[59]

Berezovsky continued buying up the rest of Sibneft, putting in bids at a cash auction for 15 percent of the company and investment tenders for a further 34 percent. No outside investor would put money into a company completely controlled by one of Russia's robber barons.

When I interviewed him on the eve of the first investment tender (for 15 percent of Sibneft sold on September 19, 1996), Berezovsky seemed remarkably relaxed. When asked whether he thought there would be a foreign bidder for the 15 percent stake, he said it was unlikely. The risks of doing business were too high for foreigners. "We control the process to a much greater degree than Western companies," he explained.[60]

The combined sum Berezovsky and Abramovich promised to invest in Sibneft in return for a 34 percent stake was $78 million, spread over three years. This sum was not even sufficient to keep the existing oil wells in good shape, let alone boost their productivity.[61]

"Government Should Represent the Interests of Business"

The prime architect of the loans-for-shares auctions, Anatoly Chubais, denied that they were rigged and that the government received absurdly low sums. "You have to remember, this was an enormous amount of money," he said, referring to the $309 million Menatep paid for 78 percent of Yukos.[62]

But the Yeltsin government clearly was not motivated by the desire to get the best possible price for the state's assets. One proof of this was the fact that no foreign bidders were allowed. "The companies were sold more cheaply than they would have been if foreign participation had been allowed," admitted Potanin. "I wanted to allow foreign participation, but my proposal was rejected in the government."[63]

How much would foreign strategic investors have been willing to pay for the stakes offered in the loans-for-shares auctions? One hint can be found in the sale of Lukoil shares in the autumn of 1995. In the loans-for-shares auctions, Lukoil bought a 5 percent stake in itself for $35 million. Only a few months earlier, the Los Angeles–based oil company Arco had bought a 6 percent stake in Lukoil for $250 million.

Another clue suggesting that the loans-for-shares auctions were rigged was that each auction had at least two recognized bids, but in almost every case, the winner paid only a few million dollars more than the starting price. This price was artificial, bearing no relation to the companies' market value, as demonstrated in the table below.

The Six Most Expensive Loans-for-Shares Auctions (in Million U.S. Dollars)[64]

Company	% auctioned	Auction price (November–December 1995)	Market cap implied by auction price	Market cap on the stock market (August 1, 1997)
Lukoil	5%	35	700	15,839
Yukos	45%	159	353	6,214
Surgut	40%	88	220	5,689
Sidanco	51%	130	255	5,113
Sibneft	51%	100	196	4,968
Norilsk	51% (voting)	170	333	1,890

Among these crown jewels of Russian industry, Norilsk Nickel seems to have sold for the most realistic price at the loans-for-shares auctions, although it still sold at one sixth of the price it would command on the open market less than two years later. The oil companies (the first five companies in the table) were the real steal, since their market capitalization would rise between 18- and 26-fold within two years after the auctions.

"Why did the state sell its property so cheaply? Because this was a corrupt clan selling to itself," asserts Colonel Streletsky, who would later write a best-selling book on the subject. "A certain portion of the government apparatus had linked up with...Berezovsky and the other [businessmen]. These people knew the best ways to steal, but to steal they needed an alliance with the government apparatus. The government apparatus, meanwhile, needed money. So, this alliance was concluded."[65]

Berezovsky took a philosophical approach to the issue. "It is my fundamental belief that, leaving aside the abstract conception of the interests of the people, government should represent the interests of business," he declared in a November 1995 interview with the newspaper *Kommersant*.[66]

For the Russian state the loans-for-shares auctions were a disaster. At a stroke, the government lost a significant portion of its revenue-generating capacity. It was a mistake that would manifest itself less than three years later, when the state financial system would collapse.

"We made a mistake in privatizing the profitable sectors, the sectors on which the government could have survived," Foreign Trade Minister Oleg Davydov would later note. "They [the banks] merely grabbed the profits from the government, grabbed them from that part of the population who do not get their paychecks. The tragedy today is that if these were still state enterprises they would be profitable, they would be paying their taxes, they would be paying their workers, they would be investing in their capital base. The enterprises would have been alive. But these so-called owners arrived and what happened? There are no profits. There are no tax payments. The plant and equipment is getting worn out. And the money escapes abroad."[67]

One of the ironies of the loans-for-shares auctions was that the funds Berezovsky and the other oligarchs used to purchase their equity stakes came from the government itself. In the earlier stages of the Gaidar-Chubais experiment with capitalism, the Yeltsin government had done everything it could to boost a handful of favored banks. These institutions were given loans by the Central Bank at negative real interest rates. They were given massive government funds to be kept on deposit at below market interest rates. They were allowed to seize the profits of Russia's trading establishment and avoid paying taxes on their windfalls. Finally, they were allowed to participate in the exclusive government bond (GKO) market, earning yields of 100 percent or more in dollar terms. Paying such enormous interest rates on its domestic debt, the Russian government was steadily going bankrupt. But such well-connected banks as Onexim, Menatep, and Stolichny grew fat on these easy profits.

Some of the money went to buy luxury goods, some to buy industrial companies in the loans-for-shares auctions.

The loans-for-shares auctions were merely another stage in the Yeltsin regime's strategy of sacrificing the interests of the nation in favor of a handful crony capitalists.

The origins of this corrupt relationship went back to Berezovsky and his 1994 book deal with President Yeltsin. As the first major Russian businessman to enter the president's inner circle, Berezovsky had paved the way for the other oligarchs. Oleg Sysuyev, who would serve as Yeltsin's deputy prime minister, would later conclude: "Yeltsin made one serious mistake: He gave Big Business the opportunity to draw impermissibly close to him and to his entourage. Nowhere did the law say how close this relation had to be. It was purely dependent on what his [Yeltsin's] instinct told him about how close this relation should be. As a result, when everybody saw how close this relationship was and how it was evolving, they concluded that these were the rules of the game set by the head of state."[68]

The loans-for-shares auctions were simply the most public illustration of those rules. The auctions may have advanced the Russian state one more step toward bankruptcy (the debt default of August 1998), but from a Machiavellian point of view, they did have one important benefit for the Yeltsin regime: They locked the oligarchs into supporting Yeltsin for reelection. By arranging the auctions as a two-stage process—in which the oligarchs loaned the government money in 1995 but were prohibited from obtaining full legal control of the industrial companies until after the 1996 presidential election— Chubais and his colleagues ensured that the winners would do everything they could to help Yeltsin's campaign for a second term. The oligarchs and the Yeltsin government were now partners in crime.

CHAPTER EIGHT

THE BLACK TREASURY
OF THE YELTSIN CAMPAIGN

Davos

Every year in late January, rich and powerful people from around the world converge on Davos, the exclusive ski resort in Switzerland. The gathering is called the World Economic Forum.

In early 1996, Boris Berezovsky arrived in Davos to discover that the forthcoming Russian presidential election was on everyone's mind. Boris Yeltsin was in trouble. Only six months remained before the elections, and the leading candidate, Communist Gennady Zyuganov, was far ahead in the polls. Surveys showed the sixty-five-year-old Yeltsin's approval ratings hovering between 5 percent and 8 percent. Among presidential hopefuls, he ranked fourth or fifth. In the parliamentary elections the previous December, Russian voters had decisively rejected his policies; the Communists and Vladimir Zhirinovsky's extreme nationalists had captured two thirds of the seats of the lower house. On the eve of the new year, Yeltsin had had a heart attack and had checked into the Barvikha Sanatorium.[1]

The world business and political community seemed to have written off his chances for a second term. Meanwhile, Gennady Zyuganov impressed Westerners with his moderate attitudes—there would be no Bolshevik-style revolution when the Communists won, he told them. "I remember clearly how the business leaders of the

West reacted to Zyuganov in Davos," Berezovsky said. "I remember the energy with which negotiations with Zyuganov were conducted, the concrete nature of the partnership arrangements that were proposed. But we [the Russian oligarchs] had no such alternative—we couldn't negotiate partnership agreements with Zyuganov."[2]

The Russian delegation at Davos included such politicians as Anatoly Chubais and Yuri Luzhkov, and businessmen like Vladimir Gusinsky, Vladimir Vinogradov, and Mikhail Khodorkovsky. Berezovsky knew that the top Russian businessmen must unite behind Yeltsin, despite their differences. His first step was to smooth relations with his archenemy, Vladimir Gusinsky. The two men met and agreed to set aside what Berezovsky termed their "fierce rivalry," and to form a united front against the Communists.[3]

"I can't say that today there is love and friendship between Boris Abramovich and myself," Gusinsky told a Russian newspaper two months later. "It's just that at a certain point you understand that by continuing to fight, you harm not only your opponent but also yourself. I don't have a chance of winning a fight with Berezovsky and he doesn't have a chance of winning against me. Therefore, we had to sit down and come to terms."[4]

Asked about the new unity of purpose that had emerged among Russia's oligarchs after Davos, Gusinsky replied: "I wouldn't talk about bankers' cooperation. Such things don't exist. Only a temporary alliance of interests is possible. A financial institution has neither friends nor enemies—only temporary allies and rivals. Time passes and everything changes."[5]

Uniting the Russian businessmen at Davos was not enough to assure sufficient support for Yeltsin back home, however. Some oligarchs—Onexim Bank's Vladimir Potanin, among others—were not even in Davos. And, for all his powers of persuasion, Berezovsky could not convince powerful Russian industrialists such as Gazprom's Rem Vyakhirev or the top oilmen that he was the figure to unite them. He turned to Anatoly Chubais. As the architect of Russia's privatization, Chubais had given something to almost every member of the new elite. He was the right man to run Yeltsin's campaign. He had always been a superb administrator—cold-blooded, clearheaded, and decisive. His loyalty to the oligarchs was strong. He also had the West's support. Both the chiefs of the multinational corporations and Western government officials considered Chubais their champion in Russia.

"[Chubais] at that time was essentially unemployed," Berezovsky told the newspaper *Kommersant.* "I met Chubais face-to-face. I asked him to unite us. I'm talking about the financial elite. We all trusted him. We knew for certain that with each of us he had an absolutely formal relationship as a government official. That was probably the main thing: We had no doubts about his propriety."[6]

After the loans-for-shares auctions were exposed as sales rigged for insiders only, President Yeltsin was forced to fire Chubais as first deputy prime minister in mid-January. But the oligarchs responded by lavishly financing the recently established nonprofit organization that Chubais took over after he was dismissed from government. Stolichny Bank, controlled jointly by Berezovsky and Aleksandr Smolensky, gave this charitable organization an interest-free loan of $3 million. The money was invested in Russian T-bills, which had an annualized dollar yield of 100 percent at the time. Later, the prosecutor-general's office would open an investigation into the misappropriation of funds at Chubais's charity. (After nine months, the investigation was dropped.) In any case, though he would work in the private sector for only half a year in 1996, he declared a taxable income of $300,000 from "lectures" and "consultations."[7]

For Russia's new business elite, Chubais was the rainmaker—he was the man who had allowed them to become fabulously rich. Berezovsky, however, retained a remarkably scornful attitude toward the architect of Russian capitalism. "Chubais is good at executing the orders given him by his master," Berezovsky would later recall. "At that time [early 1996], he was hired by those who later became known as 'the seven bankers' [Berezovsky, Potanin, and the other oligarchs]. That's a fact. He was a hired hand with very good wages. And he earned his pay well by performing the tasks set by these individuals. The task was simple: We had to win the presidential election."[8]

Berezovsky and his accomplices returned to Moscow and began to take control of the election campaign. General Korzhakov claims he was astounded by the new strategy and specifically by the new alliance between Berezovsky and Gusinsky. "When [Berezovsky] returned to Moscow, he came to me in the Kremlin and told me that it wasn't necessary to kill anybody and that it was better for everybody to be friends," Korzhakov recalls. "I was surprised. 'How can you ask to kill people one day and be friends with them the next?'"[9]

Another powerful oligarch, Vladimir Potanin, was brought into the Berezovsky-Gusinsky-Chubais clan. Among government officials, Berezovsky's trusty ally, the presidential ghostwriter Valentin Yumashev, and an old Yeltsin crony named Viktor Ilyushin, were enlisted in the cause. Now only President Yeltsin had to be convinced.

"We Will Not Give Up Power"

On February 15, Yeltsin traveled to his hometown, the industrial city of Sverdlovsk, to announce that he would run for reelection. It was a cold day and he still looked weak from his recent heart attack. His voice was hoarse and sickly, apparently from the flu. Seemingly there was little chance of a Yeltsin electoral victory. Russians were sick of the corruption, incompetence, and poverty. Though often at his best in the role of underdog, Yeltsin would need an astounding comeback to win the June elections.

The campaign was officially in the hands of one of the good old boys of the Yeltsin administration: First Deputy Prime Minister Oleg Soskovets, a former industrial manager and a stout drinking buddy of the president's. Soskovets was part of the so-called conservative faction within the Yeltsin court, which included most of the heads of the "power" ministries: Grachev of the Ministry of Defense, Barsukov of the FSB–KGB, and Korzhakov of the SBP. This same group had arranged the disastrous invasion of Chechnya a year earlier.

An old-fashioned industrial manager by training, Soskovets could speak the language of the provincial Communist bosses who still ran most of the country in Yeltsin's name. He wielded his authority in the ham-fisted manner typical of Communist Party bosses. Anatoly Chubais would later remember how, in the early years of the Yeltsin regime, when the government was preoccupied with destroying the remaining power of the Communists, Soskovets would often pigeonhole some official giving a report to the Cabinet of Ministers:

"Are you a supporter of the market economy?"

"Yes."

"Okay, then you may continue your report."[10]

Soskovets's campaign strategy was equally primitive, relying primarily on naked authority. The general attitude toward the elections was illustrated by his friend and ally, General Korzhakov. In April 1996 Korzhakov met privately with Viktor Chernomyrdin to urge the prime minister to begin exerting his influence among provincial political bosses. "The governors listen to you," Korzhakov told him. "They know that they are appointed through you, that they are fired through you and they fear you. You can simply tell them: 'Sixty percent of the votes have to be for Yeltsin.' And they will do it."[11]

That kind of strategy may have worked in the end, but it was not producing results in the early months of 1996. Korzhakov soon put on an even more brutal face. According to his own account, he was transmitting the following message to the Communists: "Watch out, lads, don't mess around. We will not give up power."[12]

It appeared that, to keep Yeltsin in power, the elections would have to be canceled.

The Gathering at Logovaz House

In mid-April, thirteen of Russia's top businessmen gathered at Logovaz House, Berezovsky's opulent mansion in central Moscow, to discuss the elections. The Group of 13, as they were called, included the heads of Russia's largest oil companies, biggest independent TV network, largest auto plant, top aerospace firm, and most of the largest banks. They drafted a petition to Russia's warring political parties, which was printed on April 27 in the top newspapers of the country.

"Society is divided," the letter began. "This schism is growing catastrophically with every day. And the fault line dividing us into reds and whites, into us and them, passes through the heart of Russia.... Ultimately victory will go not to the beliefs of one side or another, but to the spirit of bloodshed and violence. The mutual hatred of the political parties is so great that they can only take power using the means that lead to civil war and the disintegration of Russia."

The Group of 13 went on to propose their solution: "In this responsible hour, we, the entrepreneurs of Russia, make the following offer to the intellectuals, military men, representatives of the executive and legislative branches of government, members of the law-enforcement agencies, the mass media, and generally all those who have real power today and upon whom the fate of Russia depends: Combine forces to find a political compromise.... Russian politicians must be encouraged to make serious reciprocal compromises, to come to a strategic political agreement, which can be sealed by law."

The letter concluded with a threat: "Russian businessmen possess the necessary resources and will to influence both those politicians who are too unprincipled and those who are too uncompromising."[13]

The letter was interpreted by the Russian and Western media as a call to suspend the constitution and for the political parties to come to a mutually profitable arrangement among themselves. Perhaps the most extraordinary thing about the letter was how it introduced a group of top businessmen as the kingmakers of Russia. The tone of the letter was peremptory, its message a brazen call for the nation's politicians to toe the line. Subsequently, Berezovsky was quick to attribute the anticonstitutional aspects of the letter to the conservatives in the Yeltsin court: "That whole group—Korzhakov, Soskovets, Barsukov—was putting enormous pressure on the president to essentially cancel the elections," Berezovsky would tell a Russian reporter a year later. "And the path they chose was the worst and most dangerous. They were discussing banning the Communist

Party, dissolving the Duma and postponing the elections for two years...."[14]

The Group of 13's appeal for a "compromise" between Yeltsin and the Communists was not entirely genuine. The appeal seemed to be a fallback plan—a warning to the Communists that the businessmen would not permit a return to a state-run economy if Zyuganov won the election. In fact, for several months before the publication of their letter, Berezovsky and his business colleagues had been gathering repeatedly at Logovaz House to formulate a strategy for a Yeltsin election victory.[15]

Two weeks after the letter appeared, Zyuganov decided to respond to the businessmen's appeal. He proposed a television debate with Yeltsin—a "discussion" of the problems facing Russia and the means for dealing with them. Yeltsin immediately rejected the offer, fearing a spontaneous debate on Russia's decline.[16]

Boris Berezovsky, meanwhile, was busy coordinating the financing of Yeltsin's campaign. "It is no secret that Russian businessmen played the decisive role in President Yeltsin's victory," he would tell me later. "It was a battle for our blood interests."[17]

Tanya

Bolstered by the support of his new allies, Yeltsin decided against postponing the elections. Berezovsky and the other oligarchs told the president he could win in June. Yeltsin then formed a new election staff, headed by himself, with Chernomyrdin and Ilyushin as deputies. Within this new organization, Chubais headed up the "analytical group" (the common Russian euphemism for a private intelligence operation), but he was effectively the new chief of the campaign. "In this way, we gained control of the whole intellectual angle of presidential elections," Berezovsky observed. "And the president received a new channel of information, which is exactly what we were trying to do."[18]

Chubais soon took matters in hand. His headquarters were initially in the Moscow city administration building, several floors above the headquarters of the Most Group. "Very convenient," noted Korzhakov. "At his elbow is Philipp Bobkov with his professional intelligence officers and all the exclusive information of the Most security service." The placement of Chubais's headquarters next to the Most Group demonstrated how quickly the political landscape had changed—for only a year had passed since General Korzhakov's men had attacked the Most Group—"Faces in the snow!"—at the instigation of Berezovsky and President Yeltsin.[19]

Soon, Chubais's staff was transferred to the President Hotel. This heavily guarded building was closed to outsiders. It was run by the Kremlin property manager, Pavel Borodin (the focus of the 1999 bribery and money-laundering investigation by Swiss prosecutors), and served as an all-purpose repository for businessmen and officials supporting the Yeltsin campaign.

A key figure in the newly created campaign team was Yeltsin's thirty-six-year-old daughter, Tanya Dyachenko. "It was Yumashev's idea," Berezovsky remembered. "He called me at six in the morning and said: 'I have a completely ingenious idea.' He pronounced only one name: 'Tanya.' In my sleepy condition, I did not completely understand: 'What about Tanya?' He answers: 'Tanya has to work with us in the analytical group.'...I didn't understand it fully at the time, but it was indeed an ingenious idea. It opened a direct channel of communication to the president. With so little time left [before the elections], it was necessary to make decisions with lightning speed. And these were decisions that could be taken by no one else besides the president. Therefore, there had to be speed of action and trust toward this channel of communication."[20]

Berezovsky, Yumashev, and Chubais all agreed that Tanya would be a critical addition to their cause. But how would they approach her? And how would they explain her presence to the Russian political establishment? Korzhakov says that the trio portrayed Tanya's role as similar to that played by French president Jacques Chirac's daughter, Claude, in softening her father's image in the 1995 presidential election in France. In fact, Tanya would serve as much more than a public-relations adviser—she would be one of the key figures of the campaign. Around her everything would revolve and the careers of long-standing Yeltsin aides would rise and fall.[21]

Speaking about the 1996 campaign, Berezovsky gave Tanya credit for her services in the oligarchs' cause. "It is impossible to suspect that Tatyana Dyachenko had any aims apart from those that we [the oligarchs] all had," he told a Russian interviewer.[22]

Apart from being a frequent visitor at Berezovsky's mansion, Tanya appeared often at the other headquarters of the Yeltsin campaign (the official one)—the President Hotel. This was Chubais's territory. Tanya's presence in the Berezovsky-Chubais group ensured the demise of the Soskovets-Korzhakov campaign staff. Berezovsky would later recall how the "conservatives" reacted to the new development: "At first, they didn't understand fully what was going on, since the whole political balance of forces around the president had changed fairly radically."[23]

"In the campaign staff Tanya held the post of independent observer," Korzhakov would recall. "In fact...everybody knew that Yeltsin's daughter was completely dependent on the opinion of Berezovsky and Chubais."[24]

"An Insignificant Sum"

Berezovsky and his colleagues did whatever was necessary to ensure Yeltsin's victory. Government officials were not allowed to participate in the election campaign, yet virtually the entire government apparatus, especially the provincial political bosses, was mobilized. State-owned TV networks, such as Berezovsky's ORT, were supposed to remain impartial, yet they broadcast a continual stream of news programs, documentaries, and advertisements lionizing Yeltsin and attacking his opponents. By law, Party campaign spending could not exceed $3 million, yet the SBP estimated the total spent by the Yeltsin campaign at "easily over $1 billion," whereas the Washington-based think tank Center for Strategic and International Studies estimated it at $2 billion.[25]

"The money was spent on local political bosses and on bribing people," recalls Colonel Streletsky, head of the anticorruption department of the SBP. "It was spent on various fictitious parties, like the party of Ivan Rybkin [a moderately left-wing party], for instance, and a mass of various social movements, such as the Cossacks. 'Our Home Is Russia' [Prime Minister Chernomyrdin's party] also received money from this fund. Campaign headquarters were set up everywhere...."[26]

Later I asked Chubais how much the Yeltsin campaign had spent. "I have no idea," he responded. "But the officially registered election fund [$3 million]—that is an insignificant sum."[27]

"The Yeltsin campaign was financed mostly by secret contributions from Russian banks and financial-industrial groups, as well as individual businessmen," says Korzhakov. "The campaign created a slush fund into which this money was dumped; no one accounted for it. Berezovsky immediately became one of the main money managers of this slush fund."[28]

Much of the money went to pay for flattering documentaries of Yeltsin on private TV stations, billboards put up by local mayors, pro-Yeltsin rock concerts organized by the entertainment industry, leaflets and posters printed by private publishing houses. To a large extent, the flow of contributions to the campaign was recorded and controlled by campaign headquarters. The system became known as the "black treasury."

"This whole scheme was coordinated in the shadow headquarters of the election campaign, at Logovaz House," says Colonel Streletsky. "It was thought up by the same people who had thought up the loans-for-shares auctions. In the beginning of 1996, they met many times at Logovaz House: Berezovsky, Tatyana Dyachenko, Chubais, and others. At these meetings, Tatyana Dyachenko represented the interests of her father, Berezovsky represented the interests of the businessmen, while Chubais was the manager of the campaign itself."[29]

Although many commercial banks handled these funds, Aleksandr Smolensky of Stolichny Bank (in which Berezovsky would later be reported to hold a 25 percent stake) sat at the head of the table.[30]

"Among the bankers, Smolensky had the role of administrator," explains Colonel Streletsky. "Berezovsky, meanwhile, was the liaison with the government authorities. He was the closest to the government authorities and essentially performed the role of connecting link between the Kremlin and business. If there was a question that needed to be decided with the authorities—any economic question that was of interest to this group of oligarchs—Berezovsky decided it. If Smolensky was the hands, then Berezovsky was the head of the organism."[31]

Berezovsky was eager to portray himself as the chief architect of the Yeltsin campaign. He bragged publicly about his preeminence and continually denigrated the role of the only other individual who could claim the title: Anatoly Chubais. "Not a single idea during this campaign came from Chubais," he would later tell *Kommersant*. "He is not an originator of ideas, but a brilliant analyst and executive. Very often he would be opposed to our proposals, but then he would adopt them as his own. In any case, he executed our proposals and in a very capable manner."[32]

The scheme Berezovsky developed to finance the campaign was ingenious: Rather than rely on voluntary private donations, the campaign would simply recycle government funds. At a time when schoolteachers, doctors, soldiers, and workers were going for months without pay, and millions of old people were without pension checks, the Yeltsin team decided to throw billions of dollars at the president's reelection campaign. Since it was clearly illegal for Yeltsin to use budgetary funds to finance his election campaign, these funds had to be laundered through the big industrial empires of Berezovsky and a handful of other oligarchs. "The funds in the black treasury were not only illegal according to the rules of the election campaign; they were also the proceeds of black-market or illegal operations," Colonel Streletsky observed. It would be the businessmen who would donate the hundreds of millions of dollars to the black treasury. In return they would receive many times the sum of their contributions in the form of government subsidies.[33]

"The method of 'recruiting' the businessmen [by the Yeltsin campaign] had been thought up in advance," notes Colonel Streletsky. "It underpinned the privatization process. A small group of businessmen was given everything, the whole government pie.... The businessmen, including Berezovsky, who won the 1995 loans-for-shares auctions had been allowed to win, with the understanding that...the government officials were doing them a favor, were allowing them to loot, but on one condition—that the time would come when the businessmen would be expected to donate large sums of money to the election campaign."[34]

During the loans-for-shares auctions, state equity stakes worth $14 billion on the stock market in July 1997 were sold to the oligarchs for less than $1 billion. This was not the only way in which the Yeltsin government paid for the support of Berezovsky and his colleagues. The businessmen were promised additional privatization windfalls (also worth billions of dollars) after Yeltsin's reelection. Finally, according to General Korzhakov and Colonel Streletsky, Berezovsky and the other oligarchs received the opportunity to skim the campaign treasury.[35]

Thousands of different companies contributed to the Yeltsin campaign, Streletsky recalls. Virtually any company that was dependent on the Yeltsin government's goodwill could be convinced to give money for the reelection. "All these funds were thrown into a communal pot, a communal treasury, and then, from this communal treasury these funds were distributed," he says. "The money came from a broad range of enterprises and structures, but only a narrow group of people, including Berezovsky, had access to this communal treasury. This narrow group of people left part of the money for the election campaign and took part of the money for themselves....Before the money [for the campaign] got to its final destination, it passed through a chain of middlemen, each of whom kept a certain portion for himself."

The mass embezzlement was an open secret among the members of Yeltsin's entourage. On the evening of April 16, General Korzhakov and Prime Minister Chernomyrdin were drinking in the Presidential Club. It was a rare opportunity for the two men to compare notes on the progress of the campaign. Korzhakov taped the conversation.[36]

CHERNOMYRDIN: We are currently working out the financial side of the campaign.... They brought me the plans and I warned them: We're going to check everything, all the documents, analyze everything. There are huge sums coming in. These people are so untrustworthy that it's better to give the scoundrels an earful from the very beginning.

KORZHAKOV: Of course. Remember how they grew rich off your parliamentary campaign for Our Home Is Russia? I got the reports. By coincidence, I had a fellow sitting in the Olbi Club, where they were all partying. He told me: "I found myself in a sewage drainpipe. They got drunk and the conversation turned to how to divide up the cash, how to collect it, how to distribute it and to whom. Not one word about the elections."

CHERNOMYRDIN: Scum. Now even Smolensky has gotten worried.... He doesn't know where the money is going. I want to talk this over with Mikhail Ivanovich [Barsukov—the chief of the FSB–KGB].

KORZHAKOV: The fellows over there [in the campaign headquarters] have sticky palms. Chubais decides what the program is and who's in charge. The money is doled out accordingly.

CHERNOMYRDIN: We're talking four million, five, seven, fifteen...

KORZHAKOV: Yeah.

CHERNOMYRDIN: Big sums. You can earn 200 million and in hard currency.

KORZHAKOV: There is a lot to steal.

CHERNOMYRDIN: They're going to steal it anyway. The question is how much.

"It Made a Mishmash of Their Brains"

If the Yeltsin campaign was hopelessly corrupt, the Communist candidate, Gennady Zyuganov, was hopelessly dull. Moreover, it was doubtful whether the Communist Party really wanted the presidency. They seemed comfortable in their role as the opposition party. General Aleksandr Lebed, one of the candidates in the election and later Yeltsin's national security adviser, states that since 1993–94, the leaders of the Communist Party had been secretly receiving funds from the government. The Communist leaders, in other words, were less independent than their followers believed.[37]

Still, it should have been relatively easy for the Communists to win over the election. Russia was collapsing and the Yeltsin government was clearly responsible. The Communists also had a superb grass-roots organization; their corps of 500,000 activists covered the farthest reaches of the country. These loyal supporters were well suited for running an old-fashioned campaign: door-to-door solicitations, activists handing out leaflets on the street, mass rallies. But the Communists still had trouble getting their message across. They were desperately short of money and they did not have TV coverage.

The lack of a television presence was the fatal flaw of the Communist campaign. Most people in Russia received their news almost exclusively through TV. The networks—both private and public—united around Yeltsin. They unleashed an impressive advertising blitz. The president was on the news every night, visiting pensioners in the Arctic Circle, promising large funds for neglected communities, joking with collective-farm workers, shaking hands with the mayor of some forgotten industrial city. "Mr. Yeltsin totally controls the media," noted Congressman Lee Hamilton of Indiana during the April 1996 congressional hearing on Russian organized crime. "The other guy [Zyuganov] cannot even get on television."[38]

Most of Russia's mass media depended on government subsidies. Newspapers depended on cheap rates at government-owned printing presses; TV stations depended on low rates from the government broadcasting service. The biggest consumer of government largesse—Boris Berezovsky's half-state, half-private ORT television network—received more than $200 million a year in government subsidies. The fact that most of Russia's media operations needed government funds to survive clearly gave the government enormous leverage in negotiating content. The most effective weapon was bribery.[39]

Midway through the campaign, American journalists began uncovering evidence that Yeltsin's team was bribing cash-strapped journalists and their bosses to run flattering pieces about the president. Payments ranged from $100 paid to a provincial reporter for a single positive article to millions of dollars paid to the owners of the largest Russian newspapers.[40]

Colonel Streletsky, investigating the hidden money flows of the Yeltsin campaign, found that the largest payments went to the TV bosses. Among the campaign documents seized by the SBP, Streletsky says, was a budget entry for the first half of 1996 specifying $169 million payable to Berezovsky's ORT. Streletsky says that the money had, in fact, been transferred, but only $30 million ever arrived at ORT.[41]

The birth of the free press in Russia had been one of the country's few encouraging developments over the past decade. But there was a relapse in the 1996 presidential campaign. Berezovsky went so far as to admit that he "didn't believe in freedom of the press the way idealists would like to imagine this notion."[42]

Sergei Parkhomenko, editor of the *Newsweek*-style magazine *Itogi*, even declared to a *Los Angeles Times* reporter that he was willing to subordinate his journalistic ethics to prevent a Communist victory. "This is not a game with equal stakes," he said. "That is why I am willing to be unfair. That is why I am willing to stir up a wild anti-Communist psychosis among the people."[43]

The TV networks produced a number of documentaries about Boris Yeltsin, focusing on the early years, the good years. One documentary featured his wife, a cozy grandmother, in her home, talking about how happy she was with her husband. Perhaps most important was that the president almost always led the nightly news. There were some significant news events (the cease-fire in Chechnya, Bill Clinton's visit to Moscow, the Russia-Belarus customs agreement), but on slow evenings Yeltsin was shown working in the Kremlin or visiting workers in the provinces. Zyuganov, on the other hand, was rarely seen or heard. Toward the end of the campaign, after Yeltsin became ill, the Communists began focusing on whether Russians should elect a seriously debilitated president. They tried to buy advertising time on state-controlled TV, but were refused.[44]

While the Communists were handicapped by their weak public-relations strategy, the Yeltsin camp received the help of the best foreign campaign specialists. One of the first such image-makers was Tim Bell, the advertising genius behind Saatchi and Saatchi and Margaret Thatcher's election in 1979. The Yeltsin team also recruited the campaign managers responsible for California governor Pete Wilson's impressive come-from-behind election victory in 1994. The American campaign managers were installed in Yeltsin's campaign headquarters in the President Hotel. They were under strict instructions to keep a low profile and to venture from the hotel as little as possible. The California team was based in suite 1120 of the President Hotel; across the hall, in Room 1119, was Tanya Dyachenko. The professional relationship, American political strategist George Gorton bragged to *Time* magazine, was unusually close: Tanya and the Americans shared the same secretary and the same fax machines. She became a key link between the Americans and the Russian president. "The American consultants were treated like foreign royalty," grumbles Korzhakov. "After every routine staff meeting, Tanya immediately ran to them to discuss the new information."[45]

The Americans suggested such dirty tricks as trailing Zyuganov with "truth squads," which would heckle him and cause him to lose his temper. They also reinforced the more basic lessons of modern political campaigning: daily memos identifying the tasks at hand, the points to be hit, the images to be transmitted. They did simple things, such as replace a poster of a scowling Yeltsin with a smiling Yeltsin. Photograph sessions and TV appearances were strictly choreographed to seem spontaneous. A running series of public-opinion polls and focus-group sessions delved into the instinctive reactions of the Russian electorate and shifted the Yeltsin campaign accordingly.[46]

Yeltsin was sent on a grueling cross-country campaign—something that had never been done in Russia before. The Russian president put on a miner's hat and descended into the coal pits. He visited soldiers at distant army bases. He accepted the traditional

peasant gesture of hospitality of bread and salt at obscure rural settlements. At a Moscow rock concert organized by Sergei Lisovsky,
Boris Yeltsin was even inspired to get on stage and shake his belly to
the tunes. Russians had never been subjected to a direct-mail campaign
like the millions of letters bearing Yeltsin's signature sent to World War
II veterans thanking them for their service (many of the recipients
apparently thought that the letters had actually been signed by
Yeltsin).[47]

But perhaps the most sophisticated propaganda was produced
by American-trained Russian advertising men. A company called
Video International had been signed up to produce Yeltsin's official
campaign ads—fifteen different advertising spots. The strategy was
a soft sell. With Yeltsin dominating the nightly news, it was not
necessary to show him in the campaign spots. Instead, these were
one-minute biographies of sports stars and factory workers, grand-
mothers and former ministers, farmers and schoolteachers, soldiers
and artists. Accompanied by sentimental music, these people described their struggles, their hard times, their hopes and values; the
spots would end with Yeltsin's signature: "I believe. I love. I hope.
B. N. Yeltsin."

As *Washington Post* reporters Lee Hockstader and David Hoff-
man noted, "most of the subjects in the ads were closer to the profile
of Communist supporters, but all said they were voting for Yeltsin."
Most Russians had never seen this kind of sophisticated media manipulation. "It made a mishmash of their brains," noted Alexei
Levinson of the All-Russian Center for Research in Public Opinion.[48]

Coins to the People

Yeltsin proved a good candidate. He listened to his public-relations
advisers when they told him to use a teleprompter, to wear better
suits, to smile more. He knew he had to show more vigor and manli-
ness. He also responded to the campaign with a characteristically
whimsical show of largesse. He would arrive in some impoverished

industrial town and promise everybody back wages; he even promised one woman in the crowd a new car (it was delivered). Yeltsin was a czar, throwing silver coins to the people. More seriously, he even created a special "President's Fund" to pay Russia's back wages and pensions.

All this generosity was inflationary. Since the Russian economy was still shrinking, the money Yeltsin promised had to come from foreign currency reserves and foreign bank loans.

Fortunately, that spring the IMF granted Russia its largest credit ever—$10.2 billion over three years. The credit quickly disappeared. Despite the cash infusion, the Russian treasury's foreign currency reserves declined from $20 billion to $12.5 billion in the first half of 1996. The Russian government, in other words, spent at least $9 billion in foreign currency during the first half of 1996. Some of the money went to the Yeltsin campaign, some to well-connected businessmen and government officials, some to pay ordinary Russians their long-overdue paychecks.[49]

Combining sophisticated Western electoral techniques with crude pressure applied by loyal political bosses, the Yeltsin election campaign gathered steam. The president began to exhibit vitality at the office. Suddenly he was achieving notable victories in affairs of state, such as the assassination of the leader of the Chechen rebels, President Jokhar Dudayev. This operation was carried out in spectacular fashion. The Chechen president was killed in the field when a guided bomb homed in on the signal of his satellite telephone. Dudayev's death was a windfall for the Yeltsin campaign. A cease-fire with the Chechen rebels materialized two months later. Russians silently breathed a sigh of relief.[50]

Yeltsin's statesmanship received another boost on April 2, when he signed a union agreement with the neighboring republic of Belarus—an astute response to Russians' long-standing desire for the reunification of the Slavic lands of the former Soviet Union. (In fact, it was an inconsequential agreement, since it did not bring the two republics closer in any significant way.) In mid-April, Bill Clinton arrived in Moscow and indulged Yeltsin's chest-beating about Russia's great-power status—another popular stance for the Russian public.

The National Sports Fund

As the Yeltsin campaign was gaining the upper hand over the Communists, a power struggle broke out within the Yeltsin court. Boris Berezovsky had turned on his erstwhile patron, General Korzhakov. In his interviews and writings, Korzhakov does not specify when the quarrel began. As late as the summer of 1995 everything had been fine between the two men. Korzhakov had helped Berezovsky grab control of Aeroflot and establish the oil holding company Sibneft. But that autumn, when the loans-for-shares auctions took place, Korzhakov was missing. That absence from the kitchen when the state industrial pie was divided up was especially apparent in the failure of Inkom Bank (particularly close to Korzhakov) to win any loans-for-shares auctions; twice Inkom Bank attempted to enter auctions for Russia's big oil companies and twice it was unceremoniously rejected. Moreover, Korzhakov was no longer indispensable to Berezovsky as an intimate channel to President Yeltsin, since the tycoon now had Tanya Dyachenko and Valentin Yumashev. Finally, if Korzhakov's claim is true that Berezovsky asked him to kill Gusinsky, the security chief's refusal to carry out this request would have revealed him to be less than an ideal asset in the violent Russian underworld. In any case, by the end of 1995, Berezovsky had grown powerful enough to do without Korzhakov's protection.

"The entourage of the president found that Korzhakov was a big nuisance for them," says Colonel Streletsky. "Korzhakov wouldn't let [the businessmen] close to Yeltsin, wouldn't let them approach Yeltsin to advance their own mercenary interests. And the aides to the president were always worried that Korzhakov was keeping an eye on them and not letting them loot peacefully. They decided that Korzhakov had to be removed."[51]

Berezovsky moved fast. In early April he arranged to have Tanya Dyachenko initiated into some of the seamier undersides of power politics in the Kremlin: He uncovered for her the secrets of the National Sports Fund (NSF). The NSF had been founded in 1992 by Yeltsin's tennis coach, Shamil Tarpishchev. Tarpishchev exuded the same openhearted camaraderie that made Korzhakov so endearing. He was tall, with an easy smile, and his rambling gait showed off his native physical agility. Yeltsin gave Tarpishchev everything he wanted. Like the analogous foundation created by the late gangster-philanthropist Otarik, the NSF was supposed to collect money for the ruined Russian athletics apparatus.

When Tarpishchev was appointed minister of sports and athletics in 1993, he withdrew from daily management of the NSF. The new president of the foundation was a thirty-three-year-old athlete-turned-businessman named Boris Fyodorov. The NSF was a fantastically lucrative racket. It was allowed to import liquor and cigarettes duty-free and avoided paying most of the taxes due on its profits. According to Colonel Streletsky's later investigation, the NSF racked up $1.8 billion in profits within two years.[52] "This money was looted," says Streletsky. "Only trivial sums actually went to support sports. Fyodorov and his friends accumulated huge fortunes at the expense of the state budget."

In late 1994, Berezovsky chose Fyodorov as one of the founding shareholders of his TV network, ORT. Fyodorov also won the trust of another scandal-prone entrepreneur, Oleg Boiko—the man who simultaneously bankrolled the "young reformer" Yegor Gaidar and engaged in numerous shady business deals. Boiko, an associate of Berezovsky's, was an investor in the notorious Cherry Casino, and at least one top Boiko executive was seriously wounded in a gangland-style assassination attempt. In mid-1995, when Boiko's Natsionalny Kredit Bank crashed, leaving hundreds of millions of dollars in debts to the state-owned savings bank, Fyodorov was appointed president of both the bankrupt bank and Oleg Boiko's holding company, Olbi. These companies, too, became recipients of NSF largesse.[53]

NSF's highly profitable operations aroused jealousy. In March 1995, the same month that ORT general director Vlad Listyev was murdered, Lev Gavrilin, the head of foreign economic relations for the NSF, was killed by an unidentified gunman. Fyodorov, according to Colonel Streletsky, had embezzled $300 million of NSF funds. In Russia such large-scale theft almost always involved gangland violence.[54]

"Part of the [NSF] money went as bogus loans to various commercial organizations," Streletsky recalls. "Fyodorov established about eighty commercial enterprises around the NSF, and he distributed all the money that was coming in among these enterprises. Ultimately, we couldn't find the money in these commercial enterprises either. It had already gone further down the line."[55]

As one of the key parts of the crony capitalist network underpinning the Yeltsin regime, the NSF was expected to contribute to the president's reelection campaign. Sometime in late March or early April, says Colonel Streletsky, Boris Fyodorov was told to bring $10 million in cash to campaign headquarters. Fyodorov arrived at the President Hotel with the money in a suitcase and gave it to Chubais. General Korzhakov, as a dedicated sports fan and a man who still considered it his right to supervise the activities of the NSF, was angry that NSF funds were going to his rivals at the Berezovsky-Chubais campaign headquarters.

"When we heard about this, we invited him [Fyodorov] for a talk," says Colonel Streletsky. "First Korzhakov met with him in the morning in the Kremlin; then I met with him in the evening in the White House. We told him: You have to return that money to the NSF. At that time, preparations were under way for the Olympic Games and for the soccer World Cup, and there was no money [to finance the teams]. Korzhakov told him: "You have to return the money, because this money essentially belongs to the state."[56]

That same evening, Fyodorov went to Berezovsky in Logovaz House to convey his problems. Berezovsky invited his buddy, Valentin Yumashev, and the presidential daughter Tanya Dyachenko, to the interview. He also had the conversation taped. Fyodorov declared that he was being victimized by Mafiya organizations operating within President Yeltsin's administration—principally the SBP of General Korzhakov and the FSB–KGB of General Barsukov. General

Korzhakov was trying to extort $10 million from him as a personal bribe, Fyodorov said. He also accused Korzhakov's friend Shamil Tarpishchev (Yeltsin's tennis coach and the founder of the NSF) of links to organized crime. "Korzhakov and Barsukov will kill me," Fyodorov declared. "You must tell the president that he cannot surround himself with bandits."[57]

Tanya Dyachenko was shocked by this tale of corruption and criminality. "[Fyodorov's declaration at Logovaz House] was a clever ploy, thought up mainly by Berezovsky, to discredit Korzhakov, Barsukov, and Tarpishchev in the eyes of the president's daughter," Colonel Streletsky says. "The spectacle was skillfully planned and executed." The NSF affair, like so many scandals in Russia at the time, was never fully investigated, so it is difficult to assess the truth of the allegations.[58]

"We found out immediately that this spectacle had taken place and that a tape had been made," says Colonel Streletsky. "Berezovsky then became terribly frightened. The next day, he himself came to Barsukov and brought the tape with him and said: 'Listen, I have nothing to do with this, but Fyodorov came and said all these awful things. Listen to it.' Berezovsky also kept a copy of the tape to use at the appropriate time." Colonel Streletsky says that Berezovsky presented the tape to the security services because he was afraid of Korzhakov and Barsukov and wanted to show them that he was still on their side. But Berezovsky was doing more than simply showing his loyalty to his erstwhile patrons—he was letting them know that he was in possession of an extraordinarily compromising piece of information.[59]

Korzhakov forged on. On May 21, nearly two months after Fyodorov's taped interview at Logovaz House, the NSF chief's car was stopped by Moscow regional police; inside, they found a packet of cocaine. He was arrested. Though the charges were later dropped (when cocaine possession was decriminalized in 1997), analysis of Fyodorov's urine, hair, and fingernail samples confirmed that the NSF chief had indeed used cocaine. (The thirty-nine-year-old Fyodorov would die of a heart attack in 1999.) He was released almost immediately after his arrest, though he was fired as president of the NSF. His place was taken by Colonel Streletsky of the SBP.[60]

Scare Tactics

Yeltsin's campaign managers understood that if the election were to become a referendum on his record in office, he would almost certainly lose. His record as Russian president could not be defended; it was too much of a disaster even for the best spin doctors. The campaign decided to go negative. The Russian people may have hated Yeltsin and the job he was doing, but they feared famine and civil war even more. They had to be convinced that a Communist victory would be catastrophic and that Yeltsin was the only guarantee of stability.

TV networks repeatedly aired documentaries about past Communist horrors. In the closing weeks of the campaign, more than a million posters were printed with Zyuganov's face and the statement: "This may be your last chance to buy food!" The adhesive-backed posters appeared in food markets throughout Russia. At the same time, the newspaper *Kommersant*, apparently from its own funds, printed 10 million leaflets entitled "God forbid!" describing the disasters that would befall Russia if the Communists won; the leaflets were stuffed into Russians' mailboxes.[61]

On June 10, Boris Berezovsky gathered the Group of 13 (this time minus two of the original signatories) at Logovaz House to sign another open letter about the election. This letter unambiguously called for a Yeltsin victory. It fiercely criticized the Communist candidate, Gennady Zyuganov. The Communists' economic program, the Group of 13 said, "is aimed at returning the country, at best, to the situation of the mid-1980s." At worst, the letter implied, the Communists would produce famine, war, and mass terror. "If the measures set out in the [Communist] program are implemented, catastrophe will strike within four to six months," the Group warned. This terrifying prophecy, part of a fundamental shift in the Group's strategy, was Berezovsky's idea.[62]

"Today, Communist leaders are not concealing what is most important to them: to alter the form of property and to make state property dominant once again," Berezovsky told a Russian newspaper several weeks earlier. "We have conducted a superexperiment, which proved...that this form of property [socialism] generates an inefficient economy that cannot secure people's primary needs. This entails some simple consequences: If the economy can't provide for the people, then we must resort to political means. How? Build labor camps, execute a couple of million people, create an external enemy—the rest is absolutely clear."[63]

Yeltsin's campaign manager, Anatoly Chubais, still held to this line when I spoke with him eighteen months after the election. "Communism is the most evil system devised by man," he declared. "It can only work with concentration camps and terror. Those are the kinds of people we were facing in 1996."[64]

Yes, Soviet Communism had indeed been totalitarian and genocidal in the first several decades of its existence, but by the time Mikhail Gorbachev arrived, it was an entirely different animal. The rump Communist Party of Yeltsin's day was a collection of mostly elderly people, intent on protecting the social welfare system. To speak of Gorbachev's regime, let alone Zyuganov's Communist Party, in the same breath as gulags, famine, and civil war was a bit like describing today's South Carolina as a land of Ku Klux Klan gatherings and lynchings.

On June 11, the day after the Group of 13's second letter, a bomb exploded in the Moscow metro, killing four people. The Yeltsin campaign immediately blamed Communist extremists and renewed the call for Russians to "vote for civic peace and stability." No one claimed responsibility for the bombing and the perpetrators were never found.[65]

Recruiting General Lebed

While the Yeltsin team and the Communists were painting each other black, there remained the prospect of the other leading candidates' forming a bloc that would be both anti-Communist and anti-Yeltsin.

One such candidate was liberal parliamentarian Grigory Yavlinsky, co-author of Gorbachev's 500-Day Program. Yavlinsky was intelligent and perceptive (he had almost always proved right in his analysis of the situation in Russia) and a man of doubtless honesty, but his appeal to the Russian electorate was limited because he was so obviously an intellectual; in any case, he had repeatedly refused to ally his Yabloko faction (scoring about 10 percent in the opinion polls) with any other party.

Another strong candidate was the nationalist buffoon Vladimir Zhirinovsky, who won the highest number of votes in the 1993 parliamentary election campaign. This time, however, Zhirinovsky was running a lackluster campaign. He had toned down his extremist rhetoric and lost much of his TV exposure; consequently his polling numbers sank to around 5 percent. In any case, Zhirinovsky had long since proved himself to be a mere creature of the Yeltsin regime.

That left General Aleksandr Lebed. A former paratroop commander in the Afghanistan War, General Lebed had come to the nation's attention during a number of internal military operations. Under Gorbachev, he had commanded the detachments that violently put down riots in Tbilisi, the capital of the Soviet republic of Georgia. In 1991, he played a key role in swinging army support behind Yeltsin during the August Putsch. Two years later, as commander of the Fourteenth Army, he kept the peace between Russians and Romanians in the former Soviet republic of Moldova. Lebed's main attraction was his persona: tough, blunt, straightforward, and honest. Russia had long been waiting for a man on a white horse to bring order to the country, and to many, Lebed seemed to be that man.

Boris Berezovsky had already marked him out. Having cut his ties to General Korzhakov, Berezovsky decided upon General Lebed as Russia's new strongman. The hope was that, since he was relatively ignorant of Kremlin intrigues, Lebed could simultaneously scare the opposition and be easily controlled by the men behind the scenes. In the short term, the general's main value was his electoral potential. He represented a law-and-order platform and he spoke directly to that part of the electorate expected to vote Communist.[66]

On May 8, Berezovsky and other members of the Group of 13 met with General Lebed in Moscow. When the men emerged from their two-hour, closed-door meeting, no one wanted to comment on the discussions. Approached by reporters, General Lebed dismissed speculation that a deal had been struck between him and the Yeltsin camp. "They didn't buy me," he stated. "I am not for sale."[67]

They may not have bought the gruff paratroop general, but Berezovsky and his business associates threw their financial resources behind his campaign. He suddenly became a highly visible and impressive candidate, appearing frequently on television, in the newspapers, and on billboards. As Lebed's popularity rose, he pulled votes away from the Communists.[68]

On June 16, Yeltsin won a plurality in the first round of the election. He received 35 percent, while his Communist challenger, Gennady Zyuganov, received 32 percent. The big surprise was Lebed, who received a respectable 15 percent of the vote. The second round of the elections was scheduled for July 3.

Immediately after the first round, events began moving rapidly. On June 18, Yeltsin appointed General Lebed his national security adviser and secretary of the Security Council. One of the conditions Lebed set for entering the government was the dismissal of General Pavel Grachev as defense minister. Grachev was fired the next day. From a public-policy perspective, Grachev's dismissal was long overdue. Incompetent and boastful, he had presided over the disintegration of the Russian army and its blood-soaked adventure in Chechnya. But politically his dismissal was dangerous. Grachev was one of the oldest and closest comrades of General Korzhakov, and his dismissal signaled that the Berezovsky-Chubais camp was moving against the whole "conservative" faction within the Yeltsin court. This faction, now confined to three men—Korzhakov, Mikhail Barsukov (head of the FSB–KGB), and First Deputy Prime Minister

Oleg Soskovets—stood between the Berezovsky-Chubais clan and total control of the Kremlin.

A Box of Cash

For his part, General Korzhakov had already decided to expose the secrets of the black treasury of the Yeltsin campaign. That Yeltsin was now almost certain to be reelected made it easier to air the campaign's dirty laundry—there was little danger of derailing his victory. Korzhakov had to move quickly. The situation was changing daily. "During the presidential election campaign of 1996, Yeltsin ordered me not only to participate in important campaign projects, but also to control financial management of the election headquarters," he said stiffly. "The officers of the SBP discovered serious violations, whose essence can be described very succinctly: The funds of the election campaign treasury were being ruthlessly embezzled."[69]

The same day that Lebed was in the Kremlin receiving his appointment as national security chief, an operations group from Colonel Streletsky's anticorruption department of the SBP prepared to go into action. Streletsky's team felt it had little chance of getting into the President Hotel—as Chubais's base camp, this was enemy territory. Neither could the SBP force its way into Berezovsky's Logovaz House. Fortunately, the Yeltsin campaign maintained a third base of operations: Russian government headquarters (the White House)—an important repository for documents and cash relating to the Yeltsin campaign.

On the night of June 18, Streletsky's team entered room 217 of the White House. This was the office of German Kuznetsov, the deputy minister of finance and, according to the SBP, one of the key managers of the black treasury. The security men found stacks of dollar bills in Kuznetsov's safe, along with receipts for previous funds deposited in foreign banks.[70]

"The SBP officers found $1.5 million in cash inside, along with payment stubs that proved that the campaign funds were being transferred to the accounts of offshore companies," Streletsky recalls. "We didn't know who owned these firms or what services they provided. However, we did understand the mechanism by which funds were being transferred from the slush fund into foreign bank accounts.

"There were prepared money transfer orders into these offshore accounts—dummy transfer orders, we call them. The transfer orders showed, for instance, that $5 million was to be transferred to a specific account in the Bahamas for advertising services or printing services. We found twenty such accounts. On each payment stub there was a number and a date. The stubs that we were able to get were at the end of a series: numbers 21, 22, 23, 24. Multiply 24 by $5 million and you get $120 million. That gives you an idea of the size of the money transfers going on.

"Part of the funds in the black treasury was kept in cash in safes and suitcases; part was kept in banks, where special accounts had been opened. Payment was almost always made in cash in order to avoid the attention of the tax inspectorate. In order to cash the funds that were being held in Russian banks, the money was transferred abroad and cashed there. At that point the funds became black money.

"The chain could look like this, for instance: $50 million was transferred from a Russian bank to the Bahamas. Then this money was transferred to another country—in Europe, for instance, or the Baltic States. From there the money was brought into Russia in cash and deposited in the black treasury. But not all the money was brought back in. For instance, if they took $50 million out of Russia, they didn't necessarily bring $50 million back; they could bring back only $10 million. No one controlled them at all. They distributed a large portion of these funds among their own accounts, wherever they wanted. It was a large-scale fraud."[71]

Most of Russia's top banks were involved in the international money transfers: Stolichny, Menatep, Rossiisky Kredit, Most, Alfa, and so on. The Russian Central Bank did not lift a finger, perhaps because it was engaged in siphoning off at least $1 billion in IMF funds to its own black treasury offshore—the financial company, Fimaco, registered in the tax haven of Jersey.

"Every one of those people who participated in the Yeltsin election campaign, including Berezovsky, wanted to make money out of it," says Streletsky. "To do this they artificially inflated their expenses. They created fake contracts for the provision of advertising or printing services. Abroad, part of the money was distributed among the personal accounts of the bankers and the government officials, and part was returned to Russia. According to the investigations [eventually] carried out by the SBP, it became clear that about $200 million to $300 million was embezzled from the election campaign—mostly by those businessmen who were close to the campaign headquarters in Moscow."[72]

German Kuznetsov's safe proved to be a gold mine of information. After meticulously recording everything they found, Streletsky's men put the $1.5 million back into the safe. They knew that the money came from the Finance Ministry, but they wanted to see where the trail would lead.[73]

The next day, June 19, at 5:20 p.m., security guards at the White House were waiting as two top Yeltsin campaign aides came down the stairs with a heavy cardboard box—the kind that holds paper for a photocopying machine; it was full of neatly stacked dollar bills—$500,000 in cash. The two aides were arrested.[74] They were Arkady Yevstafyev, a longtime aide to Anatoly Chubais, and Sergei Lisovsky, Berezovsky's old partner from the advertising business at ORT. Both men had been assigned to top positions in the Yeltsin presidential campaign. Lisovsky had left a simple receipt in Kuznetsov's safe—a scrap of paper stating "500,000 units Lisovsky."[75]

During their interrogation by the SBP and the Federal Security Service, the two men pleaded innocent. "A. V. Yevstafyev declared that he had no connection to the confiscated foreign currency and generally did not know from where it had suddenly appeared," the Russian prosecutor-general later noted in his report. "S. F. Lisovsky, in his short explanation, stated that the confiscated foreign currency was intended for payment to artists for concerts they had performed during the election campaign."[76]

Colonel Streletsky's arrest of two top Yeltsin campaign workers with the box of cash gave the SBP the smoking gun it needed to prove massive fraud and embezzlement by the Berezovsky-Chubais campaign team.

Fyodorov Escapes Death

On June 19, the same night that the SBP was arresting Yevstafyev and Lisovsky for taking a box of cash out of the White House, in another part of Moscow the former National Sports Fund president, Boris Fyodorov, narrowly escaped death at the hands of an assassin. Fyodorov was attacked outside his home. After the first shot, the killer's pistol jammed; he then stabbed Fyodorov several times in the neck and chest. Miraculously, Fyodorov survived; he fled to Western Europe. This was the violent denouement of the strange story that had begun with Fyodorov's visit to Logovaz House nearly three months before.

The next day, Anatoly Chubais called a press conference and blamed General Korzhakov. Everyone knew about the general's conflict with Fyodorov, involving Fyodorov's dismissal from the NSF. "I think it is our common task to try to figure out what role Mr. Barsukov and Mr. Korzhakov played in the events associated with Mr. Fyodorov," Chubais declared.[77]

He pointed out that Colonel Streletsky, who took Fyodorov's place as president of the NSF, also supervised the arrest of the men with the box of cash. Both incidents, Chubais claimed, were manufactured by Colonel Streletsky and other officers of the SBP. "The so-called box of money is a traditional element of a traditional Soviet-style KGB provocation," Chubais said. "Recently we witnessed a similar situation in which drugs were planted."[78]

The perpetrators of the crime were never found. If the attack had been ordered by Korzhakov, it was a strangely amateurish job, considering that the security chief had his pick of the professional assassins of the KGB.

"My personal point of view is that this was useful only to Berezovsky, to discredit Korzhakov and Barsukov once and for all in the eyes of the president," says Colonel Streletsky. "Berezovsky had the tape of the conversation with Fyodorov and the whole country already knew about the obvious conflict between Fyodorov and Korzhakov. In other words, there were two visible enemies—the whole country could see that. There remained only to organize the assassination attempt. And it was organized. The suspicion immediately fell on us."[79]

Russia did not yet know the contents of the Fyodorov tape; the media had not yet revealed the details of the criminal intrigues Fyodorov described in the tape. But one person had heard Fyodorov's accusations against General Korzhakov: Tanya Dyachenko. The brutal attempt on Fyodorov's life seemed to confirm the terrible things she had heard that evening in Logovaz House.

Colonel Streletsky is convinced Berezovsky or someone else from his camp ordered the attack on Fyodorov. "What did this ally [Fyodorov] give to Berezovsky?" he says. "An ally is a temporary phenomenon. For Berezovsky, people are divided into two categories: a condom in its packaging and condom that has been used."[80]

On July 8, five days after the second round of the election, the newspaper *Novaya Gazeta* published a transcript of the Fyodorov tape. That night NTV also ran excerpts on the evening news. The tape was leaked to the press, apparently by the Berezovsky-Chubais group, to discredit Korzhakov before the nation.[81]

Strangely, Fyodorov himself seemed to vacillate over who was responsible for the attack. The day after the *Novaya Gazeta* revelation, he gave an interview to the *Komsomolskaya Pravda* newspaper, raising the possibility that he had been set up by Berezovsky. "The interview cited in the [*Novaya Gazeta*] article was pasted together from various fragments, including some from various other conversations," Fyodorov complained. When asked if the famous interview had taken place, Fyodorov replied: "Basically, yes. It took place at 40 Novokuznetskaya Street [Logovaz House], but someone added a lot of something else to what was spoken there."[82]

Fyodorov's wife was more direct. In an interview published in *Komsomolskaya Pravda*, next to her husband's interview, she said she believed that the people who had arranged the *Novaya Gazeta* publication were the same people responsible for the assassination attempt.[83]

In other words, while Anatoly Chubais was accusing Korzhakov's men of staging KGB-style provocations—planting cocaine on Fyodorov in May and planting a box of cash on Yevstafyev and Lisovsky in June—the interviews with Fyodorov and his wife raised the possibility that the June 19 assassination attempt itself may have been a provocation, intended to tar Korzhakov as the prime suspect.[84]

"Either They Shut Up or I'll Throw Them in Prison"

The truth about the assassination attempt on Fyodorov may never emerge. But because circumstantial evidence pointed squarely at Korzhakov and because the attack occurred on the same evening as the arrest of the men with the box of cash, those members of the Yeltsin entourage who wanted to get rid of the security chief gained enormous leverage.

On June 19, Korzhakov was driving home after a meeting with Mikhail Barsukov of the FSB–KGB, at which the two men had discussed the arrest of Yevstafyev and Lisovsky (with the box of cash). The car phone rang. It was Tanya Dyachenko, calling him from Berezovsky's headquarters at Logovaz House. "You must release them," Korzhakov remembers Tanya screaming at him. "It means the end of the election."[85]

Korzhakov refused.

That night, most of the key players of the Yeltsin campaign team—Boris Berezovsky, Anatoly Chubais, Vladimir Gusinsky, Boris Nemtsov, Yevgeny Kiselev, Tanya Dyachenko, and others—gathered at Logovaz House. The meeting lasted deep into the night. The group was highly apprehensive. No one could tell exactly how much the SBP knew about the corruption of the Yeltsin campaign or how far the investigation into the box of cash had advanced.[86]

"I spoke with [Moscow FSB–KGB director] Trofimov at one o'-clock in the morning, at the time when all this was taking place," Chubais would later recall. "He lied to me that they didn't know who Lisovsky was, while Yevstafyev had perhaps been detained a little bit, but he would be released any moment."[87]

The group at Logovaz House decided that, rather than try to cover up the scandal surrounding the box of cash, they should publicize it as a pretext for firing Korzhakov. Late-night NTV programs were interrupted by news bulletins that a coup d'état was in progress and that two Yeltsin aides, Yevstafyev and Lisovsky, had been arrested by the plotters. Most Russians heard about the incident the next morning. They were told that General Korzhakov and the FSB–KGB chief were planning to postpone the second round of the elections and destroy Russian democracy.[88]

Korzhakov found virtually all the members of Yeltsin's entourage arrayed against him. On June 20, President Yeltsin appeared on TV to announce that he was firing Korzhakov, as well as Barsukov and Soskovets.

The Western and Russian press believed Chubais's story that Korzhakov was attempting a coup d'état. Chubais, after all, was the golden boy of Russian reform, lionized in the West for privatizing the Russian economy, whereas Korzhakov was a shadowy figure, known for his advocacy of postponing the elections. Newspapers in Russia and the West reported the "box of cash" incident as a "provocation" and an "old KGB trick of planting currency." Korzhakov, Barsukov, and Streletsky, meanwhile, were denied access to the airwaves.[89]

Chubais knew that the box of cash was genuine. On June 22, two days after Korzhakov was fired, Chubais met at the President Hotel with two top campaign managers: Viktor Ilyushin, an old Yeltsin crony from the Sverdlovsk days, and public-relations adviser Sergei Zverev. Their conversation was taped by someone loyal to Korzhakov's SBP. "We have to find a means of getting in touch with Korzhakov and Barsukov," Chubais told his colleagues, "so we can explain the situation to them clearly and unambiguously: Either they conduct themselves like regular fellows or we will throw them in prison. . . . Either they shut up or I'll throw them in prison. That is absolutely certain. You can tell that to them personally from me, as a greeting."[90]

The campaign managers admitted that smuggling boxes of cash

out of campaign headquarters was standard operating procedure for the Yeltsin campaign. Viktor Ilyushin said that shortly after the scandal broke, he discussed the issue with Yeltsin.[91] "I told the Boss [Yeltsin], when I was talking with him yesterday: 'Boris Nikolaevich, right now, if you wanted to, you could catch fifteen to twenty individuals next to the President Hotel who are carrying suitcases [of cash] out of our building.... If we begin to account for the money passing through informal channels, we wouldn't be able to hold the elections....' The president answered: 'I understand.'"[92]

Still, the Yeltsin campaign could not afford any more revelations about the black treasury. That meant leaning on the prosecutor's office to hush up the affair. "Until July 3 [the final round of the elections] we don't need any noise," Ilyushin said he told the prosecutor-general, Yuri Skuratov. In the course of his meeting with Chubais, Ilyushin placed a phone call to Skuratov. "Yuri Ilyich," Ilyushin told him, "a question has come up: Is it possible to arrange things so that the documents that you will be receiving from [the FSB–KGB] not be seen by anyone but you? That they stay with you for a certain period of time, until you have had a chance to get acquainted with them and talked everything over with Boris Nikolaevich [Yeltsin].... Because we have information that if anyone else of your staff handles it, this information would very quickly find its way into the camp of our opponents.... Yes, it would be best if you keep it yourself. Don't let anyone start investigating it. And then we'll decide. Okay? Because this is what we want."[93]

Chubais and his campaign managers were concerned not only with covering up the scandal about the box of cash until the final round of the election but also with avoiding the prosecution of Yevstafyev and Lisovsky. Throughout the conversation, Chubais and Ilyushin voiced their determination not to "surrender" the two men to the law. "Our comrades were doing our job for us, they were taking on the riskiest part of it," Chubais declared. "Hell! They put their necks on the line and now we tell them: 'Excuse me, but after July 3, you'll have to fend for yourselves.' What kind of attitude is that? ... We are the ones who sent them there!"[94]

But what could the Yeltsin campaign do about the documentary evidence piling up at the prosecutor's office? In the June 22 conversation, the following exchange took place:[95]

CHUBAIS: What if, at the second stage [after the election], we were to ask Boris Nikolaevich [Yeltsin]...
ILYUSHIN: To bury it completely?
CHUBAIS: No, to ask Skuratov to send him the documents so that he could analyze them. To demand the full package of documents.
ILYUSHIN: Good idea. [Laughs.]
CHUBAIS: And then let [Skuratov] try to ask for them back...

When the transcript of this conversation was published in the newspaper *Moskovsky Komsomolets* in the autumn, the Berezovsky-Chubais camp responded that the tape was a forgery. But the prosecutor's office sent the tape off for analysis and concluded that it was genuine.[96]

Though Korzkakov, Streletsky, and the rest of the SBP men were widely portrayed as a mysterious and sinister force within the Yeltsin regime, during the presidential campaign of 1996 they appeared to have been straight enough. When Fyodorov was arrested by Moscow police for possession of cocaine, Chubais and the media speculated that the affair was a setup; yet a subsequent police and judicial investigation (when Korzhakov was out of power) revealed that Fyodorov was indeed a chronic cocaine user. When Yevstafyev and Lisovsky were arrested with the box of cash, the Berezovsky-Chubais camp was quick to declare that this incriminating evidence had been planted by the SBP; yet, a subsequent investigation by the prosecutor-general confirmed that the two men had in fact purloined the cash from the safe of the deputy minister of finance. The published transcript of the conversation between Chubais and his campaign aides was also declared a forgery; yet, the prosecutor-general's office would confirm that it was genuine. These facts suggest that throughout 1996 Korzhakov and Streletsky were telling the truth, while the Berezovsky-Chubais camp was lying.

In the subsequent investigation by the prosecutor-general's office, however, Yeltsin's campaign managers avoided being charged with covering up a crime. "A. B. Chubais...and V. V. Ilyushin both declared that...they were not attempting to obstruct the pursuit of justice in the case, but only attempting to avoid leaking information during the time of the election," the prosecutor-general noted in his report.[97]

The prosecutor's office began its inquiry into the box of cash on the basis of several charges: illegal operations with foreign currency, fraud, and theft. The charges of illegal operations with foreign currency were dropped on January 5, 1997, when this activity was decriminalized. The investigation into fraud and theft continued until April 7, 1997, when it was dropped—not because Korzhakov and his men hadn't done their jobs, but because no one could tell where the money had come from.

"Having exhausted all possibilities, the investigation failed to identify the source of the confiscated dollars," the prosecutor-general explained. "There was no proof that anyone had been harmed by the event. Neither was the legal owner of the aforementioned foreign currency identified. All these circumstances allowed the investigation to conclude that there was no sign of fraud or any other crime." If no one knew who the $500,000 belonged to, in other words, how could anyone say that it had been stolen?[98]

On July 3, in the second round of voting, Boris Yeltsin sailed into his second term with 54 percent of the votes. Western observers concluded that the election was generally free and fair.

CHAPTER NINE

OLIGARCHY

Party Town

On Halloween night in 1996 I went to a party at an abandoned theater around the corner from the Russian Foreign Ministry. Inside were several hundred revelers, mostly Americans, getting down to the blaring noise of a Russian band called Two Airplanes. Marijuana smoke hung in the air. In a side room, the crush indicated the place where the booze was. Two beleaguered bartenders were handing out large plastic cups of cheap vodka and cans of German beer. A walking corpse became entangled in the cobwebs strewn across the doorway. Prince Potemkin squeezed through, followed by Genghis Khan and Heidi the mountain girl. In a smaller room next door, a couple were making out in the corner.

"Moscow is a party town," said Mark Ames, thirty-one, grinning, noticeably the worse for drink. "Ninety-nine percent of the expats came here to make a buck, but they stay for the women—the women here are awesome!" An aspiring writer from Northern California, Ames was the editor of a raunchy expat newspaper called *Living Here*, which specialized in reviewing Moscow's restaurants and nightclubs and laughing about its politics. "This party is nothing," noted Ames. "There's a group of 100 expats traveling down to Transylvania to spend Halloween in Dracula's castle."

The parties in Moscow had always been great—intense, crazy, always surprising. In Russia there were no rules and no social order. The 100,000-strong American and European expat community was having a blast. There was danger, there was great sex, there was big money to be made. Twenty-four-year-old kids from the suburbs of New York, who would have been stuck at some pedestrian job at a Wall Street bank, suddenly found themselves the star traders for some Moscow-based outfit in Russia's booming securities market.

There was also an older generation of foreigners in Moscow. The employees of the big law firms, financial institutions, and multinational corporations—men who were there to run the rep offices or simply to clinch a deal. If they were there for a short visit, the Western executives stayed at one of the inordinately expensive business hotels. At the Metropol, just across from the Bolshoi Theater, they could decamp in opulent suites, swirling with Art Nouveau designs and Karelian birch furniture. The smallest single room, looking out at a brick wall, was $330 a night; the premier suites were $1,000 or more a night. Dinner topped $300 per person, even with mediocre wine.

Life in Moscow could be dangerous for foreigners, especially if they were trying to run their own businesses. The most famous murder of a foreigner occurred in the autumn of 1996, with the assassination of an American entrepreneur named Paul Tatum. A forty-one-year-old native of Oklahoma, Tatum had engineered one of the biggest Gorbachev-era joint ventures—a huge new luxury hotel called the Radisson Slavyanskaya. By the time the hotel was completed, in the early 1990s, it was managed by a Russian-American company (50 percent owned by the Moscow city government, 40 percent by Tatum, and 10 percent by the Radisson hotel group). The partners fell out among themselves, with Radisson and Moscow combining to push Tatum out of the project; lawsuits were filed in an arbitration court in Stockholm. Tatum, meanwhile, adopted the lifestyle of a Russian gangster, surrounding himself with bodyguards, flaunting his wealth and his numerous girlfriends, and frequenting the wilder nightclubs. The combative Oklahoman carried on an acrimonious public feud with the Moscow city government and especially with its designated representative, Chechen entrepreneur Umar Dzhabrailov. On November 3, 1996, Tatum was shot by an assassin in broad daylight as he entered a nearby metro station with his bodyguard. Despite an international outcry, Moscow police never found the killer.

The big gangster turf wars of the early 1990s might have been over, but assassinations still played an important role in Russian society. In 1996, Americans learned of the role of organized crime in Russia's once proud sports establishment; Aleksandr Mogilny and several other Russian hockey stars playing for the NHL turned to the FBI to help them deal with Russian gangsters who had followed them to North America and were trying to extort money from them. Hockey was one of the few profitable areas of Russian sport. Dozens of Russia's best players had left their impoverished country to sign multimillion-dollar contracts with NHL teams. Apart from the individual contracts, the teams paid the Russian Hockey Federation a total of nearly $10 million for the right to sign up the players. This easy source of money did not escape the attention of the gangsters. In April 1997, Valentin Sych, the venerable president of the Russian Hockey Federation, was killed by machine-gun fire as he was driving along a country road with his wife.[1]

Charitable organizations were also targeted by organized crime. Gangsters were interested in charities' lucrative foreign trade rights. By special government decree, favored charities received permission to import liquor or export oil duty-free. The huge profits made on these operations were supposed to be the government's way of subsidizing worthy social causes. But when an activity produced so much money in impoverished Russia, it was certain to attract the attention of gangsters. Boris Fyodorov's National Sports Fund was the most famous charity to be caught up in a gangster turf war. In September 1995, the chairman of the Moscow Society for the Deaf, Igor Abramov, was killed by gunmen in Moscow. A year later, the president of the All-Russian Society for the Deaf, Valery Korablinov, was also gunned down.[2]

The heavily subsidized Russian Foundation for the Invalids of the Afghanistan War was subjected to an even more horrendous assault. In November 1996, a group of Afghanistan Foundation members gathered in the Kotlyakovsky Cemetery for a memorial service in honor of their late chairman, Mikhail Likhodei, who had been killed by a remote-controlled bomb two years earlier. Someone had hidden a powerful remote-controlled bomb in a neighboring tomb. The explosion was devastating: thirteen mourners were killed and dozens injured. Police had to retrieve some of the body parts from the tops of surrounding trees.[3]

None of this dampened the spirits of Moscow's wealthy elite. The "New Russians," as they were called, had their own social circuit in the capital. They enjoyed late dinners in the expensive restaurants. A certain portion of the New Russians partied at the same nightclubs the foreigners liked, but their principal partying venues were the much wilder, more vulgar nightclubs not frequented by the foreigners. The richest New Russians soon abandoned the idea of partying in Moscow. The city was the place they made their money. As a place for relaxation, it was uncomfortable—too many people pointing fingers in the restaurants and nightclubs. The Swiss Alps or the Mediterranean coast—these were the places to have fun.

Berezovsky did not party. Everything was business, and he attended an evening event only if there were useful contacts to be made. Still, he quickly acquired the trappings of Europe's megarich, jetting around in his private plane, dropping big money at Sotheby's Russian sales, and maintaining lavish residences on Lake Geneva, in London's Kensington Palace Gardens, and on the French Riviera, where he bought one of the largest chateaux in Cap d'Antibes for a reported $27 million. He apparently did not maintain an apartment in Paris, but liked to stay in the lavish Hotel Crillon. His presence at the watering holes of Europe's billionaires allowed him not only to show visiting Russians a good time, but also to mingle with potential business partners in the West; among his friends in the international business community were important dealmakers such as former junk bond king Michael Milken and media magnate Rupert Murdoch

(Berezovsky was one of the few guests at Murdoch's 1999 wedding aboard his yacht anchored in New York Harbor).[4]

The top officials of the Yeltsin government made sure they were not left out of the party. In lifestyle (as in most other things) there was little to differentiate them from Russia's businessmen. Both the bureaucrats and the businessmen were members of the same ruling clan. Most government officials had to forgo the pleasure of wearing tailor-made suits while in Moscow; they were more careful about driving expensive cars; they dined less often in the best restaurants than they would have liked. But the top bureaucrats still could travel abroad, vacationing on the Riviera or in Florida. They could relax abroad. No one recognized them and, like Russia's businessmen, they could enjoy themselves without fear of censure. What an adventure it was—skiing in Gstaad, cruising on yachts in the Caribbean, shopping at Christian Lacroix and Chloe in Paris. Having bought their apartments, villas, castles, and chalets in the West, Russia's new ruling class increasingly made the West their home. The wives spent a good part of their time there. The children were placed in the premier European schools. The offspring of Russia's rich and powerful (including the children of Tanya Dyachenko, Anatoly Chubais, Vladimir Gusinsky, and Berezovsky) were sent to the same $40,000-a-year boarding schools in Switzerland and England.

The New Russians were a tiny minority, numbering in the hundreds of thousands at most, yet they felt they owned the mess that was Russia. It was theirs to exploit. As they traveled west, they eagerly forgot about the tens of millions of Russians they were leaving back home—that same broken mob, complaining, drinking, and dying. But the party would soon come to an end.

Berezovsky Is Appointed to the Government

At the close of the 1996 campaign, Russian voters thought they were electing a revived Yeltsin: a tough political boss, a tireless grass-roots campaigner, a man gleefully shaking his stuff at a Moscow rock concert. Voters did not know that his apparent vigor was the result of medications prescribed by the Kremlin doctors. The concoctions worked well for a while, but a week before the last round of the elections, Yeltsin finally collapsed under the strain. In late June, he suffered yet another heart attack and disappeared from public view. Only after the election did Kremlin spokesmen reveal the gravity of Yeltsin's heart condition, together with the news that the president would undergo multiple-bypass surgery. He would be confined to his sickbed for the next eight months.

With the monarch absent from the Kremlin, Yeltsin's closest aides took charge. Initially, the leader of this group was Anatoly Chubais, the architect of Russia's privatization, the coordinator of Yeltsin's election campaign, and now chief of staff. Chubais could be counted on to loyally represent the interests of the Group of 13 that had installed him back in the Kremlin inner circle. As a precaution, banker Vladimir Potanin was appointed first deputy prime minister in charge of the economy.

Now it was payback time for the businessmen who had contributed to Yeltsin's reelection. Vladimir Potanin's Onexim Bank, for instance, received huge new government accounts, while Norilsk Nickel was allowed to restructure more than $1 billion in tax debts. Vladimir Gusinsky's NTV was granted a license to take over all of Channel 4 and double its airtime. Stolichny Bank, co-owned by Aleksandr Smolensky and Berezovsky, more than doubled in size when it was allowed to take over the giant state-owned Agroprom Bank (the new, combined bank was named SBS-Agro).

Later, I asked liberal parliamentary leader Grigory Yavlinsky about Yeltsin's second term. "The government that was formed was without any clear ideology," Yavlinsky observed. "It was neither red, nor white, nor green. It was based solely on personal greed. You got a system that was corporatist, oligarchic, and based on monopolized property rights and semicriminal relationships."[5]

The one outsider in this cabal was General Aleksandr Lebed. The former paratroop general was the only genuine populist candidate of the 1996 presidential campaign and he had been rewarded with the post of chief of the Security Council. Gruff, uneducated, Lebed suddenly found himself in the center of the political arena. He spoke in a deep, gravelly voice, but it was the things Lebed said that made him a tribune of the people. "You have to remember that the president [Yeltsin] was a member of the Central Committee [of the Communist Party]," Lebed said. "So was Prime Minister Chernomyrdin and so was [the leader of the opposition] Gennady Zyuganov. They're all from the same nest. But some managed to throw away their Party cards in time and take up the democratic banner, while the others were late. You can see that these people always seem to disagree, but under the table they always manage to come to an understanding. They have always regarded ordinary people as garbage: the stones at the foot of the pyramids."

Like a good military commander, Lebed was concerned with the welfare of his men. In an interview with me, he suddenly began to speak of the fate of Russia's huge prison population. "There has never been an effort to rehabilitate people in prisons, where they are kept worse than cattle. You enter the prison system as a man and you leave it either as an animal or as excrement. And what about the specially trained men in the armed forces?" he continued. "Hundreds of thousands, with unique skills—they were simply thrown out on the street. Russia is full of a huge number of malignantly inspired people—intent on destruction and killing, full of anger, fury, hurt, and vengeance. Until this mood is changed, nothing can be done."

Lebed knew what had to be done to heal the country, but he did not know how to do it. "The key to prosperity is that the people should be not small and sickly, but big, healthy and strong, that people should consider themselves masters of their own land, that they should walk with heads held high, fearing nothing," he declared. "That's the Russian national idea."

Lebed's one indisputable achievement in office was ending the disastrous war in Chechnya. The war zone in the south had been quiet during the last months of the Russian presidential election, since Jokhar Dudayev had been killed by the smart bomb and the Chechens had agreed to a cease-fire. Only a few days after Russians voted in the final round of the presidential elections, the fighting resumed throughout Chechnya. The rebels had used the cease-fire to recoup their strength, infiltrating commandos and ammunition into the main Chechen cities. On August 6, 1996, a rebel army under commander Shamil Basayev overran the capital, Grozny. The Russian forces, as usual, were caught flat-footed. More than 500 Russian soldiers were killed; news reports carried images of armored cars burning, with their dead drivers slumped on the pavement. Within days, the battle was lost; some 3,000 Russian troops remaining in Grozny were trapped in their barracks.

In Moscow a pale, sickly Boris Yeltsin staggered through the inauguration ceremony. Someone else would have to extricate Russia from the Chechen mess. On August 12, General Lebed traveled to the town of Khasavyurt on the Chechen border to begin peace negotiations. By the end of the month, a deal had been reached and Russia had agreed to withdraw its troops. The republic was given de facto independence, although the question of Chechnya's official status was put off until 2001 (when a new man, Boris Yeltsin's heir, would be president of Russia).

Lebed says that he was impelled to make peace by information that the Chechens were planning terrorist attacks on nuclear power plants at Novo-Voronezh and other cities. "I knew I had to stop the national-level mob war in Chechnya," he says. Lebed received no public support from Yeltsin, Prime Minister Chernomyrdin, or other top government officials. No one wanted to be associated with the final act of Russia's humiliation.

While Lebed was defending his policy in Chechnya, he lashed out at the pervasive corruption and incompetence of his ministerial colleagues at home. They, in turn, were dismayed at his insistence on remaining true to his word. "I promised my supporters that I would stop the bloodbath in Chechnya and I did that," says Lebed. "I also promised them I would get rid of crime and corruption. But as soon as I fulfilled my first task [peace in Chechnya], everybody became scared; I was accused of plotting a coup, forming some kind of shadow army."[6]

On October 17, four months after he was appointed, Lebed was fired. Boris Yeltsin even made a rare appearance on TV to fulminate against the general and his "insubordination."

The autumn of 1996 was characterized by what the media termed the "sleaze war." The leading figures of Yeltsin's election campaign engaged in an extraordinary round-robin of murder allegations, accusing each other of planning assassinations. General Korzhakov, Yeltsin's former security chief, held a press conference in which he asserted that in 1994–95, Berezovsky had repeatedly asked him to assassinate Gusinsky. At the same time, the newspaper *Novaya Gazeta* published the transcript of Berezovsky's secret 1995 appeal to Yeltsin (in which Berezovsky accused Gusinsky of setting him up for the Listyev murder). Korzhakov declared further that Berezovsky and Gusinsky, now allies, had put out a contract on his life. The transcript of Anatoly Chubais's efforts to cover up the box-of-cash incident also appeared at this time. In addition, former National Sports Fund president Boris Fyodorov reappeared after his recuperation in Western Europe and declared that he feared General Korzhakov would have him killed.[7]

Amid this extraordinary spectacle, the richest businessman in Russia entered the government. On October 30, 1996, two weeks after General Lebed was dismissed, Boris Berezovsky was appointed deputy secretary of the Security Council. The tycoon had long wanted a seat in the government; as this and later government appointments would show, his preference was for an important-sounding post, with broad, vaguely defined responsibilities. The Security Council was such a post; though it had a tiny staff, it was generally charged with coordinating the nation's law enforcement and military matters. Berezovsky's appointment was supported by Tanya Dyachenko and the presidential biographer, Valentin Yumashev; Anatoly Chubais signed the papers. The nominal chief of the Security Council was a weak-willed bureaucrat and old-time Yeltsin crony named Ivan Rybkin. The fox was now guarding the chickens.[8]

Several days later, *Izvestia* revealed that Berezovsky held an Israeli passport. This information was passed to the newspaper by operatives of General Korzhakov's disbanded Presidential Security Service; they in turn had received the information in 1995 from Berezovsky's then archrival, Vladimir Gusinsky. Other Russian newspapers ran exposés of Berezovsky's past, but the tycoon reacted particularly furiously to the reports about his Israeli citizenship—the revelation jeopardized his appointment to the government, since government officials were prohibited from being citizens of foreign countries. At first, he denied that he was an Israeli citizen and threatened to sue *Izvestia.* After the Israeli government itself confirmed the information, the tycoon admitted that he had obtained the passport and said he was giving it up.[9]

In the furor surrounding his appointment and the revelation of his double citizenship, Berezovsky alienated other Russian Jews. "By Israeli law anyone who is a Jew by blood, whether he be half-Jewish or a quarter, is a citizen of Israel," he declared. "Any Jew in Russia has a double citizenship." Russian Jews reacted furiously to this slur on their loyalty to Russia, while Berezovsky complained that he was the victim of a rising tide of anti-Semitism in Russia.[10]

President Yeltsin was in no condition to weigh in on the matter—on November 5, five days after the Berezovsky appointment, Yeltsin underwent multiple-bypass surgery. Despite widespread calls in parliament and the press for Berezovsky's dismissal, the tycoon hung on.

Trading Hostages

Berezovsky's special responsibility in the Security Council was to sort out relations with Chechnya. He explained that he was in charge of helping to restore Chechnya's economy. "Business has to pay up—there's no getting around that fact—just as we paid for the fact that the Communists wouldn't take power [in the 1996 elections]," he told the newspapers. General Lebed, on the other hand, claimed that Berezovsky was interested in Chechnya because he had to "cover up his previous business deals involving Chechnya" and wanted to lay the foundation for new business deals around the restored Baku–Novorossiisk pipeline. I asked Lebed what business deals he was referring to, but the surly former national security chief refused to go into details.[11]

General Lebed told the Russian press that Berezovsky had long played a strange double game in Chechnya. "After the signing of the Khasavyurt Agreements [the Russian-Chechen peace deal]...Berezovsky came to me and started bullying me," Lebed recalled. "When he understood that I could not be frightened, he simply stated: 'What a business you have ruined. Everything was going so well. So, they were killing each other, but they've always been killing each other and always will be.'"[12]

While Russians were outraged at Berezovsky's appointment to the government—there were calls in the parliament and the press for the tycoon to be fired, for Chubais to be dismissed, and for Yeltsin to resign—the Chechen leaders were pleased. Salman Raduyev—a warlord responsible for the bloody hostage-taking raid into Dagestan a year earlier, several kidnappings, and two terrorist bombings in southern Russia in early 1997—praised Berezovsky as an "honorable man" and a man he could work with. "He [Berezovsky] has a personal interest in this oil [the Baku–Novorossiisk pipeline]," Raduyev observed. Remarkably, Raduyev's highly compromising endorsement of Berezovsky was carried in *Ogonyok*, one of the tycoon's own magazines.[13]

It was clear from Berezovsky's movements in Chechnya that he was on excellent terms with the Chechen leaders. In the aftermath of the war with Russia, the country had turned into a cauldron of violent gangs and militias. The few Russians who ventured there did so under heavy guard, for fear of being kidnapped or assassinated. Berezovsky, however, traveled often to Chechnya, without bodyguards. His old relations with various Chechen gangs in Moscow almost certainly proved useful. "He maintained these connections [with Chechen gangs] before the war, during the war, and after the war," General Lebed told me. "[It didn't matter that he was dealing primarily with Moscow-based gangs.] The Moscow-based organizations are the subsidiaries of Chechnya-based ones. When you contact the subsidiary, you contact Chechnya."[14]

The new president of Chechnya was Aslan Maskhadov. By Chechen standards, he was a moderate. He had been elected president in January 1997, after serving as the commander in chief of the Chechen forces during the war. His agenda was to enforce the rule of law in his war-ravaged land and establish relations with the outside world. But Maskhadov's authority scarcely extended beyond the capital—most of the country was controlled by independent militias and criminal gangs, usually led by the field commanders of Chechnya's rebel army. These men, having distinguished themselves in battle, were in no mood to submit to Maskhadov. Each warlord ruled his own feudal kingdom, largely defined by old clan loyalties and paid for by rogue oil refineries, drug- and gun-running operations, and various other criminal operations.

For Russia, the biggest problem in postwar Chechnya was kidnapping. Over the course of two years, more than 1,300 individuals were seized, among them several Russian generals and personal envoys of President Yeltsin. The victims were not all Russians. Dozens of Western journalists, aid workers, businessmen, and religious missionaries were kidnapped, too. Typically the victims were shunted around various Chechen towns, while their captors negotiated hefty ransoms. The hostages were a valuable commodity. Besides arms, drugs, and oil purloined from the Baku–Novorossiisk pipeline, hostages were the only valuable commodity in Chechnya. They were often traded between different Chechen militias for cash.

Foreigners commanded the highest price. The first foreigner to be kidnapped was an Italian photographer, captured in the autumn of 1996; he was released after the Italian government paid $300,000 to a Chechen counterparty. After this, the ransom demands rose steeply, typically reaching $2 million or $3 million per individual.[15]

The first famous kidnapping incident occurred six weeks after Berezovsky's government appointment. On December 15, 1996, a detachment of gunmen led by Chechen warlord Salman Raduyev seized twenty-two Russian police officers on the Chechen frontier, after the policemen had blocked Raduyev from crossing the frontier. In a bizarre gesture of intimidation, Raduyev demanded that his men be granted "freedom of movement"—otherwise the police officers would all be killed. Two days later, the menace represented by maverick warlords like Raduyev was made clear when a group of Chechen gunmen broke into a hospital in the town of Novye Atagi and killed six Red Cross workers, including four nurses. The next day, Berezovsky arrived in Chechnya for talks with Raduyev and convinced the warlord to release the twenty-two Russian policemen.[16]

This was the first of many Chechen hostage crises resolved by Berezovsky. The tycoon gloried in his humanitarian role. He would usually appear on the nightly news welcoming the victims back from captivity or boasting of his efforts to procure their release. Berezovsky's prowess in negotiating with Chechen kidnapping gangs underlay his claim to be a Russian statesman; the endless series of hostage crises, all neatly resolved by Berezovsky, served as a political platform for the tycoon to prove his usefulness to the government.

Not everybody was impressed by his role as a white knight in the Chechen hostage trade. General Lebed regarded the tycoon's actions as highly cynical political maneuvers. In January 1997, two reporters from Berezovsky's ORT television network were seized in Chechnya. Lebed flew to Chechnya to negotiate for their release. He was not successful. Upon returning to Russia, he claimed that Berezovsky had paid some Chechen warlords to block the journalists' release— in order for the tycoon to arrive a few weeks later and claim credit for their liberation.[17]

Whenever hostages were freed, the Russian government repeatedly claimed that no ransoms were paid. Chechen president Aslan Maskhadov disagreed. Chechen gangs never released their hostages because of a change of heart, Maskhadov told me; they released them because officials came from Moscow with "suitcases of cash." The primary channel for these funds to the kidnappers was Berezovsky.[18]

Berezovsky himself often boasted of his financial contributions to various Chechen groups to buy their goodwill. In the spring of

1997, he claimed that he had helped Russian businesses "donate" $1 million to build a cement plant in Chechnya; he also claimed that Logovaz had donated fifty automobiles to the Chechen Ministry of Internal Affairs. In August 1997, Berezovsky boasted that he brokered an agreement by Vladimir Gusinsky's NTV to pay a $2 million ransom for the release of three NTV journalists.[19]

Over two and a half years, Berezovsky maintained close relations with the warlords who either carried out kidnappings or were closely linked to the criminals who did. Berezovsky's favorite counterparties in Chechnya were not moderate government officials like Aslan Maskhadov. They were such terrorist leaders as Shamil Basayev and Salman Raduyev or Islamic fundamentalists like Movladi Udugov. Berezovsky's relationship with Movladi Udugov was particularly close; Udugov had been one of the leaders of the Chechen war effort and was now a deputy prime minister in Maskhadov's government. Maskhadov tolerated this man's presence for the sake of national unity, but privately scorned him as a proponent of *jihad* (Islamic holy war)—Udugov's brother was one of the leaders of the Wahabbi sect, the most fanatical Islamic fundamentalist group in Chechnya.[20]

In Moscow, former officers of the SBP and the Moscow police kept track of Berezovsky's Chechen activities. They concluded that Berezovsky served as a banker for the Chechen kidnappers, rounding up and transferring the ransom payments from the Russian side, with Movladi Udugov's Islamic fundamentalist group handling the hostage negotiations on the Chechen side.[21]

I obtained a tape of several intercepted telephone conversations between Movladi Udugov and Berezovsky, recorded either by the security services (of Russia, Chechnya, or some other country) or by some very well-equipped private intelligence agency. It is unclear exactly what is being discussed in these conversations, but it is clearly a business deal, probably hostage-related. One of the recordings is one-sided, containing only the voice of Udugov.

"Hello, Boris? It's Movladi. I tried yesterday, but some reason I was not able to get through—Okay, Boris, how are our affairs going?—Yes, I looked, but nothing has arrived. I'll have a look Monday.—Yes. Yes.

I looked on Friday. Maybe it will come on Monday. I'll look then.—Well, I don't know. Maybe it got delayed for some reason.—Yes.—What kind of documents?—Okay, I understand.—Okay, Boris, tell me, I understand from Kazbek that you will send someone on the seventh or the eighth. What's the situation there?—Okay. When can we expect him? Monday or Tuesday?—Okay, I understand. Kazbek will be there himself to meet him.—Okay. What about those two guys?—Because here they are on the verge of taking extreme measures. So far I have been able to hold them back in the hope that the situation can be resolved.—Okay, Boris. I will call you in the evening. I'll pick a better phone line and call you.—All the best. Goodbye."[22]

Often Berezovsky acted in Chechnya through Badri Patarkatsishvili, the Logovaz partner who, according to the Russian security services, had long served as the company's primary intermediary with organized-crime groups.[23]

Badri's role as Berezovsky's point man in Chechnya is illustrated in another of the telephone intercepts I obtained. The conversation between Berezovsky and Movladi Udugov is purposefully vague to frustrate potential eavesdroppers, but its thrust is clear enough.[24]

UDUGOV: Hello?
BEREZOVSKY: Hello, Movladi. Hello, my friend.
UDUGOV: Hello. Boris, how are things going with you?
BEREZOVSKY: Fine.
UDUGOV: Kazbek arrived today very upset.
BEREZOVSKY: Why?
UDUGOV: He never did understand why Badri came there.
BEREZOVSKY: Then I also didn't understand. I just spoke with Badri on the telephone and he said that everything is okay. They discussed everything and agreed on everything. What do you mean, upset?
UDUGOV: Well, the agreement was that he would bring two and a half units.
BEREZOVSKY: What?
UDUGOV: That he would bring two and a half units.

BEREZOVSKY: Movladi, you got something mixed up. I never said that. After all, I said right away that I wouldn't be able to resolve this issue. I said that I'll do what I am able to do. And I did that. I had no other objectives.

UDUGOV: Well, okay.

BEREZOVSKY: Maybe you didn't understand me.

UDUGOV: Well, I don't know. At that time, when we spoke together...

BEREZOVSKY: I always answer honestly. I said that at the moment I wouldn't be able to resolve this issue, but that I'll do what I can. And I did what I could. That's all. You must understand.

UDUGOV: I understand. Boris, I too want you to understand me. In other words, after your meeting in Tbilisi I called you and told you about the two and a half units.

BEREZOVSKY: No. You already told me about that in Tbilisi.

UDUGOV: No, it was Kazbek who spoke about that.

BEREZOVSKY: Okay, Kazbek.

UDUGOV: You spoke with Kazbek. For my part, I sent you a fax.

BEREZOVSKY: Movladi, I am telling you once more, so that there would be a full understanding: I never promised to do this. So that there would be full understanding—whatever I can do, I will do. That's it.

UDUGOV: Okay, then I didn't understand. Why did he come? To give this confirmation? This piece of paper? What's the matter—you weren't able to send it by fax?

BEREZOVSKY: No, no. They had another type of discussion. I don't know about this discussion yet, since he flew on to another destination. He hasn't come to Moscow yet.

UDUGOV: Then I understand that there has been no discussion yet. He arrived, presented this piece of paper, which was possible to have sent by fax... Tell me, please: Is this the right sum?

BEREZOVSKY: Absolutely. This is what I am able to do at the moment. I can't do anything more than that. I told you that.

UDUGOV: But I thought... That fax that I sent you—I asked for 700 or 800...

BEREZOVSKY: Movladi, okay, so you sent me that fax, but I won't be able to do that. Whatever I can do, I'll do. And whatever I could do, I did.

UDUGOV: Okay, I understand. But I counted on you to do that at my request, but now it appears that they have gotten Kazbek involved in this as well. Okay, so it turns out that we simply didn't understand each other. Boris, tell me, please, is this issue now closed?

BEREZOVSKY: No, it's not closed. It seems as if you were not listening to our conversation. I told you that right now I can't resolve this issue. Whatever I can resolve right now, I will resolve.

UDUGOV: Right now the situation over there is . . . It seems you're a bit out of touch. Okay. Today, that sum was urgently needed. But since it didn't work out, then that's it.

BEREZOVSKY: Okay, Movladi. You can reach me anytime. So long.

For the hostages themselves, the ordeal was particularly awful. As they wasted away in a pit or a basement somewhere, they were often tortured. One American religious missionary, Herbert Gregg, spent eight months in captivity and had one of his fingers cut off. A twelve-year-old girl, Alla Geifman, the daughter of a businessman in the Volga city of Saratov, was kidnapped as she walked home from school and smuggled into Chechnya. The kidnappers demanded a $5 million ransom; they underlined their demand by cutting off two of her fingers, videotaping the operation, and sending the video clip to the parents. Alla Geifman was freed after seven months in captivity.

On July 5, 1997, two British aid workers, Camilla Carr and Jon James, were kidnapped in Grozny. They had been working for a Quaker charity at a home for disturbed children in the Chechen capital. The British government resolutely stuck by its policy of never paying ransoms and discouraging others from doing so in order not to encourage further kidnappings. Fourteen months later, on September 20, 1998, the two victims were released. James had been beaten repeatedly, while Carr had been raped. Berezovsky proudly claimed credit for their release. He even sent his private plane to fly them from Chechnya to London; to the press he boasted that he had helped procure the hostages' release by donating computers and medical aid to warlord Salman Raduyev.[25]

Whether or not procuring the victims' freedom with "computers" and "medical aid" contradicted the British policy of not paying ransoms, Berezovsky's public boasts about the transaction could only serve to worsen the kidnapping spree in Chechnya. Indeed, two weeks after Camilla Carr and Jon James were released, four more foreigners working for a British telecommunications company—three Britons and one New Zealander—were kidnapped in Grozny. The men had been working on a Chechen government contract to install a cellular phone system in the city; they were seized 500 meters from the headquarters of the Chechen government's antikidnapping police unit. Two weeks later, the head of the antikidnapping unit was blown up by a car bomb.

For Chechen president Aslan Maskhadov, who was trying to turn Chechnya into a normal member of the world community, the kidnapping spree was a disaster. Soon, all foreign aid organizations—from the Red Cross and Medecins Sans Frontières to the UN High Commission on Refugees—fled Chechnya; foreign journalists and businessmen also gave Chechnya a wide berth. The country became isolated—the only foreigners seemed to be Islamic fundamentalists from the Middle East.

After the four British telecommunications workers were taken captive and his own antikidnapping chief was assassinated, Maskhadov decided he had had enough. Together with the president of the neighboring state of Ingushetia, Maskhadov declared to Russian and British newspapers that Berezovsky was financing Chechen organized-crime groups by repeatedly arranging ransom payments.[26]

"This was part of [Russian] government policy," Maskhadov told me later. "High government officials [specifically Berezovsky] would arrive from Moscow with big sums of money. These kidnappings were encouraged. They paid ransoms, maintained contact with them [the kidnappers], and ultimately attained their aim of discrediting the whole Chechen people."[27]

Maskhadov's candor about the relationship between Russian ransom payments and the kidnapping spree only made things worse. A month later, on December 8, the severed heads of the four British telecommunications workers were found lined up on the side of a highway near Grozny. The headless bodies were found a week later.

"This Was a Comedy and It Remains a Comedy"

As Berezovsky played politics in Chechnya, he continued to push his own commercial projects, despite being a high official in the Russian government. In the spring of 1997, he vigorously lobbied the Russian Central Bank to license the Andava-Aeroflot relationship. In March 1997, *Izvestia* carried detailed reports of his negotiations to buy Promstroi Bank, an institution linked to a convicted fraudster and embezzler, Grigory Lerner, who was languishing in an Israeli jail. (Berezovsky denied the *Izvestia* charges.)[28]

But the biggest prize that spring was not some shady Moscow bank, nor was it the contract to manage the finances of Aeroflot. It was the oil company Sibneft, which Berezovsky and his partner, Roman Abramovich, had incorporated and taken under management in the first round of the loans-for-shares auctions in 1995. In the second and final round of the loans-for-shares auctions Russia's prime oil and metals properties would now formally pass into the hands of the financiers, who had officially only held them as collateral. Each property would be sold to the same company that had won the auction in 1995 and which was consequently charged with organizing the auction in the second round. Sibneft was no exception.

The man responsible for overseeing the second round of loans-for-shares auctions, head of the GKI (State Property Committee) Alfred Kokh, argues that there is nothing suspicious about the organizers of the auctions ending up the winners. "This happens in any advanced country," says Kokh. "It is called underwriting." The big difference between the loans-for-shares auctions and some initial public offering underwritten by a Goldman Sachs or Morgan Stanley was that this offering was meant to fail, with no outside bidders showing up and the underwriter taking the stake for himself. But this, too, does not particularly bother Kokh. "Once we gave them [the oligarchs] the right to sell the enterprise, naturally they said: 'For two years we have been investing our brains, time, and money into this enterprise. Now what am I supposed to do—sell it to someone else? Of course I will underwrite the offering completely. I will sell it to myself.'"[29]

In the second stage of the loans-for-shares auctions, the buyers typically paid a premium of 1 percent or less over the starting price; the bids were as low as they had been in the first round. Potanin bought oil giant Sidanco for an implicit market value of $250 million—a fraction of the $5.7 billion market capitalization the company would command eight months later; Menatep's acquisition of oil company Yukos valued the company at $350 million, though it would have a market cap of $6.2 billion eight months later. "We couldn't get a better price, because the bankers who took control of the enterprises under the loans-for-shares auctions were not idiots," Kokh says. "They structured the working capital in such a way that everything was indebted to their bank. If you were to sell the enterprise to someone else, tomorrow you would get bankruptcy proceedings against it."[30]

This time Kokh is right. Having squeezed the cash out of Russia's prime industrial companies, the financial companies who won the first round of the loans-for-shares auctions ensured that none of Russia's best industrial companies were financially viable on their own. "We are a group of bankrupt companies," Menatep's Mikhail Khodorkovsky admitted cheerfully to me. "The whole country is a bunch of bankrupt companies." At the same time, the intermediaries—not only Khodorkovsky, but also Berezovsky and the other oligarchs—had grown fabulously wealthy.[31]

On May 12, 1997, the 51 percent stake in Sibneft was auctioned off. The starting price was $101 million. Berezovsky's financial vehicle, NFK (Oil Finance Company), was in charge of organizing the auction, which was structured to discourage outside bidders from participating. While in most of the other auctions bidders only had to show a bank guarantee that they had the money, the bidders for Sibneft had to deposit $190 million in cash with NFK. If the bidder were to lose, he could get his money back only forty-five days later—which gave NFK a substantial opportunity to make money off the float. Nevertheless, two serious outside bidders showed up: Alfa Bank and Onexim Bank. Alfred Kokh recounts how Berezovsky's team used the crudest possible measures to ensure that their rivals would not succeed.

"Even the address in the advertisement [for the auction] turned out to be some house under construction," says Kokh. "When they [the bidders] went there, they found the entrance to be boarded up. All the same, they managed to find [the correct] address, went there, and were simply barred [by security guards] from entering the building. Then, when they [the NFK auction organizers] finally accepted the bids, they found a whole series of infractions—that this paper was blue, but it was supposed to be yellow, and so on." Alfa Bank was disqualified because it had apparently not brought all the necessary documentation, while Onexim Bank was disqualified because of an alleged violation of an obscure bank-transfer regulation regarding its deposit. "When I raised this issue with the prime minister, Viktor Stepanovich Chernomyrdin," Kokh recalls, "he told me: 'Don't interfere.'"

The winner of the Sibneft auction submitted a bid for $110 million, just $9 million more than the starting price. The winner of the 1995 auction and the organizer of the 1997 auction was NFK (Oil Finance Company). The winner of the 1997 auction was called FNK (Finance Oil Company). No one knew for sure who was behind FNK. In public Berezovsky strenuously denied any connection to the outfit. He was already known to be linked with the organizer of the auction, NFK; to admit to being the man behind the winner of the auction, FNK, would have revealed a conflict of interest. Moreover, Berezovsky was a government official and hence prohibited from engaging in commercial activity. The public was told that the owners of FNK were two subsidiaries of Aleksandr Smolensky's SBS-Agro Bank (as Stolichny Bank was now known). But there could be little doubt that the real owner of both NFK and FNK was the same man: Boris Berezovsky.[32]

"This was a comedy and it remains a comedy," snorts Alfred Kokh. In fact, the actions of Berezovsky in acquiring Sibneft were so thinly disguised that the quote that comes to mind is Lebed's: "Berezovsky... is not satisfied with stealing—he wants everybody to see that he is stealing with impunity."[33]

New Rules of the Game

The second round of the loans-for-shares auctions was a piece of unfinished business from the first Yeltsin term. Anatoly Chubais and the rest of the government were simply keeping their part of the bargain that had been struck with the oligarchs to get Yeltsin reelected. But from now on, Chubais was determined to introduce a more civilized type of capitalism into Russia.

President Yeltsin, too, decided it was time to pay attention to his reputation; perhaps his experience with open-heart surgery made him think of his historical legacy. In March 1997, his convalescence over, he returned to the Kremlin. His first measures were designed to shake up the government and revive the "reformist" zeal of the early years of his administration; he seemed determined to rein in the crony capitalism that had stained his regime. The banker and oligarch Vladimir Potanin was dismissed from his post as first deputy prime minister. In his place, Yeltsin appointed Anatoly Chubais and another "young reformer," Boris Nemtsov, the governor of Nizhni Novgorod—a province that was a showcase of successful market reform.

Not all of the representatives of crony capitalism were dismissed. Viktor Chernomyrdin remained prime minister. Boris Berezovsky stayed on the Security Council. And Berezovsky's old-time collaborator, Valentin Yumashev, was brought in as Yeltsin's chief of staff. Still, Chubais was determined to break with the past. "Today in Russia, we have reached a crossroads between two forms of capitalism," he declared. "One form is the oligarchic capitalism of the Latin American type, in which market mechanisms function, which have free price formation and private property, but where all major decisions are made by the state, which is controlled by large financial and industrial groups. The second model can be described as people's capitalism. The economic rules in this system are the same for every participant, regardless of size. It also means that the antitrust policy of the state is applied to everyone, including the largest financial groups. Furthermore, it means that the state is separate from business and that business plays by the rules established by the state."[34]

This was indeed a new Anatoly Chubais. He wanted to change the rules according to which Russia's oligarchs had won their power and wealth. Naturally, the oligarchs found this unacceptable. "You cannot play according to one set of rules and then declare that from four o'clock tomorrow morning you are establishing a new set of rules," Berezovsky declared. "Chubais is trying to break the government and big capital. But big capital is the real support of the government today."[35]

Berezovsky repeatedly reminded the first deputy prime minister that he had been "hired" by the top businessmen to run the Yeltsin election campaign and, later, the Russian government. Chubais, however, was determined to assert his independence. "It sounds as if the board of directors of Russia has hired its chief executive officer," he noted acidly.[36]

In the brief period that Chubais and Nemtsov held the upper hand in the Kremlin, Berezovsky suffered two of the biggest business reverses of his career. The first concerned the natural-gas monopoly, Gazprom. This was Russia's wealthiest company; in natural-resource terms, it was arguably the wealthiest private company in the world. Gazprom owned a third of the planet's natural-gas reserves and dominated the gas markets of Russia and Europe. In June 1997, Berezovsky convinced Prime Minister Chernomyrdin to name him chairman of Gazprom, but the appointment was blocked at the last moment by First Deputy Prime Minister Boris Nemtsov.[37]

The second reversal for Berezovsky occurred during the privatization of the telecommunications monopoly, Svyazinvest. This holding company controlled Russia's long-distance and regional phone companies. The government had long spoken of privatizing a portion of Svyazinvest. In 1995, the Italian telecommunications company Stet was promised a 49 percent stake in the company. This deal was never consummated. Next, a consortium of Western banks— NM Rothschild, ING Barings, and MC Securities—was awarded the right to organize a $1 billion public offering for the Svyazinvest stake. But at the last minute, this deal, too, was canceled. Two Russian conglomerates, Vladimir Gusinsky's Most Group and Pyotr Aven's Alfa Group, convinced the government that Svyazinvest must remain in Russian hands; Gusinsky, in particular, was interested in Svyazinvest, since the telecom monopoly had the power to set the transmission rates for NTV. In late November 1996, the government ordered that the Most Group and Alfa Group be allowed to purchase the 49 percent Svyazinvest stake at a price determined by independent analysts. This was widely interpreted as payback for the contributions of Most and Alfa to the 1996 presidential campaign.[38]

Chubais changed the plans for Svyazinvest once again. Now only a blocking share (25 percent plus one share) would be privatized and it would be sold to the highest bidder at auction. Two rivals emerged. One was Vladimir Potanin's Onexim Bank, backed up by Renaissance Capital, Deutsche Morgan Grenfell, Morgan Stanley Asset Management, and, most important, hedge fund operator George Soros. The other bidder was Vladimir Gusinsky's Most Group, together with Alfa Bank, Credit Suisse First Boston, and the Spanish company Telefónica. Gusinsky's most important supporter, however, was Berezovsky.

In mid-July, with only a few days remaining before the Svyazinvest auction, Gusinsky and Berezovsky flew to see Anatoly Chubais, who was relaxing near St.-Tropez, at the villa of a friend—a Moroccan commodities trader who had done some big deals in Russia. Their rival, Vladimir Potanin, also showed up. The three businessmen arrived in the evening, spent four hours negotiating with Chubais, and then flew back.

"A significant part of the discussion was devoted to the rules of the game," Potanin recalls, "to the fact that these rules, which were pretty strict, but fair, were being introduced quickly. Berezovsky, as I understood him, thought that these rules...should be introduced more gradually, because they were affecting the interests of those who weren't ready for them [i.e., Gusinsky]. Berezovsky argued that [Gusinsky's] consortium had done a lot to make this deal possible, that it had worked on it, that it could expect some kind of preferential treatment."

Gusinsky, for his part, used the same argument he had once made in pushing through the privatization of Channel 4 and the creation of NTV. He had spent a lot of time overseeing the privatization, and after all those efforts, it would be unfair to let someone else walk away with the company. Gusinsky's argument was strengthened by the fact that, except for Channel 4, he had not participated in the distribution of federal assets; he had not taken part in either the voucher privatizations or the loans-for-shares auctions. In case this kind of reasoning with Chubais failed, Gusinsky and Berezovsky issued threats. "It was made clear to me and to Chubais that if the other consortium was unsuccessful in the Svyazinvest auction, they would be dissatisfied and would use various means of exerting...well, let's say, various means of influence, even to the point of unleashing a media war or destroying the deal," Potanin recalls.[39]

Gusinsky said his consortium could offer $1.2 billion; Potanin said he would pay $1.6 billion. "Gusinsky thought, as I understood him, that we shouldn't participate [in the Svyazinvest auction], that we should take on some other project," Potanin would later recall. According to some reports, Berezovsky and Gusinsky offered Potanin lucrative incentives to step aside: control over the government equity stake of Russia's electricity monopoly or victory in the privatization of the last big state oil company, Rosneft. Potanin felt obligated to bid at the Svyazinvest auction. "I didn't believe that I could refuse to participate, because that would not have been in the commercial interests of our group, our investors, and our partners," he recalls. "And Chubais didn't think he could support this kind of proposal

[Potanin's withdrawal], because as a result of competitive bidding there would be a higher price and more money for the budget."[40]

On July 25 the Svyazinvest auction took place. The starting price was $1.2 billion. Gusinsky's consortium offered $1.7 billion, while Potanin's consortium offered $1.9 billion. Potanin's company (a Cyprus-registered front called Mustcom) had offered nearly twice the sum that the government had received in all the loans-for-shares auctions. Nonetheless, the Berezovsky-Gusinsky camp immediately claimed foul play.

The day after the auction, the press war began. Sergei Dorenko, Berezovsky's star news anchor on ORT, broadcast two scathing reports on the corrupt nature of the Onexim Bank alliance. Initially, the Berezovsky-Gusinsky camp focused on the "discovery" that Potanin's bid was largely underwritten by the "speculator" George Soros. "Some people think money doesn't smell," Gusinsky declared. "For me, personally, money stinks."[41]

This line of attack failed. So Gusinsky made a public threat. "I think in the near future there will be new information, which will become public and will answer your question about whether there was an inside deal [on Svyazinvest] or not," he told the press.[42]

Within two weeks, Russian newspapers broke the story that the man in charge of the Svyazinvest auction, Chubais's friend Alfred Kokh, had taken a $100,000 "bribe" from Onexim Bank in January 1997. It was actually a book advance paid by a Swiss trading company linked to Onexim Bank. Kokh was dismissed from the government and the prosecutor-general's office began an investigation. Kokh published the book a year later and prosecutors failed to produce a conviction, but the book advance represented a clear conflict of interest.[43]

In the aftermath of the Svyazinvest auction, Chubais found himself so besieged that his life seemed to be in danger. On August 18, three weeks after the auction, Chubais's young reformers were subjected to real violence for the first time. That day, in St. Petersburg, one of Chubais's closest colleagues, Deputy Mayor Mikhail Manevich, was driving along Nevsky Prospekt with his wife when he was shot by a sniper. Manevich was killed and his wife wounded. The attack, carried out in the middle of the day on St. Petersburg's central thoroughfare, was a brazen show of force. Though police later arrested individuals suspected of having carried out the hit, it was never discovered who ordered Manevich's assassination and why. Chubais felt the threat personally. In the following months, the corridors of Russian government headquarters (the White House) buzzed with rumors of assassination plots against the first deputy prime minister.[44]

On November 4, 1997, Chubais and Nemtsov visited President Yeltsin at his dacha. "We were met by Tatyana [Dyachenko] and Valentin [Yumashev]," recalls Nemtsov. "They told me, 'You are making the biggest mistake of your career.' I was taken aback—they were so pale—and I said, 'Well, maybe, but I am acting according to my principles.'" Chubais and Nemtsov spent several hours convincing Yeltsin that Berezovsky had stepped over the line. Government officials were prohibited by law from pursuing business deals while in office, but Berezovsky had done so with relish. He had lobbied the Central Bank on behalf of his Swiss financial company, Andava. He had attempted to take over Gazprom. He had meddled blatantly in the Svyazinvest auction. In short, he was the most notorious representative of the crony capitalism that sullied the image of the Yeltsin government. Nemtsov—who remained untouched by any scandals—spoke of the need to abolish "gangster capitalism." The next day, Berezovsky was dismissed from the Security Council.[45]

A week later, Russian newspapers revealed that Anatoly Chubais had done much the same thing as his subordinate Kokh. He and four other young reformers in the government admitted to receiving book advances of $90,000 each from a publisher linked to Onexim Bank. (Again, the book was published two years later, though that didn't lessen the conflict of interest.) By Russian standards, this was minor corruption; the sums were trivial compared to the hundreds of millions that were routinely looted from state coffers by crony capitalists. Moreover, the precedent for using book publishing contracts to gain favor with top government officials had been set by Boris Berezovsky, when he published President Yeltsin's memoirs in 1994. But since the charges concerned Chubais and other young reformers and since the scandal came on the heels of revelations that Chubais had received a $3 million interest-free loan from Stolichny Bank in February 1996, Chubais's ability to serve as the standard-bearer of fairness and reform was destroyed.[46]

"I agree with Gusinsky, who said that Chubais for him is not a deputy prime minister, but a competitor," Berezovsky declared to the press.[47]

Yuksi

The war between the Chubais-Potanin and Berezovsky clans now moved to the oil industry. The government announced that it would privatize Rosneft, the holding company that owned the remaining state oil fields. Berezovsky and his partner, the young oil trader Roman Abramovich, made the first move by engineering the removal of Rosneft's general director, Aleksandr Putilov—a Lukoil man—and replacing him with one of their men. But at the last minute the privatization auction of Rosneft was postponed because of the growing financial crisis in Asia.

At Sibneft, meanwhile, Berezovsky was proving to be a mediocre oilman. He had succeeded in taking control of this company, sharing it 50-50 with Abramovich, but even with massive tax breaks, Sibneft remained a highly indebted company; it continued to sell the same amount of oil to the same clients. Berezovsky and his partner followed the same recipe (the "privatization of profits") that had been so effectively applied to Avtovaz and Aeroflot: they surrounded Sibneft with predatory intermediaries. The chief intermediary in this case was Roman Abramovich's trading company, Runicom, registered in Gibraltar, based in Geneva. This outfit sold Sibneft's oil and oil products for good cash, but delayed paying Sibneft for the shipments. The oil giant's financial statements showed that at the end of 1997 Runicom owed Sibneft $30 million; in 1998, this figure rose to $45 million. Moreover, in 1998 Sibneft gave Runicom an interest-free loan of $124 million, supposedly to import equipment, though the equipment never arrived and Runicom later returned the money. It helped that the general director of Sibneft was a longtime Runicom executive. Such self-dealing on the part of Abramovich and his deputy clearly harmed the interests of outside shareholders, but it was advantageous for the shareholders of Runicom.[48]

In the winter of 1997–98, with the prospects of privatizing state-owned Rosneft fading, Berezovsky decided to reduce his exposure to the oil business. He decided to merge Sibneft with Russia's second-largest oil producer, Yukos, controlled by Menatep's Mikhail Khodorkovsky. The resulting giant, named Yuksi, would be 60 percent owned by the shareholders of Yukos and 40 percent owned by the shareholders of Sibneft; in terms of oil production and reserves, Yuksi would be the third-largest private oil company in the world.

Despite weathering several storms in the privatization of Sibneft in 1995—the drowning of the director of the Omsk Refinery and the arrest of the owners of Balkar Trading—Berezovsky was letting the company go. At heart he was a corporate raider. He excelled at taking over companies, but he did not have much aptitude for actually running them. The oil business, with its long time horizons and huge capital investments, was not suited to his character and he was not interested in managing a company. The Yuksi merger, therefore, was a great move—not only because it was a profitable way for Berezovsky to reduce his exposure to Sibneft, but also because it united the leading anti-Chubais oligarchs (Berezovsky, Gusinsky, Smolensky, and Khodorkovsky) as shareholders in the new company. On January 19, 1998, the Yuksi merger was announced at a triumphal press conference attended by Prime Minister Chernomyrdin and Fuel and Energy Minister Sergei Kiriyenko. Berezovsky, who only a few months previously had styled himself a civil servant, flaunted his control of Sibneft and boasted of his key role in arranging the merger.[49]

Five months later, the merger collapsed. Berezovsky had not counted on the resistance of his co-owner at Sibneft, Roman Abramovich. Unlike Berezovsky, who was involved in a wide array of businesses, Abramovich was solely focused on oil. Though initially agreeing to the merger, Abramovich demanded a higher price than Yukos was willing to pay. When the deal collapsed, Abramovich was left with his controlling stake in Sibneft and his lucrative oil-trading contracts. Berezovsky, however, proceeded to reduce his stake in Sibneft (he did retain an important minority stake) and move on.[50]

The Financial Crash

By the beginning of 1998, the Russian economy was once again on the brink of disaster. The government had not fulfilled its promise to pay back wages to the army and other state employees. Plunging T-bill prices indicated that government finances were precarious. Someone had to accept the blame for Boris Yeltsin's failure to revive the country the year before. Berezovsky publicly predicted who the fall guy would be. "I am sure that Chubais has only a few weeks, if not days, left in the government," he told *Kommersant.* "He will leave the government and, of course, this will happen with the full approval of the president."[51]

Yeltsin had a predictable annual routine. Every year, he would rouse himself from his winter torpor and shake up the Russian political scene. On March 23, 1998 (ten days after Berezovsky's prediction), Yeltsin announced that he was firing Chubais. More surprising, Yeltsin also fired his longtime prime minister, Viktor Chernomyrdin. That same day, Yeltsin called in his minister of fuel and energy, Sergei Kiriyenko, and informed him that he was now prime minister. "For me it was completely unexpected," Kiriyenko recalls. "Before, in my whole life, I had met Yeltsin only twice."[52]

The decision to fire Chernomyrdin and appoint Kiriyenko had been made not so much by Yeltsin himself as by Yeltsin's closest advisers, principally his daughter Tanya Dyachenko and his chief of staff, Valentin Yumashev. Berezovsky, too, had recommended firing Chernomyrdin, although he had been opposed to Kiriyenko as the replacement.[53]

Berezovsky had reason to oppose Kiriyenko. First, he was a

young reformer, close to Chubais and Nemtsov. Second, during his time at the fuel and energy ministry, Kiriyenko had resisted attempts to rig the privatization of Rosneft in favor of Berezovsky and the other oligarchs.

Superficially, Kiriyenko's appointment seemed to be good news for Russia. The slow-moving Chernomyrdin had long epitomized the corruption and incompetence of the Yeltsin regime, while Anatoly Chubais's reputation had been irreparably damaged by the book scandal. But the dismissal of both men upset the delicate balance that had kept the different factions within the Yeltsin entourage from destroying one another and bringing down the whole regime in the process.[54]

Berezovsky's press organs led the fight against Kiriyenko. The tycoon was supported by the Communist Party; twice, the Communist-dominated Duma refused to approve Kiriyenko's appointment. Finally, on April 24, Kiriyenko managed to garner enough votes to become the new prime minister. By this time, Berezovsky had managed to convince the Kremlin to let him back into the government: On April 29, he was appointed executive secretary of the Commonwealth of Independent States, the largely symbolic trade association uniting most of the states of the former Soviet Union. Almost immediately Kiriyenko was subjected to immense pressure to appoint agents of the oligarchs to the government staff. When it became clear that he would not bend to their will, the oligarchs began to undermine his administration.

On May 12, Siberian coal miners went on strike to protest the continual delays in paying their meager wages; the miners shut down the pits and, more important, blocked the Trans-Siberian Railroad for two weeks in July, causing hundreds of millions of dollars in damage to the Russian economy. One group of miners camped out in front of Kiriyenko's government headquarters; Berezovsky and Gusinsky's TV networks featured them regularly on the evening news with their placards, meetings, and concerts. There was nothing strange about the long-suffering miners protesting their fate, but the timing of the protest was odd. In a thinly veiled fashion, Kiriyenko accused Berezovsky and other oligarchs of financing the miners' protests against his government. "The sudden organization of the miners' protest in the spring was not an accident," he told me later. "I don't want to go into details . . . but these coal miners' strikes were encouraged and financed to some degree."[55]

Kiriyenko was also facing significant financial pressure from the oligarchs. He would not have been hurt so badly by the oligarchs' opposition if the financial position of the government had not been so vulnerable, but Chubais, Kiriyenko, and other young reformers had grossly mismanaged the government's finances. The most profitable parts of Russian industry had been given away to a handful of unscrupulous tycoons, who proceeded to strip the assets, avoid paying taxes, and squirrel their newfound wealth offshore. Having built up these private business empires at the expense of the Russian state and the majority of the Russian population, the government found itself essentially bankrupt. To avoid financing its yawning budget deficit by printing money and unleashing a new bout of hyperinflation, the government simply refused to pay the army, the schools, the hospitals, and the pensioners. It also borrowed heavily from the very banks it had subsidized with government funds and assets a few years earlier.

The primary form of government debt was the zero-coupon T-bill (GKO), usually with a three-month maturity. As designed by Chubais, the GKO market represented another massive subsidy to Russian banks at the expense of the Russian treasury. Only Russian banks were permitted to buy GKOs; foreign institutions were excluded. With competition at the GKO auctions thus limited to a few Russian institutions, the annualized yield on these securities ranged from 60 percent to 200 percent in dollar terms in 1995–98. If foreign banks, with their huge funds, had been allowed into the GKO market at this time, the yields would have been a fraction of this astronomical rate. The government's T-bill policy ruined Russia's domestic credit market—what company could borrow funds at dollar-based interest rates of 100 percent?—and bankrupted the government with high debt-service costs. The banks that collected these huge interest earnings—Onexim, Menatep, Alfa, and Stolichny, among others—put almost nothing back into the Russian economy. The most dramatic increase in GKO emissions came in 1996, when the Yeltsin regime needed a quick infusion of cash to boost the economy during the elections, but the emissions continued to rise with every year, reaching $70 billion two years later. By the time Kiriyenko became prime minister, all the proceeds of new GKO emissions went to pay off the old GKOs. The GKO operation had put the Russian government on a treadmill and now the treadmill was moving too fast for the government to keep up. With the government forced to sell more GKOs at ever higher interest rates simply to service its debt, it was clear that something would have to give. The GKO operation had become a pyramid scheme.[56]

In April and May, Russian banks began unloading their GKOs en masse. Ironically, it was not the foreign institutions that led the flight out of the GKO market, but Russia's supposedly patriotic, home-grown banks. The giant Berezovsky-linked bank, SBS-Agro, and other well-connected financial institutions were desperate for new government subsidies, and when they did not get them, they had no choice but to liquidate their government securities portfolio.

"Why was there such an acute conflict between Berezovsky and SBS-Agro on the one hand and the government on the other?" observes Kiriyenko. "These banks had grown up at a time of hyperinflation, when it was possible to benefit from state monetary outflows, and when state property was divided up [at minimal cost]. These banks were unable to survive without the constant inflow of state funds. SBS-Agro was basically bankrupt by June or July."[57]

In late June, the oligarchs briefly reunited in a meeting reminiscent of the Group of 13 and placed their demands before the government. "The representatives of big business told us that it was imperative to create a council of economic mutual aid," Kiriyenko recalls. "This would be a council of oligarchs, which would give the government recommendations and advice. They made us understand that if the government agreed to the creation of such a council, they would support [the government]. Otherwise the oligarchs would pursue policies to bring this government down."

Kiriyenko refused. He continued to insist that Russia's top financial-industrial groups pay their taxes and he rejected their pleas for a bailout. Berezovsky and his allies continued to talk down the Russian market, while fleeing the GKOs as fast as they could; at the same time, their media outlets portrayed the Kiriyenko government as an ineffectual, fly-by-night administration. The crisis of confidence in the Kiriyenko government and the Russian economy deepened.

To avert the looming catastrophe, President Yeltsin recalled his disgraced former economics czar, Anatoly Chubais. Chubais was still the official most trusted by Western financial institutions; if anyone could negotiate a new loan with the IMF, he could. By falsifying some financial figures and hiding the catastrophic state of the Russian economy, Chubais managed to convince the IMF and several commercial banks to put together a $23 billion loan package. (Later, Chubais would tell the Russian press how the government had misled foreign creditors: "We ripped them off.") The first IMF installment ($5 billion) disappeared within days, sold to well-connected Russian banks desperate to unload their rubles.[58]

On August 17, 1998, the GKO pyramid collapsed. The Russian government announced that it was dropping its support for the ruble. Within weeks, the ruble lost three quarters of its value and inflation increased to more than 100 percent on an annualized basis. Western investors were hit in several ways at once. First, the ruble was devalued massively; second, the government suspended its service of the GKOs and most other government securities; finally, the government declared a ninety-day moratorium on Russian banks honoring their foreign currency commitments. According to some estimates, foreign banks and investors lost $40 billion in the Russian debt market and $8 billion in the equities market; other estimates put the combined losses at more than $100 billion. In the aftermath of the crash, stock markets around the world plunged 5 percent or 10 percent or more; such institutions as Citibank, Chase, BankAmerica, and Credit Suisse declared that most of their profits that year had been wiped out by the collapse. Russian losses caused a major hedge fund, Long-Term Capital Management, to collapse spectacularly—so big was the threat to the world monetary system from this collapse and other aftershocks of the Russian crisis that the U.S. Federal Reserve felt obliged to organize a bailout of the hedge fund. The world's most famous hedge fund operator, George Soros, saw his investments in Russia, including the $1 billion invested in the telecoms holding Svyazinvest, almost completely destroyed by the financial collapse. "I was fully aware that the robber-capitalist system was unsound and unsustainable and I was quite vocal about it; nevertheless I allowed myself to be sucked into the Svyazinvest deal," Soros would later write. "It was the worst investment of my professional career."[59]

As most of Russia's commercial banks collapsed during the devaluation, ordinary Russian citizens once again saw their savings destroyed. The financial crash symbolized the final breakdown of Russia's post-Communist economic reform. In the midst of all the failures of the Yeltsin era, Chubais and the young reformers could at least claim one success: stabilizing the ruble and taming inflation. This monetary stabilization was the reason the young reformers had subjected the Russian population to such suffering—why the economy had been reduced to half its former size, why salaries were left unpaid, why social services were allowed to break down. Now the monetary policy, like the rest of the Russian economy, lay in ruins.

On August 23, less than a week after the crash, Yeltsin fired Prime Minister Kiriyenko and his entire cabinet. Berezovsky and the other oligarchs had attained their goal of bringing down Kiriyenko, but in the process they had hastened the collapse of the Russian economy and destroyed much of their own personal wealth. Now that the young reformers had been routed, the political momentum seemed on Berezovsky's side. At his urging, Yeltsin chose the old stalwart, Viktor Chernomyrdin, as his next prime minister. It was a distinctly uninspired choice. Chernomyrdin was widely regarded as incompetent and corrupt; moreover he was being returned to a post from which he had been fired five months earlier. This time both Berezovsky and Yeltsin were blocked by unprecedented opposition from the parliament. Twice the Communist-dominated Duma refused to ratify Chernomyrdin, despite heavy pressure from the Yeltsin camp. Finally, facing the prospect of a constitutional crisis in addition to the economic disaster, Yeltsin withdrew Chernomyrdin's candidacy. On September 10, he nominated a man acceptable to the parliament, former foreign intelligence chief and foreign minister, Yevgeny Primakov. The sixty-eight-year-old Primakov was regarded as clean, patriotic, and conservative. His nomination was ratified by the Duma the next day.

CHAPTER TEN

VLADIMIR PUTIN
TAKES POWER

Eavesdropping on the President

The alarmist predictions about Primakov's ushering in Communist economic policies or printing money and presiding over a bout of hyperinflation never materialized. The ruble stabilized, inflation decreased, business confidence began to recover. Primakov's diplomatic manner helped create a new political consensus in Russia while his plodding, old-fashioned way of speaking reassured the public.

For Berezovsky, however, Primakov represented a threat. The financial crash had discredited Yeltsin and his entourage. With Primakov's appointment, political power shifted to the prime minister and the Duma. Both the Duma and Primakov's government were staffed largely with either Communists or patriotic liberals, both equally hostile to crony capitalism. There was talk of Yeltsin's serving out his term in a strictly symbolic role, with real power left in the hands of Primakov and his ministers. There were also rumors about bringing the more rapacious businessmen to justice.

Berezovsky's influence on the Yeltsin family was weakening. He no longer held exclusive claim to being what the press discreetly called the family's "financial adviser." Yeltsin's daughters and sons-in-law had developed relationships with other businessmen, including Berezovsky's partner, Roman Abramovich, whose nickname inside the Yeltsin family, according to General Korzhakov, was "the Cashier." While these and other financial relationships would be exposed only later, top Russian government officials such as former Prime Minister Sergei Kiriyenko began to notice that Berezovsky was losing his control over the Kremlin inner circle.[1]

"After spending five months at my job, I got the impression that all the stories that Berezovsky tells about his ability to influence the president directly is pure fiction," Kiriyenko told me. "Berezovsky is able to project a demonic image quite brilliantly. He does this thanks to the fact that he has access to good information, and as soon as he finds out that someone, somewhere, is trying to do something, he immediately appears in public with a similar suggestion, so that the next day, everyone will think that it was done because Berezovsky mentioned it."[2]

Berezovsky got his information on the inner workings of the Yeltsin administration in two ways. He continued to maintain close contact with key members of Yeltsin's staff and he employed surveillance technology to listen in on important people's phone calls. He used the services of a powerful private intelligence agency called Atoll, which he had set up several years earlier with the help of General Korzhakov's SBP; Atoll was equipped with state-of-the-art surveillance technology. While many big businesses in Russia maintained their own intelligence agencies, Berezovsky's outfit was special. It was engaged, according to Russian prosecutors' allegations, in spying on the Yeltsin family.[3]

"When the press began to write about the financial relations of Tanya Dyachenko and Berezovsky [in 1998], the president's daughter decided to terminate her relationship with Berezovsky," remembers General Korzhakov. "But he began to blackmail her, promising that he would tell the press the whole truth about the finances of the president's family."[4]

In the autumn of 1998, several fly-by-night Websites posted the transcripts of intercepted telephone conversations between top government officials. Among the transcripts were numerous discussions between Berezovsky and Tanya Dyachenko. Although the pages were withdrawn after a few days and the transcripts seemed to contain little incriminating information, the Russian prosecutor's office suspected that Atoll was behind the operation and that the Website was meant as a warning to Dyachenko and others in the Yeltsin entourage that dirty laundry could be aired very quickly.[5]

The telephone intercepts forced the hand of the Primakov government. It decided to move against Berezovsky.

Arrest Warrant

As I prepared for a trip to Russia in the winter of 1998–99 to continue my investigations on Berezovsky, I was warned by former Foreign Trade Minister Oleg Davydov that the situation in the country was tense. Galina Staravoitova, a liberal woman parliamentarian and a friend of Chubais's, had just been assassinated in the lobby of her apartment house in St. Petersburg. This was not a good time to start digging up stuff on Berezovsky, Davydov said.

"With a guy like that, you have to be very careful," Davydov told me. "He has contacts with the criminal world, so to speak. Therefore you have to conduct all these matters with him carefully. Over there we don't talk—we shoot. Or there are compromising materials that appear in the papers. I think that... if you try to collect information about [Berezovsky] in Russia, you will run into problems.... They can do anything. What they did with Staravoitova is only one thing. With a foreign journalist they aren't going to think twice about it."[6]

In fact, Russia was becoming a safer place. The government of Yevgeny Primakov was launching the first concerted crackdown on organized crime since the fall of Communism. Investigations were begun of numerous famous people; the prosecutors seemed determined to uncover the links between the Yeltsin entourage and the bandit capitalists. During my stay in Moscow that winter, I unwittingly witnessed the beginning of the crackdown on the empire of Boris Berezovsky.

On the morning of February 2, 1999, I was standing by the window of my room in the Hotel Baltschug, enjoying a cup of coffee and looking at the pink dawn over the roofs of Moscow. Suddenly, three white vans screeched up to the building across the street. A dozen men in face masks, camouflage uniforms, and automatic rifles jumped out. Then, accompanied by other men in brown leather jackets, carrying briefcases and video cameras, they entered the building.

It was the headquarters of Sibneft. The raiders were members of the police and the prosecutor-general's office. Officially, the police were looking for evidence connected with Berezovsky's private intelligence service, Atoll. The next day the prosecutor-general's office announced that it had opened a criminal case against Andava and Forus Services for alleged money-laundering and currency-law violations connected to Aeroflot. Again, the prosecutor-general's office was looking for much more—it was investigating suspected fraud, embezzlement, and tax evasion perpetrated by a network of Berezovsky-linked companies. Aeroflot's offices were raided, as were those of NFQ (advertising company) and FOK (financial company). On the same day that police searched Aeroflot's offices Yeltsin's son-in-law Valery Okulov fired nine top managers and directors—including the remaining Berezovsky appointees. He also announced that he was terminating the special relationship between Andava and Aeroflot.[7]

Berezovsky's primary TV network, ORT, was cut off from government subsidies and subject to bankruptcy proceedings. ORT's advertising chief, Sergei Lisovsky, was raided by the tax police. Berezovsky fought back by obtaining the promise of Australian media magnate Rupert Murdoch to invest in the network. Meanwhile, Atoll, the Berezovsky-linked intelligence agency, was shut down, pending an investigation. The first half of 1999 witnessed numerous criminal cases against various individuals and companies—all of them officially unrelated, but all of them somehow linked to Berezovsky.

The outfit investigating organized crime at automaker Avtovaz was the Samara division of the MVD. After the headquarters of the Samara MVD were destroyed in a fire in February 1999 (consuming all the documents inside and killing at least sixty people), the MVD announced that it was opening a criminal case against fraud and money-laundering at the automaker. At the same time, the prosecutor-general's office began criminal proceedings against Aleksandr Smolensky of SBS-Agro Bank. Smolensky was Berezovsky's closest ally among the big Russian bankers; SBS-Agro, in which Berezovsky reputedly held a 25 percent share, was the third-largest bank in Russia and the authorized bank of the Presidential Administration. Smolensky himself was accused of embezzlement in connection with the Chechen bank fraud of 1992–93, though this case would later be dropped. Prosecutors also opened an embezzlement and money-laundering case against Anatoly Bykov, a Berezovsky ally in the aluminum business.[8]

The prosecutor-general, Yuri Skuratov, was also presiding over an investigation of 780 top government officials on the suspicion that they had used inside information to play the GKO market. Among the suspects Skuratov named publicly were Anatoly Chubais and several other young reformers. The inside knowledge of the government's financial strategy had apparently enabled government officials to cash out of the government securities market in the summer of 1998 and to transfer their profits to foreign currency accounts abroad just before the GKO market was suspended and the ruble crashed.[9]

The Russian government's newfound zeal in going after the big crooks and criminals in early 1999 was echoed abroad. The collapse of the ruble the previous summer had convinced the West of the folly of supporting a Russian government that connived with a handful of crony capitalists to loot the country. Western governments were finally curbing the money-laundering resulting from the oligarchs' depredations in Russia. In the autumn of 1998, the U.S. Treasury Department, together with the FBI and the Justice Department, began a secret money-laundering investigation into at least $7 billion funneled out of Russia through the Bank of New York. Federal officials disclosed that at least part of this money came from criminal activities such as kidnapping. The source of the rest of the funds was less clear. It was now apparent that even the most venerable American financial institutions had been infected by Russian organized crime.

The most suspicious portion of the Bank of New York money flow originated with a group of obscure Moscow institutions: Flamingo Bank, Sobin Bank, DKB, and MDM Bank—all at least partly owned by SBS-Agro. Among the managers of these banks were several close friends of Berezovsky and Roman Abramovich. American investigators found no immediate evidence that either Berezovsky, Abramovich, or Aleksandr Smolensky was aware of the alleged money-laundering activities of their obscure affiliates, but the fact that SBS-Agro was named in the Bank of New York investigation posed potential problems for Berezovsky in the future.[10]

In Switzerland the federal prosecutor's office began investigating several cases of alleged embezzlement and money-laundering perpetrated by top Russian officials. The most famous of these involved bribes allegedly given by the Swiss-Albanian construction company Mabetex to Pavel Borodin (the Yeltsin crony who managed the Kremlin's real-estate holdings) and to members of the Yeltsin family. The Swiss also began to focus on Berezovsky and his shenanigans with Aeroflot's foreign cash flows. In the early summer of 1999, Swiss prosecutors raided Andava, Forus, and a number of other Swiss enterprises linked to Berezovsky; they also froze these companies' bank accounts, as well as the personal bank accounts of Berezovsky, his Aeroflot partner, Nikolai Glushkov, and numerous other individuals connected with the operation. Later that summer, when Berezovsky requested entry into Switzerland for medical treatment (he was rumored to have hepatitis), he was denied an entry visa.[11]

With Berezovsky being pressed by law-enforcement agencies at home and abroad, President Yeltsin had no choice but to jettison his longtime companion once again. On April 2, 1999, Berezovsky was dismissed from the post of executive secretary of the Commonwealth of Independent States. On April 6, because both Berezovsky and Glushkov failed to present themselves to answer questions on the Aeroflot case, the prosecutor-general's office issued warrants for their arrest. The charges were "illegal business activities" and "money-laundering." The arrest warrants were transmitted to Interpol, obligating foreign governments to arrest Berezovsky and Glushkov on sight.[12]

Berezovsky remained remarkably cool. When the warrant was announced, he was in Paris, staying at the Hotel Crillon on the Place de la Concorde, across the street from the American Embassy. He granted an interview to Britain's *Sunday Telegraph*, in which he denied the charges against him, labeling them a political intrigue, and declared his intention of going home to Russia. At this moment a key government official spoke out in Berezovsky's defense. The minister of internal affairs, an old Yeltsin loyalist named Sergei Stepashin, declared that if Berezovsky came back to talk to the prosecutors, he would not be arrested.[13]

Berezovsky returned to Russia and spoke to the prosecutors on April 14; the arrest warrant was duly revoked. Berezovsky declared that he was being persecuted by Primakov and the former KGB for political reasons. Perhaps. But the fact remains that the Primakov administration was the first government of the Yeltsin era to make a determined effort to hold Berezovsky and other crony capitalists accountable to the law. It would also be the last Yeltsin government to do so.

Sidelined and discredited though he was, President Yeltsin still retained the constitutional power to dismiss the prime minister and his government. Back in January 1999, Primakov had attempted to broker a nonaggression agreement among the different branches of government to guarantee political stability for the next eighteen months. The Duma, which had been gathering votes to impeach Yeltsin, would promise not to take action against the president. Yeltsin, in turn, would promise not to dismiss either the Duma or the Primakov government until the end of his term. Yeltsin, on the advice of his entourage, refused to sign the agreement.[14]

From Yeltsin's point of view, the Primakov administration was spinning out of control. Primakov may have brought stability and trust back to the Russian political arena, but the corruption investigations at home and abroad were extremely dangerous for Yeltsin—

they threatened not only to bring down crony capitalists like Berezovsky, but to tarnish Yeltsin's daughter Tanya. Moreover, Primakov did not offer enough of a guarantee that Yeltsin himself would not face prosecution once he left office. For all his diplomatic assurances to President Yeltsin, Primakov was overly concerned with restoring the rule of law and punishing the guilty. Finally, Primakov's government evoked loud protests from the United States and other Western governments. On March 24, the United States and its NATO allies began bombing Serbia to force it out of the province of Kosovo. Primakov prepared to take a hard line against the NATO attack and the world suddenly faced the prospect of a Cold War–style confrontation between Russia and NATO. President Clinton called Yeltsin to express his concerns. For Yeltsin, always sensitive to the approval of the West, this was another black mark against his prime minister. The task before Yeltsin and his entourage was clear: They had to head off the investigations and they had to assure the accession of a friendly politician to the Russian presidency in June 2000.

On May 12, Primakov was fired. His replacement as prime minister was Sergei Stepashin, the man who, as minister of internal affairs, had promised to protect Boris Berezovsky from arrest.[15]

"The dismissal of Primakov was my personal victory," Berezovsky stated in an interview with *Le Figaro* several months later.[16]

Berezovsky Triumphant

With Primakov gone, the last vestiges of political consensus evaporated. The Duma pressed ahead with impeachment measures against Yeltsin. This was the chance for the Duma to assert itself and defy the president. In the past such direct legislative challenges to President Yeltsin had been foiled with the help of wide-scale bribery of Duma deputies. On May 15, the Duma collected a large majority on all five charges—that Yeltsin had unconstitutionally broken up the Soviet Union in 1991, that he had illegally shelled the parliament in 1993, that he had permitted the deterioration of the military, that he was guilty of genocide in allowing the massive rise of Russian mortality rates, and that he was guilty of starting the illegal war in Chechnya in 1994. But on no count did the parliament manage to obtain the necessary two-thirds support, mostly because some 100 parliamentary members were absent from the proceedings. (Zhirinovsky's large parliamentary bloc, notorious as the most corrupt in the parliament, was absent en masse.)[17]

In arranging Primakov's dismissal, Berezovsky overcame one of the biggest threats he had ever faced. Even for such a master of political intrigue, it was an amazing feat. Just a few months previously, he was a wanted man, hounded by Russian prosecutors. With his businesses the focus of money-laundering investigations in Switzerland and the United States, a life of exile promised no comfort. If Berezovsky was to survive as a free man, he would have to triumph politically in Russia. And this is what he did, by reasserting his influence in Yeltsin's inner circle and becoming, once again, the chief power broker in the Kremlin. Now he began to strengthen his hold on the government.

The new prime minister, Sergei Stepashin, was an old Yeltsin loyalist. A career MVD bureaucrat, Stepashin had seen his influence soar under Yeltsin; he had served as head of the FSB–KGB, minister of justice, and then minister of internal affairs. Stepashin may have been a proven team player, but he was not an unprincipled creature of Berezovsky or the rest of the "family." To ensure that the new government would obey his wishes, Berezovsky pushed other, subservient politicians to the fore.

The new first deputy prime minister (number two in the government and the man in charge of managing the Russian economy) was Nikolai Aksyonenko, former head of the massively inefficient and corrupt Ministry of Railways. Aksyonenko was a protégé of Berezovsky and Roman Abramovich. Together with the new fuel and energy minister, Viktor Kalyuzhny, Aksyonenko helped Sibneft get lucrative oil-export contracts; Aksyonenko also began to take control of the most important natural-resource monopolies, such as the oil pipeline monopoly Transneft. The new minister of internal affairs was Vladimir Rushailo, former head of the organized-crime squad of the Moscow police. Rushailo had worked closely with Berezovsky in the hostage negotiations in Chechnya and was widely rumored to be "Berezovsky's man." Another important Berezovsky agent in the government at this time was Aleksandr Voloshin, Yeltsin's chief of staff. Five years earlier, Voloshin had helped Berezovsky set up the Avva investment scheme; his appointment as Yeltsin's chief of staff in February 1999 marked a reassertion of Berezovsky's control over President Yeltsin's entourage. Subsequently, Voloshin proved cynical and effective in spearheading the defeat of the Duma's impeachment initiative, intimidating the independent press, and pulling the strings behind important government appointments.[18]

Berezovsky might have placed several of his men in key government positions, but this did not mean he controlled the government completely. Although the investigations into Berezovsky's operations by Russian prosecutors lost steam, the criminal cases against him and his allies remained open. Abroad, the inquiries actually gathered momentum. The Bank of New York matter went public. The boldest measures of the Swiss federal prosecutor's office in the investigation of Berezovsky and Kremlin property boss Pavel Borodin were undertaken not during the Primakov era, but when Stepashin was prime minister. In Hungary, meanwhile, the most obviously thuggish individual targeted by Russian law-enforcement authorities, Krasnoyarsk Aluminum boss Anatoly Bykov, was arrested and extradited to face murder charges in Russia. Stepashin, in other words, may have been loyal to Yeltsin, but he refused to take the necessary action to derail the criminal investigations against Berezovsky and other members of Yeltsin's entourage.[19]

On August 9, Stepashin was fired. His replacement was Vladimir Putin, the head of the FSB–KGB.

It was a strange choice. Until recently, the forty-six-year-old Putin had had a distinctly unremarkable career. He had joined the KGB in 1975, after obtaining a law degree at Leningrad State University. He finished the Andropov Academy (the KGB training institute), was assigned to the First Chief Directorate (the KGB's primary foreign espionage arm, then under the command of General Kalugin), and spent seven uneventful years in his hometown, Leningrad. Only in 1984 did Putin finally get his coveted foreign posting, but instead of being assigned to one of the key Western capitals, he was stationed in a Soviet satellite country, East Germany. And he was not even posted to East Berlin, but to a provincial capital, Dresden, which boasted only ten KGB members on staff. Putin worked as a member of this staff for six years, collaborating closely with the Stasi (the East German secret police) and traveling occasionally to the West. In 1990, once the Berlin Wall had fallen, he was recalled to Russia; again he was posted not in the center of the action—Moscow—but off to the side, in Leningrad. The KGB by this time was suffering from budget cutbacks and a massive reduction of personnel. Lieutenant Colonel Putin, like hundreds of thousands of other secret-service men, was demobilized into the KGB reserve.[20]

Living in a humble communal flat with his wife and two daughters, Putin functioned under a standard KGB cover, as foreign affairs assistant to the dean of Leningrad University, working with foreign students and foreign teachers. In August 1991, when the hard-liners' coup against Gorbachev failed, Putin knew that his old career was finished and he resigned his commission from the KGB. Fortunately, he was picked up by his old law school professor, Anatoly Sobchak, the eloquent democrat who had been elected mayor of Leningrad two months before the coup. Sobchak appointed Putin deputy mayor and placed him in charge of the Committee for Foreign Relations (responsible for dealing with foreign diplomatic missions, foreign humanitarian aid, foreign trade, and foreign investment). In 1994, the former KGB man was appointed first deputy mayor and placed in charge of privatization and tax collection from the private sector.[21]

Under Sobchak and Putin's rule, St. Petersburg did not prosper. Although the city was largely spared the violence of the mob war and other excesses of the Yeltsin era, it sank steadily into debt and poverty. On June 16, 1996, Sobchak was voted out of office. His loyal associate, Vladimir Putin, left with him.

Again, Putin found a political patron from the distant past: Anatoly Chubais, a Leningrad native, who had known both Sobchak and Putin from the early days of the democratic movement. In August 1996, Chubais was President Yeltsin's chief of staff and head of the Presidential Administration. This institution was responsible to Yeltsin alone (unlike other government institutions, which were responsible to the prime minister); it was based in the former Central Committee complex on Old Square. Within a few years, the Presidential Administration had ballooned to 2,400 employees. These officials covered the gamut of government affairs in Russia—from overseeing the oil industry to setting media policy. The Presidential Administration was an extraordinarily powerful institution: Like its predecessor on Old Square—the Central Committee of the Communist Party—the Presidential Administration was both omnipresent in Russian life and so secretive that it was known to most citizens by rumor only.

Just as Soviet leaders had been chosen inside the secretive Central Committee, so the succession to Yeltsin was determined by murky intrigues within the Presidential Administration. At first, Putin was posted as deputy to the Kremlin property chief, Pavel Borodin. Borodin's department seemed to epitomize the corruption of the Yeltsin government (as of the publication of this book, the investigation into Borodin's activities in Switzerland is still open). Putin apparently kept his nose clean; his particular task was to manage Russian state property abroad. In March 1997, when Berezovsky's longtime ally, presidential ghostwriter Valentin Yumashev, took over from Chubais as head of the Presidential Administration, Putin was promoted to deputy chief of staff and, in July, first deputy chief of staff. Putin's new responsibility was to head up the Control Commission, serving as a kind of presidential auditor checking on the work of Russia's provincial governments. He developed a reputation for industriousness and severity. The governors, so brazen in defying the Kremlin up to now, began to fear the surly technocrat vetting their affairs for the Kremlin.

Berezovsky's camp quickly sized up Putin as a man who would go far. In September 1997, Berezovsky's newspaper *Nezavisimaya Gazeta* published a long analysis of how Anatoly Chubais was trying to take control of the levers of power in the Kremlin (this was a month after the Svyazinvest auction). Among Chubais's ambitions, the newspaper alleged, was to take control of the FSB–KGB by firing its current director, Nikolai Kovalev, and replacing him with Vladimir Putin. Chubais had brought the former KGB lieutenant colonel to Moscow, and Berezovsky's newspaper was worried that at the head of the FSB–KGB Putin would loyally execute Chubais's orders.[22]

Less than a year later, Putin did indeed take over from Kovalev as the head of the FSB, but it was Berezovsky, not Chubais, who was responsible for the change. Berezovsky had his own reasons for replacing Kovalev. When Kovalev headed up the FSB, Berezovsky was hit by the publication of compromising documentation from the Aeroflot-Andava relationship. Though the FSB's role in exposing this scandal is unclear, Berezovsky came to regard Kovalev as a threat. (Later, Kovalev would make his political orientation public by joining the Fatherland-All Russia movement of Berezovsky's archrival, Moscow mayor Yuri Luzhkov.) Kovalev's career ended when the FSB chief was hit by a scandal of his own. In April 1998, Aleksandr Litvinenko, a lieutenant colonel in the FSB's organized-crime squad, and several colleagues announced that they had been ordered to kill Berezovsky. It was a strange revelation. Litvinenko himself, while a career KGB man, had moonlighted as Berezovsky's bodyguard; his FSB department had worked with Berezovsky on numerous Chechen kidnapping crises. Most knowledgeable observers concluded that the alleged FSB plot to assassinate Berezovsky was a fabrication. (A year later, Litvinenko would be arrested on charges of illegal searches and other violations of duty.)[23]

"My personal opinion is that this was some sort of a trick," says Sergei Kiriyenko, who served as prime minister at the time the scandal first broke. "I doubt that the FSB ever planned any action against Berezovsky. I think that Berezovsky was just trying to play up his own importance and attract attention to himself."[24]

In any case, the scandal was used as an excuse to fire Kovalev and install Vladimir Putin as head of the FSB in July 1998. The appointment came as something of a shock for the FSB. Putin had reached only the rank of lieutenant colonel before retiring nearly a decade earlier, and now he was running the entire secret service. He moved quickly to consolidate his position, firing more than three dozen top FSB generals and division heads and replacing them with loyal subordinates. He soon proved his political loyalty to his boss, President Yeltsin, by destroying the career of prosecutor-general Yuri Skuratov.

"Prosecutor Skuratov, who was working on corruption and repeatedly declaring that he intended to root out corruption in Russia, unexpectedly stumbled on a link to the Kremlin," notes former KGB general Oleg Kalugin. "Instead of ordering the investigation to be stopped in time, he let the investigation move ahead. As a result, the names of the members of the Yeltsin family, and of Yeltsin himself, as well as of his financial sponsors and underwriters, all entered into the historical record. Skuratov was told to stop the investigations, but he had slipped from [the Kremlin's] control."[25]

Putin initiated an investigation of Skuratov. In early February 1999, when the prosecutor's office announced that it had opened criminal cases against Berezovsky and other Yeltsin cronies, Russian government TV showed a videotape of a naked man identified as Skuratov frolicking in bed with two prostitutes. Immediately, the Kremlin announced Skuratov's resignation. The prosecutor-general, however, found supporters in the upper house of the Russian parliament, the Federation Council, which had the right to ratify the dismissals and appointments of the prosecutor-general, and refused to accept Skuratov's resignation. Over the next year, Yeltsin would repeatedly attempt to fire Skuratov and evict him from his office, but the Federation Council would repeatedly retain him in his post.

"Still, Yeltsin was convinced that Putin was a completely reliable, loyal member of the family [the Yeltsin clan]," notes Kalugin. "Putin was a man of Prussian-style obedience: Once the president orders something, Putin will do it. This made him irreplaceable for President Yeltsin."[26]

With Putin appointed prime minister in August 1999, Berezovsky and the rest of the Yeltsin clan could be confident that the corruption investigations would be curtailed. But what would happen after Yeltsin left office in June 2000? The family decided that Putin was the man to guarantee that there would be no retribution after the change of power. Yeltsin officially designated Putin his heir to the Russian presidency. The Yeltsin clan also established a pro-Kremlin political party called Unity—an ad hoc collection of loyal officials and celebrities—to contest the December 1999 parliamentary elections. Berezovsky and his Kremlin allies mobilized all their resources to win votes for Unity in the parliamentary elections and to get Putin elected in the subsequent presidential contest. The Kremlin had an overwhelming array of power at its disposal. It had the armed forces and the police. It had control of the government budget and the cash flow of Russia's wealthiest companies. It dominated the airwaves with its TV stations. The one thing the Kremlin did not have was a viable presidential candidate. That Putin was endorsed by such widely hated individuals as Yeltsin and Berezovsky could only hurt him; Putin's popularity ratings in the summer of 1999 hovered between 2 percent and 5 percent—far behind the leading political party, Fatherland–All Russia, led by anti-Berezovsky politicians Yevgeny Primakov and Yuri Luzhkov.

Pressed by money-laundering investigations abroad and almost certain to lose both the parliamentary and the presidential elections in the coming months, the tight-knit group of kleptocrats who constituted "the family" were facing a grim future. Berezovsky knew that he would not be able to duplicate the feat of 1996, when the mobilization of billions of dollars of cash and a crushing propaganda campaign in the media enabled Yeltsin to stage a come-from-behind victory. This time the Kremlin did not have a complete monopoly of the media, since a number of national newspapers and at least one national TV network (Gusinsky's NTV) supported the Primakov-Luzhkov alliance. Moreover, the Primakov-Luzhkov alliance was a much more attractive alternative than the Communists had been in 1996. In any case, the Russian people could not be fooled twice—they would not vote for the designated Kremlin heir just because they were told to do so. A victory for Putin and Unity would require some dramatic external event. It would require a war.

The Second Chechen War

A year earlier, Berezovsky had warned the Russian political establishment about Chechnya and reminded everybody of his central role in keeping the militants there at bay. "With respect to the war in Chechnya . . . it is a dangerous delusion to believe that this question has been resolved," Berezovsky told the press in the spring of 1998. "When Nemtsov and Chubais demanded that I be fired [in the autumn of 1997], they should have thought about who would work on this problem in the future. It is clear to everybody that with my departure, the situation [in Chechnya] has not improved. It is slowly getting worse. From this moment, I believe that Chubais and Nemtsov are fully responsible for what is going on in Chechnya."[27]

It was less than two months after this prediction that Berezovsky was reappointed to the Russian government as executive secretary of the Commonwealth of Independent States. His relationship with various Chechen warlords and terrorists again assumed an official character. Berezovsky had correctly identified Chechnya as a key factor influencing Russia's political process. Chechnya was his trump card. Over the years, he had stayed afloat in the Russian political game by wielding his influence in this gangster-ridden state. Now, in 1999, with the end of the Yeltsin era approaching, Berezovsky played his Chechen hand to the full.

In early March 1999, a personal envoy of President Yeltsin, MVD general Gennady Shpigun, was kidnapped at Grozny Airport. Sergei Stepashin, then serving as minister of internal affairs, declared that the act was an insult to the Russian government that could not be overlooked. Stepashin, first as minister of internal affairs, then as prime minister, devised a plan to punish the Chechens by bombing the terrorist training camps, and to establish a security zone by occupying the northern (pro-Russian) part of the republic up to the shores of the Terek River. This military operation was supposed to be carried out in August and September. The question was how to get the Russian population behind the new war—in order not to end up with the same kind of mass opposition that had sapped the war effort during the First Chechen War of 1994–96.[28]

The Chechens themselves provided the solution. In the first days of August 1999, a force of several thousand Chechen fighters crossed into Dagestan, occupied several villages, and proclaimed an Islamic Republic. The Chechen force was led by two men. One was Shamil Basayev, the terrorist commander famous for his bloody hostage-taking raid on a hospital in Budyonnovsk in 1995. The other was a mysterious man named Khattab—an Islamic fundamentalist from Saudi Arabia who had taken part in numerous wars in Afghanistan and Central Asia and who was reportedly linked to the Saudi terrorist Osama bin Laden (the prime suspect in the U.S. embassy bombings in Africa in 1998). Khattab was a member of the Wahhabi movement, an Islamic fundamentalist sect financed by Saudi donations and expanding rapidly in Chechnya and Dagestan; another man linked to the Wahhabi movement was Berezovsky's longtime hostage-negotiations partner, Movladi Udugov. The villages in Dagestan occupied by Basayev and Khattab were well known for their sympathies to the Wahhabi brand of Islam, although the rest of the Dagestani population was fiercely hostile to the incursion by the Chechens.

The invasion of Dagestan provided the Russian government with an excuse to deploy massive force in the Northern Caucasus: The Chechens were the aggressors, the Kremlin was fighting to defend its territory, and the Dagestani population was overwhelmingly behind the Russians. The Chechen force attacking Dagestan, though large and well-equipped, was a private operation—the largest of a number of independent militias that had sprouted up throughout Chechnya with the help of outside donations, ransom payments, black-market trading, and other criminal operations.[29]

A month after the Dagestan incursion, I called the president of Chechnya, Aslan Maskhadov, who strongly denied any connection with the fighting. "Basayev [the commander of the militants] is an ordinary citizen of Chechnya," he declared. "He can go to Dagestan, to Kosovo, to Bosnia, but he represents only himself, Shamil Basayev. In no way does he represent the Chechen people, let alone the government. These people, who have grown their beards long and are preaching the Great Jihad are controlled and financed by someone else, including the financial oligarchs who surround Yeltsin in Moscow."[30]

Maskhadov was referring primarily to Berezovsky. Berezovsky had donated $1 million in cash, possibly more, to Shamil Basayev. Berezovsky would later admit to having donated the money, although he claimed it was for the "reconstruction of a cement factory." Berezovsky, of course, had a more consistent, long-term financial relationship with Movladi Udugov, whose brother was one of the leaders of the Wahhabi sect. On several occasions the tycoon transferred money to Udugov, apparently as part of hostage negotiations. The recorded phone conversations between Berezovsky and Udugov, filled with discussions of money-transfer arrangements, apparently occurred as late as July 1999—less than a month before the Chechen attack on Dagestan. The newspaper *Moskovsky Komsomolets*, which reprinted portions of telephone intercepts, speculated that the negotiations between Berezovsky and his Chechen counterparts concerned the attack of Basayev and Khattab on Dagestan, but there is nothing in the conversation to prove that—it could well have been yet another routine hostage negotiation.[31]

To the extent that Berezovsky represented the interests of the Yeltsin regime in Chechnya, the Kremlin had been undermining the moderates, supporting the extremists financially and politically, and consequently sowing the seeds of conflict. At best, it was simply a misguided policy: Berezovsky and his Kremlin allies were simply too clever for their own good and ended up with the double game exploding in their faces—much like the KGB's double game in financing private entrepreneurs and organized-crime bosses in the late 1980s ended up hastening, not preventing, the disintegration of the Soviet Union. The worst-case scenario is that the Berezovsky strategy with the Chechen warlords was a deliberate attempt to fan the flames of war. Why would the Kremlin (acting through Berezovsky) want to support the Islamic fanatics that later ended up shedding so much Russian blood? Maskhadov argued that it was all about the desire of the Yeltsin clique to hang on to power.

"Soon there will be elections for the Russian parliament and for the presidency," says Maskhadov. "For the Russian government, this is a big problem. The president [Yeltsin] is sick. The people surrounding him are interested in who will take power afterward. If someone from the opposite camp takes power, they won't be able to avoid arrest. Therefore, at any cost, they must bring one of their own people to power. Do you understand? And what better excuse than Chechnya could they find to declare a state of emergency and postpone the elections?"[32]

It took only a month for the Chechen war to spread from the remote mountains of Dagestan to the heart of Russia. On September 9, at 5 a.m., a bomb went off under an apartment building in a shabby working-class neighborhood of Moscow. Inside, more than 100 people were sleeping. The building itself was a standard eight-floor block—one of the thousands of cheap residential units thrown up in the days of Leonid Brezhnev. The building collapsed entirely; the inhabitants were crushed under the rubble. Preliminary estimates put the death toll at ninety-four. A few survivors were pulled out in their ripped and bloodied nightclothes.

Four days later, shortly after midnight, an even more powerful explosion destroyed another Moscow apartment building, killing 118 people. Russia had been hit by an unprecedented wave of terrorism; in just over three weeks, five blasts in Moscow and two provincial cities claimed the lives of at least 300 people.

The blasts transformed the Russian political landscape. Prime Minister Putin declared the nation besieged. Paranoia swept Russia's cities. Police and FSB operatives imposed a regime of heightened security. Horror and grief gave way to fury against the Chechens, who were officially blamed for the bombings. Within days, Russian popular opinion rose overwhelmingly in support of the war in Chechnya. Putin's approval ratings surged. The young prime minister talked tough, voicing ferocious determination to wipe out the terrorists even if he caught them in the outhouse. He was now the leader of a war effort against a terrorist state that threatened peaceful Russian citizens in their homes. No one claimed responsibility for the blasts. The Kremlin blamed the same Islamic militants who had invaded Dagestan, but the top Chechen warlords, including the terrorist commander Shamil Basayev, denied that they or their cohorts had anything to do with the bombings. No evidence surfaced linking any individuals or groups to the bombings. Arriving on the eve of the crucial 1999–2000 elections, the blasts were reminiscent of the mysterious bombing of the Moscow subway a week before the presidential elections of 1996— an incident the Kremlin had blamed on "Communist extremists" and that conveniently reinforced the Yeltsin regime's claim to being the only force guaranteeing peace and stability in Russia.

The fact that Berezovsky, together with other members of the Yeltsin inner circle, had long maintained a secret relationship with Chechen extremists gave rise to the suspicion that the 1999 apartment bombings had been organized by the Russians themselves. The French newspaper *Le Monde* carried reports of the Russian arms-export monopoly providing Shamil Basayev's men with weapons and of a meeting in the resort town of Biarritz in the summer of 1999 between Berezovsky, Aleksandr Voloshin (Yeltsin's chief of staff), and Chechen rebel commanders. Later, *Le Monde* quoted Berezovsky as saying that the man he met in Biarritz was not a Chechen commander, but Vladimir Putin, on the eve of Putin's appointment as prime minister. The mystery surrounding the apartment blasts deepened. On September 22, six days after the last explosion, police in the provincial town of Ryazan found a bomb in the basement of a local apartment house; the local FSB unit treated the find as a terrorist incident and prepared to make arrests. However, FSB headquarters in Moscow declared the next day that the bomb had been a fake and that it had been planted as part of a security exercise to test the vigilance of local authorities. Because of contradictory statements by local and federal authorities and the lack of a proper investigation into the incident, what actually happened in Ryazan remains unclear. A week later, the French newspaper *Le Figaro* asked former Security Council chief Aleksandr Lebed if the Russian government had organized the terrorist attacks against its own citizens. "I am almost convinced of it," Lebed responded.[33]

Lebed's pronouncement created a sensation—it was the first time a top Russian official had publicly voiced a suspicion that had merely been hinted at in the national press. Lebed's public-relations staff later claimed that the general had been quoted out of context. Berezovsky flew to Krasnoyarsk, where Lebed was serving as governor, to talk to the maverick general. It is not known what the two discussed. But after Berezovsky's visit, Lebed fell silent; he remains conspicuously absent from the Russian political scene to this day.[34]

Lebed's accusations were echoed several days later by Chechen president Aslan Maskhadov, who stressed the tremendous boost the terrorist incidents and the subsequent war hysteria gave to the Kremlin clan's political hopes. "Chechnya [is] a testing ground for all these political games, a pawn in all these preelection maneuvers [in Moscow]," Maskhadov told me. "Today there is indeed the smell of war. We will probably have a big war here, a big conflagration. But the fault is wholly with the Russians. All this fundamentalism, extremism, and terrorism has been artificially created."[35]

Maskhadov alternately blamed Berezovsky and the Russian "secret services" for the terrorist wave. The Chechen president did harbor an animus against Berezovsky, if only because of the tycoon's repeated support of the extremist elements in Chechnya at the expense of Maskhadov's own presidential authority. Moreover, now that he had been dragged into a new war with Russia, Maskhadov had an interest in demoralizing the Russian side by portraying the Kremlin as scheming behind the backs of Russian soldiers. Still, the allegation should be treated with respect—Maskhadov was an experienced observer of how Berezovsky and other members of the Yeltsin clan operated in Chechnya over the years.

It is hard to believe that Prime Minister Putin was behind the bombings. While it is true that the blasts, more than any other event, ensured Putin's victory at the polls, there is nothing in the man's past to indicate that he would commit such a monstrous crime to gain power. On the contrary, Putin's past career betrays an unusual dedication to a fixed code of conduct (albeit an authoritarian one); there is nothing to suggest the bottomless cynicism necessary to massacre one's own people to promote one's career. If the explosions were organized by the Russians themselves, it is likely that they were the work of maverick members of Putin's camp. The new prime minister, after all, did not control all the levers of power in September 1999. He relied largely on the independent political bosses such as Berezovsky, acting for his benefit or in his name.

Some Russian newspapers speculated that Berezovsky may have been behind the bombings. If Berezovsky had arranged the September blasts, the crime would bind Putin to him forever. Even if Putin had known who was behind the explosions, he would not have been able to say anything at the time. The new prime minister was just finding his bearings in the Kremlin; a shattering revelation such as that the terrorist bombings had been organized by the Kremlin clique would have destroyed both Putin's political hopes and the Russian war effort against the Chechens. Neither could Putin reveal Berezovsky's role in the bombings (if in fact the tycoon was involved) later, after winning the presidential election, since he would have to admit that he had initially covered up the crime. Hence, if Berezovsky was in any way involved in the bombings, this secret would remain an iron bond attaching Putin to Berezovsky.[36]

But all of this is speculation. There is simply too little evidence either way. The most likely explanation is that the attacks were in fact carried out by Chechen militants or by Islamic extremists acting on behalf of their embattled coreligionists. Both Shamil Basayev and other commanders such as Salman Raduyev had carried out terrorist assaults against the Russian civilian population in the past and had boasted of their exploits. The Wahhabi commander, Khattab, was linked to the notorious international terrorist Osama bin Laden. Chechen field commanders gloried in a murderous ferocity toward their foes. These men publicly executed Russian prisoners of war and civilian hostages by cutting off their heads with large hunting knives, and videotaped the procedure. Clearly, there were plenty of candidates in the bowels of the Chechen underworld capable of carrying out the 1999 apartment bombings.

In the aftermath of the blasts, the Russian war effort swung into high gear. No longer was it a matter of a border war or mere punitive strikes against terrorist training camps. The old Stepashin plan of a limited war, with Russian forces waging a quick, relatively bloodless campaign to secure a defensible border on the Terek River, gave way to an all-out drive to subdue all of Chechnya. More than 100,000 Russian troops converged on the small country. Within weeks, they had passed the Terek River and had surrounded Grozny.

The Russian war effort was characterized by several idiosyncrasies. Although the stated enemies of the Russians were Shamil Basayev, Khattab, and other extremist warlords, it was that part of Chechnya loyal to the moderate president, Aslan Maskhadov, that was attacked first. The extremists were left mostly untouched. Two obvious targets for Russian warplanes, for example, were the Grozny homes of Shamil Basayev and Khattab, which the warlords had converted into command centers. These houses should have been targeted on the first day of war, yet Russian airplanes ignored them in favor of other objectives such as an old propeller plane sitting on the tarmac at Grozny airport or the central marketplace of the city. A *New York Times* correspondent even visited Basayev in his home at this time and published an article in which she described the house as swarming with the warlord's lieutenants. Yet, it was only several weeks later that Basayev's house was bombed, by which time the terrorist commander had long since disappeared into the bunkers of Grozny.[37]

For all the claims by Russian political and military leaders that this would be a "smart" war, with minimal casualties, the conflict soon degenerated into the same kind of war of attrition that had characterized the First Chechen War. The rapid Russian advance through northern Chechnya was followed by a horrific battle for Grozny—a relentless series of Russian assaults against well-fortified

Chechen positions in the ruined city. The carnage in Grozny lasted for months, before the rebels finally abandoned the city and fled to the mountains of the south; thousands of people on both sides were killed, but the war continued.

The Second Chechen War only solidified the hatred felt by many Chechens toward the Russians. Russia was now facing a terrorist threat for the foreseeable future—a threat that would force the country to tighten domestic security arrangements and become more of a police state. I asked Oleg Sysuyev, a former deputy prime minister and an adviser on Putin's campaign staff, whether the continuing problem of terrorism would harm Russia's nascent democracy. "Undoubtedly," Sysuyev replied. "Such strong support of Putin, mostly connected with the war, is a worrying phenomenon. It will be difficult to avoid the recurrence of terrorist incidents and it will be difficult to avoid very harsh countermeasures by the authorities, which will undoubtedly harm democracy."[38]

Aluminum Magnates

Berezovsky felt sufficiently at ease under the new administration to embark on one of the biggest business takeovers of his career. In the autumn of 1999 he began preparations for assuming control of some of Russia's largest aluminum companies. The purchases were announced on February 11, 2000. The oil company, Sibneft, disclosed that certain "shareholders of Sibneft" had purchased controlling stakes in the Bratsk and Krasnoyarsk aluminum smelters—the two largest in the country. On the same day, a spokesman for the owner of Russia's fifth-largest aluminum smelter, Novokuznetsk, announced that a controlling stake in the smelter had been bought by Logovaz. At a stroke, Berezovsky, Abramovich, and certain other partners had taken over two thirds of the Russian aluminum industry. It was a huge prize. Russia was the second-largest aluminum producer in the world, after the United States. Aluminum was one of the country's top foreign exchange earners. It was also one of the most gangster-infested parts of Russian industry.

All three of the aluminum smelters bought by Berezovsky and his allies had been controlled by the Trans-World Group, an international trading company that had risen from obscurity in 1991 to dominate Russia's aluminum industry. Like Logovaz in its various commercial forays, Trans-World somehow always managed to find itself in the center of the roughest spots of the Russian metals business and emerge victorious.

The key figure in the Trans-World Group was Lev Chorny, a trader from Tashkent, Uzbekistan. Crippled from a childhood bout with polio, Lev Chorny joined his older brother, Mikhail, in a number of commodities trading deals in the early 1990s, including timber export. In 1991, the Chorny brothers hooked up with David Reuben, a small international metals trader based in London. The men set up a loose partnership. The Chornys would manage the procurement of aluminum in Russia; their interests abroad would be represented by a Monte Carlo–registered company called Trans-CIS Commodities. David Reuben, through his London company, Trans-World Metals Ltd., would sell the aluminum to foreign buyers. Overall, the partnership would become known as the Trans-World Group. Its business was founded on barter arrangements. Trans-World would provide aluminum smelters with raw materials (principally refined alumina) and would lend the smelters money to produce the aluminum; it would be paid with finished aluminum for export. Trans-World managed to obtain most of the best barter contracts in the Russian aluminum industry. By 1994, it owned majority or blocking stakes in Russia's biggest aluminum smelters and controlled two thirds of the country's aluminum production; it also owned significant stakes in other large Russian metals companies. In terms of metal produced, Trans-World could claim to be the third-largest aluminum company in the world.[39]

The key to Trans-World's success was the Russian side of the operation, or, more specifically, Lev Chorny and Trans-CIS commodities. Chorny had good political protection in the Kremlin. His partner in Russia was Vladimir Lisin, former deputy to First Deputy Prime Minister Oleg Soskovets. Top Kremlin patronage did not prevent the

Chorny brothers from being dogged by allegations of organized-crime associations, however. In 1995, for example, the MVD announced that it was investigating Trans-CIS for involvement in the Great Chechen Bank Fraud of 1992–93; the MVD claimed that Trans-CIS got its first capital by using fictitious banks in Chechnya to defraud the Central Bank. No corroborating evidence was found (the Central Bank archives in Grozny were destroyed during the First Chechen War) and no charges were ever brought against either of the Chorny brothers.[40]

The one prize that seemed to evade the Chornys was the giant Krasnoyarsk Aluminum Smelter. Trans-World had had a barter contract with the smelter and had bought a 20 percent stake in it, but in the autumn of 1994 the smelter's general director turned against the brothers, canceled the barter contract, and attempted (unsuccessfully) to erase their stake from the share register. Faced with the prospect of losing one of their most lucrative export contracts, the Chornys fought back. The key personality at Krasnoyarsk was not the general director, but a thirty-three-year-old entrepreneur named Anatoly Bykov. A former athletics trainer, Bykov had formed a much-feared association of sports clubs in Krasnoyarsk in the late 1980s. He soon went into business, taking control of local hotels, casinos, and car dealerships. But Bykov's biggest prize was the Krasnoyarsk Aluminum Smelter. The sportsman had started a metals trading company in the summer of 1992 and over the next five years would relentlessly proceed to gain total control of the smelter. The battle for Krasnoyarsk Aluminum was exceptionally bloody, even by Russian standards; at least five top executives were assassinated gangland-style.[41]

Some of the most famous murders connected to Krasnoyarsk occurred in 1995. That year, Yugorsky Bank, based in an imposing new skyscraper in central Moscow, decided to move beyond its traditional oil-finance business and enter into a partnership with Krasnoyarsk Aluminum. In early 1995, Yugorsky Bank hired Vadim Yafyasov as vice president. Yafyasov had worked in the Ministry of Metallurgy, in a department charged with granting metals export licenses. Shortly after joining Yugorsky Bank, Yafyasov was appointed deputy general director of Krasnoyarsk Aluminum. On the evening of April 11, 1995, Yafyasov was shot dead in his car as he was arriving home. Three months later, it was the turn of Yugorsky Bank's president, Oleg Kantor; hitmen ambushed Kantor's car, first killing his bodyguard with a pistol shot to the head and then murdering Kantor himself, stabbing him repeatedly and finally eviscerating him. It is unknown to this day who killed the Yugorsky Bank executives and why.[42]

Krasnoyarsk had another relationship that promised independence from the Chorny brothers—an association with a New York–based metals trading firm called AIOC. This outfit was run by a thirty-four-year-old South African named Alan Clingman; it rose from obscurity to become one of the most successful traders in Russian aluminum, copper, nickel, zinc, steel, ferroalloys, coal, and precious metals. "I'm making close to 100 percent on my equity," Clingman boasted to me in 1994. By 1995, AIOC had a particularly fruitful relationship with Krasnoyarsk Aluminum. It owned a small equity stake in the smelter and, more important, it had the barter contract for Krasnoyarsk's exports. Late in the summer of 1995, AIOC's point man in Russia, a trader named Felix Lvov, booked a flight abroad. He arrived at Sheremetevo Airport, passed through customs and passport control, and waited to board his flight. He was approached by two men who identified themselves as law-enforcement agents and asked him to follow them. Several days later, a man waiting for a bus on the highway a few miles from the airport stepped into the woods to relieve himself and discovered Lvov's decomposing body. The murder remains a mystery. But AIOC never recovered. It lost the Krasnoyarsk barter contract and less than a year later declared bankruptcy and disappeared.[43]

Trans-World, meanwhile, regained its position at Krasnoyarsk. By 1998, it controlled the smelter in partnership with Anatoly Bykov. Still, the Chorny brothers continued to be hounded by law-enforcement authorities. On February 21, 1997, Minister of Internal Affairs Anatoly Kulikov made an extraordinary appearance in front of the Duma and declared that Russia should take steps to protect the strategically important aluminum industry from organized crime. He stated that almost all the deals concluded by the Bratsk and Krasnoyarsk smelters are controlled by mobsters and reported that he had asked the prosecutor-general's office to forward the results of its investigations of Lev and Mikhail Chorny. "Mafiya structures are monopolizing the market, destroying it, and turning economic relations into relations between criminal syndicates," Kulikov declared three weeks later in a letter to *Kommersant.* Again, as in 1995, the MVD's investigations into the Chorny brothers' activities went nowhere and no charges were ever filed.[44]

After such a tumultuous but profitable foray into the Russian aluminum industry, why did Trans-World Group decide to get out in 1999? Neither Lev Chorny nor David Reuben commented to the press. But it appears that Trans-World's cozy arrangement in Russia was coming under pressure. Its primary partner at the Krasnoyarsk Smelter, Anatoly Bykov, was in jail in Hungary. Trans-World had lost control of the crucial Achinsk Alumina Combine, and the Putin government was raising the tax burden on the barter arrangements that had been the key to Trans-World's profits in the past. There had long been talk of a rift between Trans-World's London-based chief, David Reuben, and the Russian chief, Lev Chorny. (Lev's brother, Mikhail, had already gone off on his own with the Sayansk Aluminum Smelter.) Reuben's operation in London, striving to stay clean, had never delved into the details of how operations in Russia were run—it was merely the Western marketing arm. In this sense, it was probably Chorny who decided to split, exchanging an increasingly ineffective partner (Reuben) for more politically connected ones (Berezovsky and Abramovich).[45]

Partnership with Berezovsky promised to give Lev Chorny a new lease on life. Chorny had lots of liquid capital but little political influence (especially after Soskovets was fired in the summer of 1996). Berezovsky, for his part, was perpetually strapped for cash (his great wealth lay primarily in nonliquid assets), but he could provide the political protection Chorny desperately needed. Berezovsky and his partners thus gained control of one of Russia's biggest sources of foreign exchange revenues. Berezovsky also had the added pleasure of striking a blow against his old rival Anatoly Chubais, who was now chief of Russia's electricity monopoly, UES. Chubais was allied with Russia's second-largest aluminum company, Siberian Aluminum. Helped by rate hikes initiated by Chubais's electric companies, Siberian Aluminum had long been moving onto Trans-World's turf. Now Siberian Aluminum was beaten to the punch.[46]

While the aluminum business seemed to be an unexpected detour for Berezovsky, there were certain signs that he had been involved with aluminum before. A 1993 oil-and-gas yearbook, compiled by a Los Angeles–based trade publication and the big accounting firm Ernst & Young, listed the major commodity export contracts fulfilled in the first years of the Yeltsin regime; the yearbook listed Logovaz as having exported 840,000 tons of aluminum in 1991–92. This was a massive shipment—worth about $1 billion at the time. I was never able to confirm this shipment with any other source; if Logovaz really did export such a large amount of aluminum, it seems likely that it was merely lending its name and Special Export status to a deal concocted by someone else. In any case, Logovaz's Special Export status was withdrawn two years later and Logovaz was nowhere to be seen in the aluminum business for the duration of the 1990s.[47]

Berezovsky, however, maintained close relations with the key leader of the Trans-World Group, Lev Chorny. Over the years, the career paths of Berezovsky and Lev Chorny seemed to pass at the same time through the same places. In 1992–93, for instance, when Berezovsky was establishing contact with Chechen organized-crime groups for protection in the auto market, Trans-CIS Commodities was allegedly using Chechnya as an offshore registration center to perpetrate bank fraud. In 1994–95, Berezovsky and Chorny shared the same political patrons: First Deputy Prime Minister Oleg Soskovets and security chief General Aleksandr Korzhakov. In 1997, Logovaz and Trans-World found themselves allies in the fight against Vladimir Potanin's Onexim Bank in the loans-for-shares auctions. In 1998, Lev Chorny was said to have participated in Berezovsky's bid for a telecommunications company, PLD Telecom. In 1999, both men were widely mentioned as collaborating in the takeover of the influential newspaper *Kommersant*. Still, while the career paths of Berezovsky and Chorny seemed remarkably synchronized in many respects, it was impossible for outsiders to know the exact relationship between the two men.

After he'd taken control of the automobile, television, airline, and oil industries, Berezovsky's entry into the aluminum business represented his fifth major foray into business. Evidently he felt he had a friend in Vladimir Putin. In many ways, Berezovsky had served as the new prime minister's patron. He had played an important role in Putin's elevation as head of the FSB in the summer of 1998. He had played an even more important role in Yeltsin's choice to replace Stepashin with Putin as prime minister a year later. He was also playing a key role in the media campaign building up Putin's popularity in the run-up to the elections. But none of these factors guaranteed Berezovsky immunity from prosecution.

Legal Immunity

The wave of nationalism unleashed by the Chechen War changed the political landscape for the 1999 parliamentary elections. The pro-Kremlin party, Unity, grew from complete obscurity into one of the most popular political organizations in the country. At the same time, Luzhkov and Primakov's Fatherland-All Russia Party, which a few months earlier seemed guaranteed to win the parliamentary elections (and the presidency afterward), sank relentlessly in the opinion polls.

Berezovsky played a key role in these developments. His TV network, ORT, may not have been the only one in the country, but it was still the dominant source of news and information for most of the population. Generations of Russians had been conditioned to swallow the official Party line served up by Channel 1. Berezovsky made sure that the network supported the Chechen War, glorified Putin, and promoted the prime minister's party. At the same time, ORT's star news anchor, Sergei Dorenko, began a weekly roundup of the news, broadcast every Sunday evening, which he used to excoriate the Luzhkov-Primakov alliance. Luzhkov, the pugnacious Moscow mayor, was accused of incompetence, corruption, and murder. It was more difficult to find dirt on the former prime minister, Yevgeny Primakov, so the Kremlin publicity managers attacked him on his age— one set of billboards in Moscow and other cities pictured nothing more than a wheelchair. The pro-Putin propaganda campaign was extremely effective and, as in 1996, it was Berezovsky who masterminded the key strokes (with the help of the anchorman Dorenko). "Dorenko alone was able to determine the results [of the parliamentary elections] in December," Putin's campaign aide Oleg Sysuyev would tell me later. "This is an exaggeration, perhaps, but it has a large measure of truth."[48]

Berezovsky had decided to run for a seat in the Duma himself. Perhaps it seemed strange that the master of Kremlin intrigue wanted to dedicate time and money to become an elected member of parliament, but Duma membership had one very alluring advantage: It conferred immunity from prosecution. Rather than holding its elected officials to higher standards of behavior than the average citizen and subjecting them to even greater legal scrutiny, the Russian constitution placed the parliamentarians above the law.

As a result, the most outrageous criminals found shelter from the law inside the Duma. There were gangsters like Sergei Skorochkin, who had actually killed people while serving as a parliamentary deputy. Skorochkin admitted to gunning down two men in his constituency (he claimed they were gangsters putting the squeeze on him); but he had parliamentary immunity and was never investigated. Several months later he was murdered. And there was Sergei Mavrodi, who had been jailed for defrauding millions of investors with his MMM pyramid scheme. During the 1995 parliamentary elections, the MVD announced that at least 83 candidates standing for the 450 seats were ex-convicts. Among the elected members of parliament were murderers, pimps, casino magnates, and drug dealers. Not surprisingly, the Duma became the scene of mob-related violence: at least three parliamentary deputies were murdered, as were more than a dozen staffers.[49]

Alongside earnest and legitimate parliamentarians, there was a tremendous array of charlatans and criminals. The Duma became a huge barn, beneath whose roof all sorts of disreputable figures bayed before the TV cameras. No man epitomized the spirit of the parliament more than the nationalist buffoon Vladimir Zhirinovsky. The Russian economy may have been in ruins and Russian society torn apart by lawlessness, but at least the country had freedom of speech. Zhirinovsky was a great speaker. Russians loved watching his antics— throwing a glass of water at the young reformer Boris Nemtsov, scuffling with a female parliamentarian, pretending to throw punches over the shoulders of friends restraining him. The next day he would be threatening Germany and Japan with nuclear annihilation or accusing the head of the FSB–KGB, Sergei Stepashin, of being an agent of the Mossad. Zhirinovsky's role in Russian politics was only to entertain. He was there to show all Russians that parliamentary democracy was a joke. During the 1999 campaign, when Zhirinovsky was asked why he had so many reputed gangsters on his party's electoral list, he answered that he was glad to have them, because "the leaders of the shadow economy represent the real power in the country."[50]

Not surprisingly, with rare exceptions, such as its refusal to ratify Chernomyrdin as prime minister in September 1998, the Duma proved a remarkably malleable institution. Often, it failed to assert even the meager powers it was entitled to under the Yeltsin constitution. In most of the key votes, such as the ratification of Yeltsin's prime ministerial candidates or the defeat of the impeachment measures, the Kremlin managed to obtain a majority by bribing or blackmailing dozens of lawmakers. The Duma may have been useless in defending the interests of Russia, but it proved highly effective in defending its core constituency: Russia's new criminal class. Over the years, the parliament did nothing to fight against organized crime. It was not until 1997—four years after the beginning of the mob war— that the Duma finally passed an organized-crime bill that gave the authorities the legal weapons to fight racketeering. A money-laundering bill was passed only in the summer of 1999, but was vetoed by President Yeltsin.[51]

In his run for the Duma in the autumn of 1999, Berezovsky chose an obscure constituency: the autonomous republic of Karachaevo-Cherkessk—an impoverished corner of the North Caucasus, with 300,000 inhabitants. The native Karachai people had long retained fraternal relations with the Chechens. Like the Chechens, the Karachai had been deported en masse by Stalin because of their alleged collaboration with the Nazis. The governor of Karachaevo-Cherkessk was Anatoly Semyonov. An ethnic Karachai, Semyonov had served as the Russian land forces commander during the First Chechen War. Several months after the end of that war, General Semyonov was cashiered, pending an investigation that he had approved the sale of Russian armaments to the Chechen militias. Though Russian commanders were commonly accused of selling arms to the Chechens during the First Chechen War, the specific charges against General Semyonov were never proved and the case was dropped.[52]

On December 19, with the support of General Semyonov, Berezovsky easily won his seat in the Duma and received his long-coveted parliamentary immunity. The Russian prosecutor-general's office continued to pursue its case against Berezovsky, but the tycoon could now be certain that the case would not come to trial unless the Duma voted to lift his parliamentary immunity. Berezovsky's partner in Sibneft, Roman Abramovich, had also won a seat in parliament, representing the constituency of Chukotka, a frozen wasteland opposite Alaska, notorious as Russia's most impoverished and primitive region.

The December parliamentary elections were a gigantic triumph for the Kremlin. The recently formed Unity Party garnered a surprising 23 percent of the vote, just behind the Communists, who won their usual 24 percent. The only two real opposition parties, Luzhkov and Primakov's Fatherland-All Russia and Yavlinsky's Yabloko faction, performed terribly, winning 13 percent and 6 percent, respectively.

Putin's triumph over his political foes encouraged President Yeltsin to take the step his entourage had long been considering. On

December 31, in his traditional New Year's Eve address to the nation, Yeltsin announced that he would not serve out the remainder of his term; he was resigning the presidency immediately and handing over his powers to Vladimir Putin. The presidential election would be moved up by three months, to March 26, 2000.

Yeltsin's abrupt resignation made a mockery of the spirit of Russia's democratic constitution. He had not violated the law, but the cynical political calculation behind the move was clear. The premature ending of his presidency caught the opposition off-guard, while it was still reeling from its defeat in the December parliamentary elections. The war in Chechnya had not yet degenerated into a bloody mess, harming Vladimir Putin's popularity. Finally, Yeltsin's resignation meant that Putin had the advantage of running for president as an incumbent. The transfer of power seemed less democratic than monarchic—a reigning king resigning in favor of his designated heir.

Yeltsin's farewell speech was marked by an extraordinary apology. "I want to ask for your forgiveness," he told his audience, "for the fact that many of the dreams we shared did not come true and for the fact that what seemed to us so simple turned out to be tormentingly difficult. . . . The pain of each of you has called forth pain in me, in my heart—sleepless nights, tormenting worries about what needed to be done so that people could live better and more easily."

Several days later, in an unusual display of Christian piety, Yeltsin and his family traveled to Jerusalem to celebrate Orthodox Christmas in the Holy Land. Berezovsky went along as well.

The same evening that Yeltsin made his resignation speech, Vladimir Putin, the new acting president of Russia, signed a decree granting Boris Yeltsin and his family an array of benefits and privileges, including immunity from criminal prosecution.

EPILOGUE

Berezovsky's most destructive legacy was that he hijacked the state for his private interests. In other countries, powerful businessmen lobby the government to advance their interests. Berezovsky took control of the men who ran the government and forced the state to feed his business empire.

He and other crony capitalists produced no benefit to Russia's consumers, industries, or treasury. No new wealth was created. Berezovsky's business ventures involved taking over existing enterprises that already were either highly profitable or exceptionally well-endowed with resources. He did not make these companies healthier or more competitive; on the contrary, they stagnated or declined under his tutelage. His takeover of the TV network ORT did not produce any improvements in programming or production efficiency; it remains today an uninspired channel that most probably would be bankrupt if not for massive state subsidies. His takeover of oil giant Sibneft did not produce significant improvements in productivity or financial management. Aeroflot, in spite of its monopoly position in a booming market, expanded lethargically and was beset by cash-flow difficulties. The grandiose investment project Avva collapsed ignominiously. The only major bank with which Berezovsky was involved—SBS-Agro—collapsed as well, leaving depositors with no means of withdrawing their money. Berezovsky's business career has been dependent on the corruption of government officials or state industrial managers. The only significant business enterprise he created from scratch was the network of Logovaz car dealerships. But even here, his profits were based not on providing superior service or finding a new market niche, but on the collusion of Avtovaz managers in providing him with cars at a price lower than their production cost.

If Berezovsky was incompetent as a business manager (in the Western sense of the term), as a master of political intrigue he was second to none. From the time he first entered Boris Yeltsin's inner circle, in 1993, Berezovsky proved unstoppable in his rise to power; no other operator in the Kremlin matched his combination of creativity and ruthlessness. Berezovsky masterminded two successful presidential election campaigns (in 1996 and 2000) to put his chosen candidate on the throne. He survived corruption scandals, a car bomb, murky Kremlin intrigues, a murder investigation, even an arrest warrant. Time and again, his opponents breathed a sigh of relief, believing that he was finally finished, but he outmaneuvered them and saw them destroyed.

Berezovsky's success both in making money and in claiming to be a valuable statesman in the government's service was due in part to his relationship with some of Russia's strongest gangsters. On several occasions, he made his way between warring organized-crime groups and somehow emerged the winner. In the end, except for the need to maintain relations with the Chechen gangsters who had emerged as that country's rebel leaders, Berezovsky managed to separate himself from the sordid world of mob turf wars. He was a man of big ideas, drawing up a grand strategy and leaving his subordinates and intermediaries to deal with the execution.

"It is truly impossible in many instances to differentiate between Russian organized crime and the Russian State," observed U.S. Congressman Benjamin Gilman, opening the special hearings on Russian organized crime in April 1996. "Many aspects of the Russian State are virtually a full-fledged kleptocracy—dedicated to enriching those in power and their associates."[1]

When FBI director Louis Freeh was asked in those same hearings whether a "permanent criminal culture" was emerging in Russia, he replied: "The jury is probably still out on that. I think they are at an obviously critical time... as to who is really in charge—the criminals or the democratically elected officials."[2]

Freeh was speaking diplomatically. Organized crime in Russia was not some kind of shadow government—it lay at the core of the Yeltsin regime. And what of Boris Yeltsin—the giant who slew Communism and ended decades of Soviet imperialism? His legacy is one of almost unmitigated failure. The elections that he presided over were free but not fair. The market was free but not open to outsiders. The freedom Yeltsin brought was the freedom enjoyed primarily by a handful of political bosses and crony capitalists. For the Russian people the Yeltsin era was the biggest disaster (economically, socially, and demographically) since the Nazi invasion of 1941.

To the extent that Russia has free speech and democratic accountability, these principles were largely established in the late Gorbachev years. Yeltsin was supposed to further the aims of glasnost and perestroika, but neither freedom of speech nor democratic accountability was expanded in any significant way. Democracy became a curse word—to be called a democrat became synonymous with being labeled a crook. The two concepts that were supposed to lead Russia to a Western-style future—privatization and democracy—were discredited. On the streets of Moscow, people began to speak of privatization as "grab-it-ization" (*prikhvatizatsiya*) and of democracy as "shitocracy" (*dermokratizatsiya*).

The failure of democracy and capitalism in Russia in the 1990s will have far-reaching consequences. Russia has always been one of the world's great ideological battlegrounds. In the nineteenth century, it was the standard-bearer for the principle of absolute monarchy and the divine right of kings against the principle of republican government. In the early twentieth century, Russia was the birthplace of Communism and totalitarianism. In the 1990s, the "American model" was tested on Russian soil. After the fall of Communism, the United States was left as the world's only superpower and faced the historic challenge of exporting its ideology to its former geopolitical rival. The American model had political, economic, social, and cultural components. Could it work in a country as large and as old as Russia? The history of the Yeltsin regime suggests that it could not. It is unlikely that the United States will again have such an opportunity to establish its system of values in Russia. In 1991, when Yeltsin came to power, Russia was virtually a blank slate in ideological terms. Communism had been discredited. No other ideology had surfaced to take its place. Russians were looking to America. They wanted to be allies, friends; they wanted to emulate America. The introduction of American values was the underlying, unspoken justification for the reforms of the Yeltsin regime. When these reforms failed, the infatuation with America died as well.

Perhaps the problem is that the American model, both as presented by American policy makers and as understood by the Russians, was a perverted version of the real thing. Western policy makers and advisers mistakenly believed that all it would take to propel Russia from Communism to a Western-style future was to dismantle the old state command system and open up the economy to free markets and private ownership. They overlooked the need to prepare the state and society for this change.

Private property or free markets alone do not guarantee a high level of civilization. Even the most impoverished countries have private property and free markets. What they lack is a healthy state and a healthy society. Today, these are the two essential preconditions for civilization.

There are several salient characteristics defining a healthy state: a good legal code and the means to enforce it; the equality of all citizens before the law and the state; a sound financial basis allowing for the provision of such public goods as national defense, law enforcement, transportation, education, medical care, and pensions; an efficient and effective government apparatus. A healthy state is uncorrupted by wealthy individuals, powerful businessmen, or special-interest groups; it is an honest broker for all the conflicting interests of society. Finally, a healthy state protects the weak from predation by the strong.

A healthy state should not be confused with a powerful state. The U.S.S.R. was powerful but not healthy. Its power was based on fear, arbitrary rule, bureaucracy, corruption, lawlessness, and the absence of independent local governments or civic organizations. The Soviet Union's terminal illness was the result of its failure, despite massive propaganda, to encourage a sense of duty or civic responsibility among both the masses and the elite. It failed to produce citizens. Those who believe that a healthy state means a strong central government forget that the central government is merely the top of the pyramid. The foundation is the network of local governments and independent civic associations competing with the central government in addressing local and national needs. Without a broad foundation of such local and civic institutions, a strong central government is a fragile structure—a tall tower on a shallow base. The Soviet Union was such a structure. Over the course of seven decades, the Communist dictatorship destroyed churches, elected local governments, independent trade unions, professional associations, charitable organizations—all independent institutions, in other words, that could challenge the Communist Party's monopoly on power. In the end, the U.S.S.R. collapsed because of the hypertrophy of state power.

A healthy society can be defined by the strength of its system of values—a factor as important as it is difficult to measure. The nation that Boris Yeltsin inherited lacked the values that are the foundation of prosperity and democracy. How can private enterprise flourish when society is suffused with envy? How can the economy grow when the value of honest work is universally derided? How can democracy flourish when no one wants to take responsibility for the common good? The pervasive nihilism in Russia is the result of the Communist regime's destruction of such key building blocks of a healthy society as family, religion, and independent civic association. Boris Yeltsin's Russia was not a nation of citizens, but a mass of fractured families and isolated individuals. Russians were subjects, not citizens.

The fact that Boris Yeltsin inherited an unhealthy state and an unhealthy society made it difficult for his reforms to succeed. Yet, Yeltsin and his ministers did little to address the problem. The Russian state grew more corrupt, more inefficient, more arbitrary in its exercise of power. Russian society grew even sicker than it had been under the Communists; there was a decline in both family values and the sense of civic responsibility. The disregard for the value of other human beings, already rampant under Communism, deepened under Yeltsin's watch. Often it seemed that Russia's perverted value system rewarded any activity that victimized one's neighbor; this criminal mind-set became so dominant that adhering to principles of honesty, decency, or law-abidance became equivalent to moral dissent.

In the absence of either a healthy state or a healthy society, the application of such Western liberal principles as privatization and free prices could only precipitate Russia's destruction. While Chubais and other young reformers naively followed a lopsided version of the American model in macroeconomic policy (neglecting the role of good government and healthy social values in breeding America's success), Russia's businessmen were guided by a perverse understanding of American capitalism on the microeconomic level. Whenever I asked Russia's business magnates about the orgy of crime produced by the market reforms, they invariably excused it by pointing to the robber barons of American capitalism. Russia's bandit capitalism was no different from American capitalism in the late nineteenth century, they argued.

Communist propaganda had always maintained that making money in a free market was a purely predatory and criminal activity. Soviet schoolchildren had been taught that the United States, as the paragon of capitalism, was controlled by a ruthless, superrich elite; they were taught that all the great financial and industrial empires powering the American economy had immoral origins—behind every fortune was a legacy of theft, lies, even murder. The American captains of industry were little more than crooks and criminals. Russia's new business magnates had all absorbed this image of Western capitalism in school; when they went into business, they acted accordingly.

"Perhaps our criminals are the most powerful people in the country today, but this is just a phase," they would say. "Just like America. Look at all your great capitalists—Rockefeller, Ford, Carnegie, Morgan: They all started out as criminals as well."

The old American robber barons may have bent the rules occasionally, but they were neither criminals nor looters. On the contrary, the robber barons, whatever their moral flaws, helped turn the United States into the strongest economic power in the world. They built the railroads that opened up the country. Carnegie built the world's largest steel industry. Rockefeller created the world's largest oil industry. Ford invented a way to mass-produce automobiles for the American middle-class consumer. Morgan financed America's industrialization and turned Wall Street into a market where small investors would not be defrauded. No, Berezovsky and his colleagues could in no way be compared to the robber barons of American history.

In its scale and rapaciousness, the looting of the state that took place during the Yeltsin regime was unprecedented—it was, perhaps, the robbery of the century. At the root of the disaster was the Russian penchant for playing the double game, for pursuing an essentially dishonest policy. Rather than consistently following a set of clearly defined principles, the Russian mentality was to say one thing and do another. This tendency to play the double game was evident in the KGB's willingness to finance organized-crime groups and the new commercial banks in the 1980s in hopes of controlling them and prolonging the existence of the Soviet Union. It was also evident in the Yeltsin regime's sponsorship of a handful of crony capitalists in hopes of using them to create a genuine free-market economy. It was evident in Russia's long, tangled relationship with Chechnya.

The West, too, betrayed a penchant for the double game. The fact that the Yeltsin regime had turned into a gangster state was often blithely dismissed; Russia's lawless market was described as "raw capitalism" or "frontier capitalism," with the implicit analogy to the American nineteenth century. The Clinton Administration, in particular, while trumpeting the principles of democracy and the free market, repeatedly ignored evidence that the Yeltsin regime was a kleptocracy. In 1998, a top Russia analyst at the CIA told *The New York Times* that the Clinton Administration routinely discouraged reports about the corruption of the Yeltsin regime. One such report about Prime Minister Chernomyrdin (said by the CIA to have amassed a personal fortune of $5 billion by 1996) was returned by Vice President Al Gore with a "barnyard epithet" scribbled across it. This self-delusion on the part of the Clinton Administration was confirmed in a 1999 article by Fritz Ermarth, a veteran CIA Russia hand and former chairman of the National Intelligence Council. Ermarth spoke of American policy makers' "disdain for analysis about the corruption of Russian politics and their Russian partners." Ermarth attributes this primarily to the "warping of intelligence analysis to fit political agendas" and to "a cynical Washington habit of...preserving the image of a foreign policy success."[3]

The U.S. government's repeated praise of the Yeltsin regime as "democratic" and "reformist" damaged the liberal principles on which Western societies are based. The issue came to a head during the 1996 Russian presidential elections, when the Clinton Administration was faced with a choice of supporting either Yeltsin or the Communist candidate, Gennady Zyuganov, who was said to represent a return to the Cold War past. There was no reason for the United States to favor either one. When one is faced with a choice between two evils and not compelled to choose either one, the correct course of action is to abstain. But the Clinton Administration abandoned the stated U.S. policy of staying aloof from other countries' political processes and threw its weight behind Yeltsin, promoting his campaign with both rhetoric and money.

Berezovsky's career in the 1990s undoubtedly was very exciting: All around him, history was being made—Communism was destroyed, the Soviet Union fell apart, democracy and free markets were proclaimed, huge fortunes were acquired. But what was left at the end of it all? Russia was ravaged and destroyed. Millions of Russians died premature deaths. Most of Berezovsky's companions of the road ended up as nonentities, despised by their survivors. Eventually, of course, Russia's era of self-destruction will draw to a close and the nation will undertake the difficult task of rebuilding. Vladimir Putin may well be the man to accomplish this task. But first he will have to deal with the corruption and crony capitalism epitomized by Boris Berezovsky.

APPENDIX I

Boris Berezovsky Appeals to the President

Immediately after the murder of Vlad Listyev, on March 1, 1995, Moscow police fingered Boris Berezovsky as a possible suspect. Concerned that he was on the point of being arrested, Berezovsky went to the offices of General Korzhakov's SBP to record a personal videotaped message to President Yeltsin. He was accompanied by Irina Lesnevskaya, a top TV producer at ORT and a close friend of the Yeltsin family. This is a transcript of the videotape, which was recorded sometime in the first days of March 1995. The document is important not only for the facts it reveals, but also as an illustration of the subtle ways in which both Berezovsky and Lesnevskaya attempt to win the president's trust and support.

LESNEVSKAYA: Boris Nikolaevich [Yeltsin], I am turning directly to you.

I know who killed Vlad [Listyev].

After you signed the decree on the establishment of NTV—I believe it was in November of 1993—I was appalled. How could this be? Speaking with you on November 7, at your dacha, during the interview with [inaudible], you promised me that we would have an alternative television, an independent television, where all independent TV companies could come and produce independent programming. And suddenly, one TV company, supported by one bank, is given a whole TV channel.

I called Naina Iosifna [Yeltsin's wife]. I cried, I wept, I explained to her that you have again been misled. I don't know if this was the cause, or if something else was, or if someone said something to you, but you stopped your decree before it had a chance to be ratified by all the interested parties.

Within eight hours, I got a call from Igor Malashenko, the director of NTV, who said:

"Did you do this?"

I said: "I don't know if it was me or not, but I did have a part in it."

"Gusinsky wants to see you."

I said: "I am not acquainted with him. If he wants to, he can come to REN-TV."

Gusinsky snatched up the telephone and very politely, almost gently, said that he could not come to visit me and would be very grateful if all the independent TV companies could come to him.

That's how all of us—Vlad Listyev, I (Irina Lesnevskaya), my son, Dimitri Lesnevsky, Andrei Razbash, [unintelligible], all those who today present the programming of ATV, REN-TV, and Vid on Channel 1—arrived at Gusinsky's.

Over the course of forty minutes he ran around the room, shouting and screaming, shouting that he hadn't taken into account one stupid wench, that the institutions [of government] are working [on this], that he had been pushing through this decree for half a year and now the decree has been stopped.

He tried to frighten us. He insulted us. He told me: "You have one son. What could be more important for a mother than that her son should be alive and well? But on the 101st-kilometer mark of the Moscow highway, there are always some kind of incidents. He could be accidentally hit by a car, his car could turn over, burn up..."

In short, he frightened us so badly that we signed a joint agreement, our whole TV company, that we would be producing our programs for NTV and giving them to this channel.

It is true, our programs are actually transmitted on this channel, just as they are on NTV.

Exactly in a week's time, the decree on NTV was signed and entered into force.

Two or three weeks later, we got a call from a journalist acquaintance of ours who said: "Folks, you need a security service." I have to mention names, yes? Ivan Konnov, who works at NTV. Ivan Konnov called and said: "Folks, you need a security service. You never know, but Gusinsky will not forgive you. I will send you some lads."

Two guys came and told us that they are currently without work, that they used to work for RIA, that RIA has collapsed, that everything there has been looted, that they are ready to create a ready-made security organization for journalists, that this would not be expensive—$300,000 or $400,000. We gasped. We don't have this kind of money. They started painting scenarios: liquidate the liquidators, fatal injections, umbrellas [the Bulgarian dissident assassinated in London], and so on. Then he [*sic*] said: You have partners, I heard you're planning to incorporate Channel 1. Gusinsky has such an organization, while you are all undefended. How can this be? You have Berezovsky, you have someone else there, banks—put us together and we will establish a good organization.

Well, we said our goodbyes, apologized, said that we don't need this, that we don't play these kinds of games, that we're journalists, after all, and that we don't intend to establish an organization similar to Gusinsky's.

He came again, two weeks later—two of them, Sergei Sokolov and Sergei Kulishev—and said, folks, how can this be? and began again to tell us all kinds of things and asked us, if we didn't want to do this thing, if we could introduce them to Berezovsky, because Berezovsky would probably be interested in such an organization.

We did a very stupid thing, Boris Nikolaevich. We are gullible people. We brought these people over to Boris Abramovich Berezovsky, told him about them, introduced them, and left them to sort things out.

Two weeks later there was the explosion of the car at Logovaz. I don't remember the date, it's not important, it could have been three weeks later. In any case, Boris Abramovich took them on a temporary basis, to look at their work, study them, what kind of lads are they, what do they offer, what are they all about, and they started hanging around Logovaz. They were not signed up permanently.

Then the explosion took place. The perpetrators were never identified. The driver was killed, the bodyguard was crippled, Berezovsky was wounded. Our incorporation [of Channel 1] was stopped; all the documents gathered dust on the shelves. We thought that Berezovsky would never return after such a blow . . . These lads disappeared; we didn't see them anymore.

In September Berezovsky returned—it could have been August. We went to Logovaz and saw the two Sergeis, who are next to Berezovsky, already permanently employed in his organization. The haphazard process of the incorporation came to its logical conclusion and finally you signed the decree on the incorporation of Channel 1.

After Vlad Listyev was chosen . . . Often meeting with Sokolov and Kulishev at Berezovsky's, they were always complaining to me that he doesn't let them close to him, that he tells them to leave the room when he has visitors, that Berezovsky isn't giving them access to all the information and how are they supposed to protect him if he conducts himself in this way and is always visible. [They complained] that at Gusinsky's the organization was up and running, that they were trying to create one here, but that they were being blocked. And he [one of the security agents] was surprised that Vlad had been chosen general director and not me. Vlad was a such and such . . .

Well, to make a long story short, in order not to exhaust you, after the murder of Vlad, we swore that we would do everything to, first, find out who did it and, second, to take our revenge. How can we take our revenge? Only by calling a press conference, making a film, writing an article.

Now we have come to Korzhakov, because we understand that what has happened is a coup d'état.

This is worse than the White House [the battle over the parliament in 1993]. This is worse than the GKChP [the 1991 Putsch]. This is inside the city. An enormous organization has been created. It controls everything—all the Mafia organizations, all the criminals—decides who should live and who should die...

In short, I am certain that this [Listyev's murder] was done by these people [the two security agents], on orders from the Most Group [Gusinsky]. These people were infiltrated in our midst so that we could introduce them to Berezovsky.

The facts speak for themselves. Boris Nikolaevich, now everything must be done so that the case would be handled personally by Korzhakov and the FSB, above all not the police. Because currently a theory is being prepared and witnesses are already on hand to the effect that Vlad was killed by Berezovsky, and this will continue...

This theory appeared tonight.

[Aside] Wait, Sasha, turn the camera here. It's not important. This is for Boris Nikolaevich—he will understand that we are nervous.

When we were called into the prosecutor's office and asked questions, we understood that they have only one theory: that the murderer is either Berezovsky or his chief deputy, Badri Shavlovich [Patarkatsishvili], that all the evidence points only to Berezovsky.

But we have spent the whole past year all together! We understand that this is unrealistic and impossible! Therefore... Please continue.
BEREZOVSKY: Good day, Boris Nikolaevich.

You know that this assassination attempt that was carried out on the seventh of June of last year still has not been solved, although you gave a clear order to the minister [of internal affairs] to seriously undertake this task.

All the same, from time to time, I was visited by officers of the Regional Administration of the Moskvoretsky Region [RUVD—the regional administration for internal affairs], who proposed various theories.

The last time they came was not long ago. They warned me ahead of time that they would be coming—this was about three weeks ago. They came to me and brought one other person. My visitors were Sergei Lvovich Kozhivin and his assistant. They brought with them a person whom they introduced as a thief-professing-the-code [a gangster boss], and said that this person knew who had ordered my assassination the last time and that currently a new assassination was being planned.

In return for telling me who did this, who is planning it currently, they demanded a guarantee—not they, but this person who was the thief-professing-the-code—a guarantee in the form of payment and a guarantee in the form of protecting him against future actions that could cause problems for him.

All this was taking place in the presence of officers of the Moskvoretsky RUVD and they were the initiators of this meeting.

I said that I would have to think about it. I asked them what sum of money they were thinking about. They mentioned $500,000 and said it could be paid in installments. We agreed to meet once more, at which time I was supposed to give them their answer.

I had asked my security agents, Sergei Sokolov and Sergei Kulishev, to record the conversation onto an audiotape and a videotape, in order to have proof that I was being blackmailed. So this conversation was recorded on audiotape and videotape by Mr. Sokolov and Mr. Kulishev and we agreed to a follow-up meeting at which we would discuss the terms of the agreement and the sum of money I would pay.

For the following meeting they once again came to visit me. I had collected the money for the meeting—$200,000—and in the course of the bargaining session, also recorded on tape, in the presence of the RUOP [the regional police administration for organized crime] officers, Sergei Lvovich Kozhivin and one other officer, I handed over a sum of $100,000 to this person by the name of Nikolai who had been characterized by the RUOP officers as a thief-professing-the-code.

The transfer of the money took place on the twenty-fifth of February. A day before the murder—the first of March—early in the morning, I flew to London as part of our delegation led by [Prime Minister] Viktor Stepanovich Chernomyrdin. There, in the evening, I received news that Vlad Listyev had been murdered. I immediately took a charter flight and returned home.

Of course at that time I had absolutely no idea about the preceding events and Vlad's death.

On Saturday... Was it Saturday or not when you were flying off with Aleksandr Vasilievich [Korzhakov] and Mikhail Ivanovich [Barsukov]? Was it Friday? When was the holiday?

On Friday, when you... After the memorial service... While I was attending the memorial service for Vlad at Ostankino, I was called by my assistants and told that Logovaz is due to be searched and visited by the OMON [paramilitary police].

I was terribly surprised and turned to Oleg Nikolaevich Soskovets, who happened to be right there, asking him to help me contact the Administration of Internal Affairs or the Ministry of Internal Affairs so that I could understand what was happening.

Oleg Nikolaevich got in touch with Viktor Feodorovich Yerin [the minister of internal affairs], and Viktor Feodorovich assured Oleg Nikolaevich that this was all a misunderstanding and that nothing would happen.

When I arrived, after the memorial service, at about three o'clock in the afternoon, at the office of Logovaz, or rather the visitors' center of Logovaz, I found there armed men from OMON and officers who introduced themselves as the officers of RUOP. They presented a search warrant and a warrant empowering them to interrogate me in my capacity as a witness.

I asked them what they were planning to find in the Logovaz visitors' center and why they needed a search warrant.

To this question, after approximately one hour of discussions, they answered that they needed only the corporate charter of Russian Public Television [Channel 1]. Of course, this charter of Russian

Public Television was immediately presented to them, but all the same, it remained completely unclear why this had to be accomplished by a group of armed men and why it necessitated a search warrant.

After this, in the course of nine hours, they engaged in outright intimidation and blackmail. Ultimately, they demanded that we go to RUOP to give our statements there. Well, this whole affair was already smelling like a provocation.

I contacted [acting prosecutor-general Alexei] Ilyushenko, through an assistant of Aleksandr Vasilievich Korzhakov, since he was absent at the time—he was with you... I contacted Ilyushenko, Ilyushenko gave a direct order to his first deputy, Uzbekov, and Uzbekov immediately telephoned Logovaz and gave a direct order to the RUOP officers there not to take me and my deputy, Badri, to the police station, but to take our statements here, in place.

Still, they failed to execute this order of Mr. Uzbekov and this intimidation continued for nine hours.

LESNEVSKAYA: Logovaz was surrounded by armed...

BEREZOVSKY: In the course of all this time, Logovaz was surrounded by a group of armed men, and so on and so forth.

Now, yesterday, NTV aired a program, *Itogi*, where essentially or, shall we say, in thinly veiled fashion, I was identified as the person who both masterminded and executed the murder of Vlad Listyev. And correspondingly it became clear that this theory had been developed not recently—it was all prepared ahead of time. Already, according to other sources, there was a thief-professing-the-code sitting in jail who had made a declaration that my deputy, Mr. Patarkatsishvili, contracted for Vlad's murder and that only due to the fact that this thief-professing-the-code had been arrested by the law-enforcement agencies was he unable to fulfill the contract.

From this it becomes absolutely clear that all of this has been planned ahead of time, since this man had been arrested before Vlad's murder, and therefore the source of all these actions now becomes completely clear.

Boris Nikolaevich, I personally have no doubts that all this is a provocation.

Boris Nikolaevich, I would like to draw your attention to the fact that when you are absent from Moscow, when there is no one in Moscow who is close to you, Moscow sinks into absolute lawlessness. Nothing could have stopped these people from their unquestionably lawless actions.

At that time, Radio Liberty... the representatives of Radio Liberty... were present in the offices of Logovaz and generally they too witnessed this whole shameful affair.

At that time, in the course of that whole time, Valentin Yumashev was present as well; he saw it all with his own eyes and undoubtedly will tell you all about it.

But the most important thing, Boris Nikolaevich... The most important thing, Boris Nikolaevich, I would like to say that realistically it is very difficult to tell nowadays who is the real authority. I am not telling you this to egg you on in that action that you have already decided to take. By now this is completely insufficient. Boris Nikolaevich, you no longer rule in Moscow. Moscow is ruled by scum!

Boris Nikolaevich, I do not doubt your intentions, your endless dedication to the tasks which you perform. But your people are being set up. One by one, your people are being eliminated from under you.

Thank you.

LESNEVSKAYA: I would like to continue.

Boris Nikolaevich, we, the assembled bankers and independent TV companies, did not unite in vain. We wanted to make Channel 1 into a channel that would be inaccessible to the Mafiya, into a presidential channel, your channel, Boris Nikolaevich, because we believe in you.

I know that two years ago, when I made a documentary about you and then a second one, I know that I got on all the lists. On the first day as soon as the power in the country changes, we will be eliminated.

Four times I refused to make a documentary about Luzhkov. The orders and various telephone calls continue to this day.

I have no doubts that this theoretical construct, this jesuitical plan to murder Vlad was thought up and executed by the Most Group, by Mr. Gusinsky, by Mr. Luzhkov and that organization that is supporting them—the enormous pyramid with all its branches—the former KGB.

Who benefits from all this? Very simple. To whom, Boris Nikolaevich, did you give that TV channel? These are people who killed the standard-bearer, the best journalist in the country! And you, meanwhile, fire Pankratov, fire Ponomarev... It's these people who are your enemies.

And so Luzhkov at last...

The country is outraged. "Get rid of all the power ministers! Get rid of Yeltsin!" And you no longer have a leg to stand on. They have identified the murderer—Berezovsky and Badri Shavlovich [Patarkatsishvili]...

The theory is fantastically jesuitical, but it's convincing. This is more than a coup d'état.

Therefore, I turn to you, Boris Nikolaevich, a person whom I trust, notwithstanding the many disappointments of the past several years. I love your family and your wife, and I know that she, too, is partial to me. Therefore, I am turning to the family—I have one son... We did a very interesting thing. We wanted to create a good TV channel, in order to free it from the filth, from the Mafiya and from the stench that is now boiling up in Ostankino.

I ask you: Listen to the end to everything we have told you and take measures.

Thank you.

APPENDIX II

The Yeltsin Campaign Covers Up

On June 19, 1996, two workers from the Yeltsin campaign, Arkady Yevstafyev and Sergei Lisovsky, were arrested by members of General Korzhakov's SBP as they were trying to smuggle a box with $500,000 in cash from Russian government headquarters (the White House). The next day, General Korzhakov was fired.

With only two weeks to go before the final round of the presidential elections, the Yeltsin campaign was anxious to cover up the box-of-cash incident. The campaign was terrified that Korzhakov would expose the mass corruption that had surrounded their campaign. The following conversation took place in the President Hotel (the headquarters of the Yeltsin campaign) on June 22, 1996, among campaign managers Anatoly Chubais, Viktor Ilyushin, and Sergei Zverev. The conversation was secretly taped by someone close to General Korzhakov's SBP.

The tape, which was later deemed authentic by the prosecutor-general's office, illustrates not only the mentality of the Yeltsin entourage during the election campaign, but also the ease with which they ordered the law-enforcement authorities to derail the investigation and toe the Party line. The following is an excerpt from the conversation.

CHUBAIS: I'm telling you. The language here must be completely unambiguous. We have to tell [Korzhakov] to his face: Either your guys shut up or we'll throw them in prison. That's all. We have so much evidence, together with documentation, that there will be enough for each of them to get fifteen years. All the blood, all the looting that they're responsible for. To the last detail. And it [the evidence] is in pretty secure locations. Many locations. If anything happens to any of us, we will publish this material immediately. I personally drew up the plan two months ago, worked it out to the last detail, because I knew whom I was dealing with. Now the picture is the following: Either they shut up or I'll throw them in prison. That is absolutely certain. You can tell that to them personally from me, as a greeting.

ILYUSHIN: I do not fit this role and therefore I won't do it. Will Berezovsky be able to do it? After all, he is sort of in contact with them. He's friendly with them.

ZVEREV: No, no. We shouldn't pick on Berezovsky. Over there he's well [unintelligible].

ILYUSHIN: I understand it this way: If I am fired tomorrow, I will be beyond the Kremlin wall and won't be able to enter into any doors. But what forces do they still have after they have been relieved of their duties? Who is helping them operate? Are they doing it just by themselves? Or do they have this whole apparatus still functioning under their command?

CHUBAIS: First of all, of course, it is still functioning without their leadership. Secondly, Aleksandr Vasilievich [Korzhakov]. There is a rumor that in July he is going to be returned to his post. The same Krapivin [head of the Federal Guard Service] . . .

ILYUSHIN: Under the orders of Chernomyrdin?

CHUBAIS: I doubt it. According to the witness statement . . .

ILYUSHIN: Yes, but what happened there . . . What do we do with the Boss [Yeltsin]? We have to talk to him.

ZVEREV: We have to finish the job.

CHUBAIS: Arkady Yevstafyev understood the situation perfectly. He understood perfectly well the risks he was taking. Absolutely and clearly. He didn't even pose any questions. A completely reliable guy. [Unintelligible] yesterday with a box of a million dollars.

ILYUSHIN: Lisovsky?

CHUBAIS: No, Arkady. About Lisovsky [unintelligible] in this scheme. It is precisely Lisovsky who should withdraw into the shadows. Yevstafyev was carrying out...It was decided: Forward! If we consider [unintelligible] that person. It's clear that he is obligated to sacrifice himself, but in such a way that all this...

ILYUSHIN: I told the Boss, when I was talking with him yesterday. I told him: "Boris Nikolaevich, right now, if you wanted to, you could catch fifteen to twenty individuals next to the President Hotel who are carrying out—carrying suitcases [of cash] out of our building." He was sitting with a stone expression. I told him: "Because if we begin to account for the money passing through informal channels, we wouldn't be able to hold the elections. Therefore, we haven't had any interruptions so far, but to organize [unintelligible] is simple enough." The president answered: "I understand."

CHUBAIS: We, the people who support the president, would lay our heads down for him. Literally.

ILYUSHIN: This is what I have to suggest. You, Anatoly Borisovich [Chubais], have a talk with him, keeping in mind certain details of yours. I will talk with the Boss on Monday as well. While, with respect to the basic...I will report to him, first of all, that I had the meeting. After all, I went to him for permission to meet with the representatives of the power ministries [the security services, the armed forces]. I will tell him that the meetings took place and will say that, in our opinion, we have to have a directive from Skuratov [the general prosecutor] not to surrender these guys [Yevstafyev and Lisovsky]. And to support them, naturally, to control the course of events, so they wouldn't end up ruining everything.

CHUBAIS: At what time are you going to see him?

ILYUSHIN: I usually talk to him at nine o'clock.

CHUBAIS: I had better make my report afterward.

ILYUSHIN: On Tuesday.

CHUBAIS: If it were during your meeting, I would make my appointment later.

ILYUSHIN: I have an appointment with him at nine o'clock, either by telephone or I go to see him...

CHUBAIS: I will [unintelligible] call him at the same time. The main thing is to take it away from Skuratov [unintelligible] and put it under the president's control.

ILYUSHIN: The twenty-third for me is completely open...

ZVEREV: Korabelshchikov [an assistant to Yeltsin] said that...

ILYUSHIN: He [Yeltsin] is going on vacation, yes?

ZVEREV: Yes.

CHUBAIS: Is he going to be at the office? You're going to meet him at the office, right?

ILYUSHIN: If he's not going to be at the office, I will try to visit him at his dacha. For you, it will be even easier to meet with him at the dacha, right?

CHUBAIS: No. He bawled me out the last time. That time with all that Chechen business.

ILYUSHIN: Okay. I will talk it over with him on Monday, either by telephone or I'll go see him.

CHUBAIS: That's very important—whether he's at the office or at the dacha. At the office I'll contact him directly, but at the dacha I won't contact him. So let's coordinate our actions, Viktor Vasilievich. What you wanted to say was [unintelligible] to protect these guys. The FSB must be given an order to protect them. Krapivin must be given an order to protect them, so that they would know that it is the president who is ordering it. With respect to the prosecutor-general—a request for him to hand over the complete set of documents for [unintelligible] of the president.

ILYUSHIN: And we keep them ourselves...

CHUBAIS: Our comrades[1] were doing our job for us, they were taking on the riskiest part of it, put their skins on the line...

ILYUSHIN: I am talking in general. Because they had a question: What are we going to do with Sobchak? I said—exactly the same thing.

CHUBAIS: Who exactly had this question?

ILYUSHIN: Skuratov. I said: "Until July 3, we don't need any noise."

ZVEREV: We have to divide our tasks into two parts. There's noise here because of the elections. And the personal security of these toilers[2] also probably needs to be guaranteed in some way.

ILYUSHIN: I raised this question, but only having the preelection period in mind. I have to tell you honestly that I did not discuss what will happen after the elections, since I am convinced that we will all have to make our own way. I can't offer you much help there.

ZVEREV: If the elections turn out well, we'll have a chance to make our way out.

CHUBAIS: Let's keep in mind the simple things. Hell! They put their necks on the line and now we tell them: "Excuse me, after July 3 you'll have to fend for yourselves." What kind of attitude is that? I am categorically opposed to this kind of approach. These people are threatened under criminal statutes. Yes, so it turned out that Ilyushin and Chubais are here and they are over there. But we are the ones who sent them there! And not just anybody.

ILYUSHIN: So, we will work in this direction.

CHUBAIS: Yes, we must answer for this with our heads. How am I going to look them in the eye otherwise? Are you crazy?

ILYUSHIN: I agree with this formulation of things.

CHUBAIS: Whatever happens...So, you did your job and afterward we split up, so to speak. What happens afterward—you get five years in prison. Sorry, but it happens. It happens to everybody.

ILYUSHIN: No, maybe I didn't focus the conversation on this, but I fully support this position and perhaps in this case I didn't figure out my point of view well. Of course, I will certainly continue my discussions with Skuratov about this. That's the right thing to do.

CHUBAIS: There is a key question: Should we prevent the transfer of documents to Skuratov?

ILYUSHIN: But we won't be able to do anything. When I got a call from Trofimov [head of the Moscow FSB] yesterday, he told me that I [Trofimov] am obliged to transfer the documents.

CHUBAIS: I don't believe Trofimov. Not a single word he says.

ILYUSHIN: When I spoke with Skuratov today I didn't ask any questions. He said that today all the documents will be transferred to him.

CHUBAIS: Trofimov organized it all personally. I spoke with Trofimov at one o'clock in the morning, at the time when all this was taking place. He lied to me that they didn't know who Lisovsky was, while Yevstafyev had perhaps been detained a little bit, but he would be released any moment.

ILYUSHIN: I can tell you more about Trofimov. As far as I know, he was fighting with Barsukov [former head of the FSB] almost from the beginning. But Trofimov, as a military man, was obliged to carry out the orders of his superior. I wouldn't yet strike Trofimov from the list of the people who can smooth out the situation, although, of course, I can't guarantee anything.

CHUBAIS: I have no doubt that Trofimov is on the other side of the barricades. I don't know what kind of relations he had with Barsukov, but I have no doubts that he is an enemy who wants to destroy us. This was completely obvious from his attitude toward us. I had a personal conversation with him at one o'clock in the morning. It was completely clear what position he was taking.

ILYUSHIN: But we won't be able to block the transfer of the documents to the prosecutor's office if the prosecutor's office requests them.

CHUBAIS: Why? First there is the [unintelligible] of the president to Skuratov. Second, the directive to Kovalev [the director of the FSB] to string things along. That's all.

ILYUSHIN: The fact that it will be strung along—that's without a doubt. But the transfer [of the documents from the FSB to the prosecutor's office] will already take place today.

CHUBAIS: But in the prosecutor's office, there is Ilyukhin [Viktor Ilyukhin, one of the leaders of the Communist Party, chairman of the Duma Security Committee and a former member of the prosecutor's collegium] and he has the run of the place there.

ILYUSHIN: What if I ask Skuratov to keep the documents to himself? [Ilyushin calls Skuratov on the phone.]

ILYUSHIN: Yuri Ilyich, a question has come up: Is it possible to arrange things so that the documents that you will be receiving from Trofimov not be seen by anyone but you? That they stay with you for a certain period of time, until you have had a chance to get acquainted with them and talked everything over with Boris Nikolaevich [Yeltsin].

[Skuratov answers.]

ILYUSHIN: It must be done precisely in this way. Because we have information that if anyone else of your staff handles it, this information would very quickly find its way into the camp of our opponents.

[Skuratov answers.]

ILYUSHIN: Yes, it would be best if you keep it yourself. Don't let anyone start investigating it. And then we'll decide. Okay? Because this is what we want.

[Telephone conversation ends.]

CHUBAIS: What if, at the second stage, we were to ask Boris Nikolaevich [Yeltsin] . . .

ILYUSHIN: To bury it completely?

CHUBAIS: No, to ask Skuratov to send him the documents so that he could analyze them. To demand the full package of documents.

ILYUSHIN: Good idea. [Laughs.]

CHUBAIS: And then let [Skuratov] try to ask for them back . . .

ILYUSHIN: You have to understand that my relations with him [Yeltsin? Skuratov?] are also, so to speak, formal. I can't tell him what I would tell to any old person.

CHUBAIS: I understand.

ILYUSHIN: Moreover, I know that Skuratov has been very heavily influenced by the guys [Korzhakov and Barsukov].

CHUBAIS: Yesterday, Barsukov's guys tried all day to get through on the various TV networks.

ZVEREV: Television has been closed to them, but the radio . . .

CHUBAIS: I'll ask Boris Nikolaevich [Yeltsin] as well.

ILYUSHIN: That's the best. Because he has been informed of everything. You have a different level of trust with him. I am somewhat on the sidelines in your relations.

CHUBAIS: Generally, Skuratov is reasonable enough. He's not the problem. The problem is in the prosecutor's office and generally.

ILYUSHIN: All the better that we worked all these things out ahead of time.

CHUBAIS: We have to find a means of getting in touch with Korzhakov and Barsukov, so we can explain the situation to them clearly and unambiguously: Either they conduct themselves like regular fellows or we will throw them in prison. Because this continuation of the conflict at such a critical time will simply lead to ...

ILYUSHIN: They haven't calmed down, have they?

CHUBAIS: See for yourself. The information is coming to the surface. Where else could it be coming from?

ILYUSHIN: How would they organize a channel for this information? Let's think about it. Who are they using?

ZVEREV: What about Georgi Georgievich [Rogozin, the deputy to Korzhakov]?

CHUBAIS: That's for sure.

ILYUSHIN: That's for sure.

CHUBAIS: He's gotten so cocky that now he's started hinting that the FSB should be given over to them ...

ILYUSHIN: I will say that he [Rogozin] is conducting himself ... Of course, he is working for several intelligence agencies at the same time, but he is judging things too harshly ...

[End of tape.]

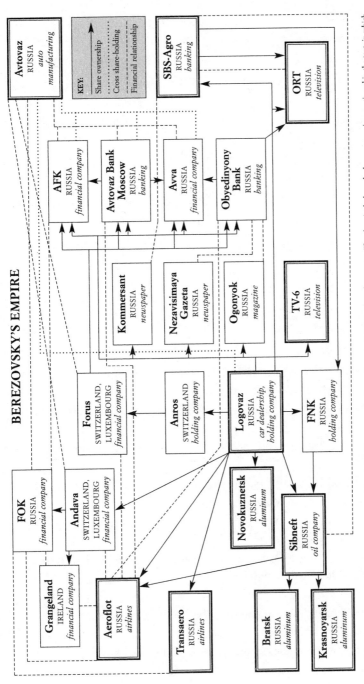

BEREZOVSKY'S EMPIRE

NOTE: This chart is simplified and is best read as a schematic that illustrates a complex and constantly shifting business empire. The chart includes only those companies mentioned in the text, in which Berezovsky is believed to have an ownership stake—either directly or indirectly. Not all of the shareholder relationships are proven. Moreover, the chart does not distinguish between the corporation and the individuals who own the corporation. Logovaz, for example, does not own Andava; Berezovsky and Glushkov own that company (two of the founding shareholders of Logovaz). It is not Sibneft that owns ORT, but certain major shareholders of Sibneft. Finally, the chart does not reflect any specific time frame—these relationships were established over a period of at least eight years (1992–2000).

LIST OF SOURCES

Aeroflot. *1996 Annual Report* (Russian version).

———. *1997 Annual Report* (English version).

———. *1997 Annual Report* (Russian version).

———. *1997 Annual Report* (United States SEC filing).

———. *1998 Annual Report* (Russian version).

———. "Contract No. 249/1 (4/1) of 10 June, 1996" (between Aeroflot and FOK).

——— -FOK-Grangeland payment invoices and receipts for December 1996.

——— list of shareholders, provided to the author by general director Valery Okulov in Moscow in September 1998.

Aksyuchits interview. Viktor Aksyuchits, head of the Christian Democratic Movement and political adviser to First Deputy Prime Minister Boris Nemtsov, interviewed by the author in New York in June 1990 and in Moscow in August 1997.

Alekperov interviews. Vagit Alekperov, chairman of Lukoil, interviewed by the author in Moscow in December 1995.

Aleksandrov interview. Mikhail Aleksandrov, investment strategist for Alfa Capital, interviewed by the author in Moscow in December 1993.

Aluminum executive source. A top executive of one of Russia's largest aluminum companies, interviewed by the author by phone in March 2000.

American Petroleum Institute. *Basic Petroleum Data Book* (Washington, D.C., September 1994).

Andava Group. "Contract No. 9705—AF01" (May 29, 1997).

———. "VEK filing." (June 1997 report to the Foreign Currency Export Control board (VEK) of the Russian Central Bank.)

Andava S.A. *Procès Verbal de la Séance du Conseil d'Administration* (February 8, 1996).

——— registration documents. (Registre du Commerce in Lausanne, Switzerland.)

Anros S.A. registration documents. (Registre du Commerce in Lausanne, Switzerland.)

Aslund, Anders. *How Russia Became a Market Economy* (Washington, D.C., 1995).

Aven interview. Pyotr Aven, minister of foreign trade in 1992, interviewed by the author in Moscow in May 1992.

Barsukov memo. A memorandum written by Mikhail Barsukov, director of the FSB, in the late spring of 1996 about alleged embezzlement and criminal ties at ORT, reproduced in Korzhakov, pp. 436–37.

Basayev interview. Shamil Basayev, Chechen terrorist commander and former prime minister of the Republic of Chechnya, whose interview was posted on the Chechen rebel Web site, www.qoqaz.net, on February 21, 2000.

Bekhoyev interview. Shah-Mirza Bekhoyev, president of the Bureau of High-Speed Ships and prominent Chechen businessman, interviewed by the author in Moscow in September 1998.

Berezovsky interview. Boris Berezovsky, interviewed by the author in Moscow in September 1996.

——— interview with *Itogi* (NTV, November 18, 1996).

——— interview with *Kommersant-Daily* (October 5, 1995).

——— interview with *Kommersant-Daily* (November 16, 1995).

——— interview with *Kommersant-Daily* (June 17, 1997).

——— interview with *Kommersant-Daily* (March 13, 1998).

——— interview with *Kommersant-Weekly* (November 12, 1996).

——— interview with *Le Figaro* (September 22, 1999).

——— interview with *Moskovskie Novosti* (May 16, 1996).

——— interview with *Obshchaya Gazeta* (December 3, 1998).

——— interview with *Sevodnya* (November 14, 1996).

——— interview with *Sunday Telegraph* (March 11, 1999).

———— et al. "Vyiti iz Tupika!" printed in *Izvestia* (April 27, 1996).

———— and Lesnevskaya appeal. A personal videotaped appeal by Boris Berezovsky and Irina Lesnevskaya to President Yeltsin, made in the first days of March 1995 in the offices of the Presidential Security Service. The author obtained a copy of this videotape in 1998 from a former member of the Presidential Security Service, viewed it, and recorded it on audiotape.

———— press conference (Moscow, May 13, 1997).

———— press conference (Moscow, August 20, 1997).

———— -Udugov Conversation 1. A tape of a conversation between Boris Berezovsky and Chechen deputy prime minister, Movladi Udugov, recorded sometime in July 1999, obtained by the author in the autumn of 1999. Partially reprinted in *Moskovsky Komsomolets* (September 22, 1999).

———— -Udugov Conversation 2. A tape of a conversation between Boris Berezovsky and Chechen deputy prime minister and warlord, Movladi Udugov, recorded sometime in July 1999, obtained by the author in the autumn of 1999. Partially reprinted in *Moskovsky Komsomolets* (September 14, 1999).

Boettcher interview. Dieter Boettcher, director of the metals trading division of Metallgesellschaft A.G., interviewed by the author by phone in April 1992.

Borovoi interviews. Konstantin Borovoi, chairman of the Russian Commodities and Raw Material Exchange, interviewed by the author in Moscow in May 1992 and July 1993.

Brunswick Brokerage. *Russian Equity Guide—1997/98* (Moscow, 1997).

———— analyst report. "Russian Automotive Industry: Industry Analysis" (Brunswick Brokerage, January 31, 1996).

Bull interview. James Bull, managing director of international operations for IP Network, interviewed by the author by phone in January 2000.

Bykov interview. Anatoly Bykov, chairman of Krasnoyarsk Aluminum smelter, interviewed by the author in New York in July 1998.

Chebotarev interview. Gennady Chebotarev, former deputy chief of the organized crime division of the MVD, interviewed by the author in Moscow in May 2000.

Chernomyrdin transcript. The tape-recorded conversation between Prime Minister Viktor Chernomyrdin and SBP chief Aleksandr Korzhakov, reprinted in Korzhakov, pp. 362–86.

Chubais interview. Anatoly Chubais, head of the State Property Committee and first deputy prime minister, interviewed by the author in Moscow in January 1998.

————interview with *Kommersant-Daily* (March 5, 1998).

————transcript. The tape-recorded conversation between Anatoly Chubais, Viktor Ilyushin, and Sergei Zverev, reprinted in *Moskovsky Komsomolets* (November 15, 1996) and Streletsky, pp. 249–56; the transcript was authenticated by the prosecutor-general's office. [See Skuratov letter.]

Cilluffo interview. Frank Cilluffo, task force director of the Global Organized Crime Project of the Center for Strategic and International Studies in Washington, D.C., interviewed by the author by phone in May 2000.

Clingman interview. Alan Clingman, chairman of AIOC, interviewed by the author in New York in August 1994.

Cockburn, Andrew and Leslie. *One Point Safe* (New York, 1997).

Committee on International Relations. *The Threat from Russian Organized Crime: Hearing before the Committee on International Relations, House of Representatives, One Hundred Fourth Congress, Second Session: April 30, 1996* (U.S. Government Printing Office, Washington, D.C., 1996).

CS First Boston. "Gazprom Voucher Auction Completed" (October 7, 1994).

————."Russian Capital Markets" (June 1994).

CSIS. Center for Strategic and International Studies, *Russian Organized Crime: Global Organized Crime Project* (Washington, D.C., 1997).

Cuendet interviews. Yves Cuendet, a director of André & Cie., interviewed by the author by phone in October–December 1998 and in Lausanne in March 1999.

Davydov interviews. Oleg Davydov, former foreign trade minister, interviewed by the author in New York in April 1998, in Helsinki in September 1998, and by phone in November 1998.

Derr interview. Ken Derr, chairman and CEO of Chevron, interviewed by the author by phone in February 1998.

Dickens interview. Chris Dickens, head of Chris Dickens Communications Ltd., interviewed by the author by phone in November 1996.

Dixon interview. Mark Dixon, president and editor of *East European Business Review*, interviewed by the author in New York in July 1993.

Eberstadt, Nicholas. "Russia: Too Sick to Matter?" in *Policy Review* (June–July 1999).

Ermarth, Fritz. "Seeing Russia Plain: The Russian Crisis and American Intelligence" in *The National Interest* (Spring 1999).

Ferrero interviews. William Ferrero, general director of Andava S.A., interviewed by the author by phone October–December 1998 and February 1999 and in Lausanne in March 1999.

FOK registration documents. Finansovaya Obyedinyonaya Korporatsiya certificate of registration and bylaws, obtained by the author in the winter of 1998–99.

———. *1996 Annual Report* (Moscow).

Forus Services S.A. registration documents (Registre du Commerce, Lausanne, Switzerland).

Freeh press conference. A November 1997 press conference in Moscow by Louis Freeh, director of the FBI.

Fyodorov transcript. Boris Fyodorov's conversation with Berezovsky and Dyachenko in Logovaz House, transcribed and partially reprinted in *Novaya Gazeta* (July 8, 1996).

Fyodorov (Boris) interview with *Komsomolskaya Pravda* (July 10, 1996).

Fyodorov (Olga) interview with *Komsomolskaya Pravda* (July 10, 1996).

Fyodorov interview. Svyatoslav Fyodorov, world-famous eye surgeon and 1996 presidential candidate, interviewed by the author by phone in August 1993.

Gaidar interviews. Yegor Gaidar, former acting prime minister (1991–92) and architect of Yeltsin's market reforms, interviewed by the author in Moscow in January 1998 and by phone in February 1998.

Gazprom analytical department. "Gusinsky, Vladimir Aleksandrovich," a thirty-page report, obtained by the author in August 1997.

———. "Potanin, Vladimir Olegovich," a thirty-page report, obtained by the author in August 1997.

———. "Ekonomicheskie i Politicheskie Perspektivy Svyazki Chubais-Potanin," a twenty-page report, obtained by the author in August 1997.

Glazyev interview. Sergei Glazyev, minister of foreign trade in 1993, interviewed by the author by phone in May 1993.

Gorbachev interview. Mikhail Gorbachev, former general secretary of the Communist Party of the Soviet Union, interviewed by the author in New York in May 1992.

Goskomstat. *Rossiisky Statistichesky Ezhegodnik—1997* (Moscow, 1997).

———. *Rossiya v Tsifrakh—1998* (Moscow, 1998).

Gouliev interviews. Rasul Gouliev, former deputy prime minister and Speaker of Parliament of Azerbaijan, interviewed by the author in New York in February 1998.

Grangeland Holdings registration documents. Annual reports and registration documents filed with the Companies Registration Office in Dublin, Ireland.

Gray, Francine du Plessix. *Soviet Women: Walking the Tightrope* (New York, 1990).

Greenspan interview. Alan Greenspan, chairman of the Federal Reserve, interviewed by the author by phone in December 1989.

Griffiths interview. Susan Griffiths, partner of Global Cash Management, Ltd., interviewed by the author by phone in February 1999.

Gusinsky interview. Vladimir Gusinsky, head of the Most Group, interviewed by the author by phone in September 1994.

——— interview with *Obshchaya Gazeta* (April 18, 1996).

——— interview with *The Boston Globe* (March 13, 1994).

——— interview with *Ekho Moskvy,* a Moscow radio station (August 15, 1997, 14:15).

Hamilton interviews. Thomas Hamilton, former BP executive and executive vice president of Pennzoil, interviewed by the author in December 1995.

Handelman, Stephen. *Comrade Criminal: Russia's New Mafiya* (New Haven, 1997).

Heffner interview. Robert Heffner, chairman of oil services company GHK Corp., interviewed by the author by phone in January 2000.

House of Lords. *Opinions of the Lords of Appeal for Judgement in the Cause "Berezovsky (Respondent) v. Michaels and others (Appellants), Glouchkov (Respondents) v. Michaels and others (Appellants) (Consolidated Appeals)"* (London, May 11, 2000).

Hughes interview. Lou Hughes, head of international operations for General Motors, interviewed by the author in Geneva in February 1997.

Human Rights Watch. *Abandoned to the State: Cruelty and Neglect in Russian Orphanages* (New York, 1998).

Jordan interviews. Boris Jordan, head of equities trading for CS First Boston Russia division and subsequently president of MFK-Renaissance, interviewed by the author in Moscow in December 1993 and October 1997.

Kalugin interviews. General Oleg Kalugin, former head of the First Chief Directorate of the KGB, interviewed in Moscow in September 1990, in Washington, D.C., in March 1999, and by phone in March 2000.

Kiriyenko interview. Sergei Kiriyenko, Russian prime minister in 1998, interviewed by the author by phone in New York in December 1998.

Kharshan interviews. Mikhail Kharshan, director of the First Voucher Investment Fund, interviewed by the author in Moscow in December 1993, January 1994, and September 1994.

Khatsenkov interview. Georgi Khatsenkov, Russian jewelry trader, interviewed by the author in Moscow in August 1993.

Khodorkovsky interviews. Mikhail Khodorkovsky, founder of the Menatep Group and chairman of Yukos, interviewed by the author in Moscow in September 1994 and September 1996. A third interview with Khodorkovsky was conducted and taped by *Forbes* staff writer Kerry Dolan in New York in spring of 1998.

Kislinskaya interviews. Larisa Kislinskaya, organized-crime expert for *Itar-Tass* and *Sovershenno Sekretno,* interviewed by the author in Moscow in August 1993 and January 1998.

Kokh, Alfred. *The Selling of the Soviet Empire: Politics and Economics of Russia's Privatization—Revelations of the Principal Insider* (New York, 1998).

——— interviews. Alfred Kokh, deputy prime minister and privatization chief, interviewed by the author in New York in June 1998 and in Moscow in December 1999.

Konstantinov, Andrei, and Malcolm Dikselius. *Banditskaya Rossiya* (St. Petersburg, 1997).

Korostelev interview. Yuri Korostelev, head of financial department of the Moscow City Government, interviewed by the author in Moscow in September 1998.

Korzhakov, Aleksandr. *Boris Yeltsin: Ot Rassveta do Zakata* (Moscow, 1997).

——— interviews. General Aleksandr Korzhakov, head of the Presidential Security Service, interviewed by the author in Moscow in October 1997, December 1998, and January–February 1999; witness statement signed February 1999.

——— interview with *Moskovsky Komsomolets* (October 30, 1999).

——— interview with *Moskovsky Komsomolets* (November 3, 1999).

——— press conference (Moscow, October 11, 1996).

Kozlov interview. Aleksandr Kozlov, Aleksandr Lebed's public relations chief, interviewed by the author by phone in October 1999.

Kulikov letter. Anatoly Kulikov, minister of internal affairs, in a letter to *Kommersant-Daily* (March 12, 1997).

Kuppers interviews. Rene Kuppers, general director of Forus Services S.A., interviewed by the author by phone in October–December 1998.

Kvantrishvili interview. Otari Kvantrishvili, president of the Lev Yashin Foundation for the Social Rehabilitation of Athletes, interviewed by the author by phone in August 1993.

Lebed interviews. General Aleksandr Lebed, former head of the Security Council of the Russian Federation, interviewed by the author in Moscow in October 1997 and September 1998.

—— interview with *Le Figaro* (February 17, 1997).

—— press conference (Moscow, October 17, 1996).

Lesnevsky interview with Radio Liberty. Dimitri Lesnevsky, cofounder of REN-TV, interviewed on *Radio Liberty* (October 5, 1995).

Lieven, Anatol. *Chechnya: Tombstone of Russian Power* (New Haven, 1998).

—— interview. Anatol Lieven, Institute for Strategic Studies (London), interviewed by the author by phone in September 1999.

Lis's registration documents. "Founding Agreement of Open Joint-Stock Company Lis's (Moscow 1992)," obtained by the author in October 1996.

Luzhkov interview. Yuri Luzhkov, mayor of Moscow, interviewed by the author in Moscow in September 1998.

—— interview with *Moskovsky Komsomolets* (November 5, 1999).

Maret interviews. Christian Maret, chief executive of André & Cie. (Vostok), interviewed by the author in Moscow in September–December 1998.

Maskhadov interview. Aslan Maskhadov, president of the Republic of Chechnya, interviewed by the author by phone in September 1999.

Mavrodi interview. Vyacheslav Mavrodi, cofounder of MMM trading and financial company, interviewed by the author in Moscow in May 1992.

Maximov, Aleksandr. *Rossiiskaya Prestupnost: Kto est Kto* (Moscow, 1998).

Mayor interviews. Alain Mayor, former head of Russian operations for the FINCO department of André & Cie., interviewed by the author by phone in October–December 1998 and January–March 1999.

MC Securities. "Russian Companies: Monthly Valuation Analysis" (August 5, 1997).

——. "Siberian Oil Company: New Kid on the Block" (May 13, 1996).

Medaas interview. Sveinung Medaas, head of the Moscow office of Phibro Energy Inc., interviewed by the author in Moscow in May 1992.

MFK. "International Company for Finance and Investments—1997" (brochure).

Modestov, Nikolai. *Moskva Banditskaya* (Moscow, 1996).

Moody interviews. Jim Moody, chief of Organized Crime/Drug Section of the FBI, interviewed by the author in Washington, D.C., in July 1993 and by phone in May 2000.

Mordassov interview. Vladimir Mordassov, vice director of foreign economic relations for the ministry of fuel and energy, interviewed in Moscow in July 1993.

Morgenstern interview. Claudia Morgenstern, senior investment officer at the International Finance Corp., working on Russian privatization, interviewed by the author by phone in January 1994.

Nemtsov interview. Boris Nemtsov, first deputy prime minister 1997–98, interviewed by the author in Moscow in May 2000.

Nikolaev interview. Alexei Nikolaev, president of Avtovaz, interviewed by the author in Togliatti, Russia, in June 1996.

Nurnberg interviews. Andrew Nurnberg, London-based literary agent, interviewed by the author by phone in October–November 1999.

Okulov interviews. Valery Okulov, general director of Aeroflot, interviewed by the author in Moscow in September 1998 and by phone in November 1998.

———— press conference. Valery Okulov, general director of Aeroflot, press conference of February 8, 1999, in Moscow.

Onexim Bank analytical service. "Berezovsky" (a 250-page dossier obtained by the author in January 1998).

————. "Gusinsky" (a 200-page dossier obtained by the author in January 1998).

————. "Smolensky" (a 200-page dossier obtained by the author in January 1998).

Onexim Bank. "The United Export-Import Bank Open Joint-Stock Company" (brochure).

Orekhov interviews. Andrei Orekhov of Grant Brokerage, interviewed by the author in Moscow in December 1993 and September 1994.

ORT registration documents. "Protokol No. 1 of the founding meeting of the Closed Joint-Stock Company Russian Public Television (AO ORT)—Moscow, 24 January, 1995," obtained by the author in October 1996.

Pashkevich interview. Andrei Pashkevich, director of public relations for Moscow RUOP (the organized-crime squad of the Moscow police), interviewed by the author in Moscow in October 1996.

Petrakov interview. Nikolai Petrakov, economic adviser to President Gorbachev, interviewed by the author by phone in July 1993.

Platts-Mills interview. Jonathan Platts-Mills, director of Lonrho, Plc., interviewed by the author by phone in April 1992.

Polevanov interview. Vladimir Polevanov, chairman of the State Property Committee in 1994–95, interviewed by the author in Moscow in September 1998.

Politkovsky interview. Aleksandr Politkovsky, member of the board of directors of TV Company VID and longtime Listyev partner, interviewed by the author by phone in November 1996.

Potanin interviews. Vladimir Potanin, president of Onexim Bank, interviewed by the author in Moscow in September 1996 and October 1997.

Premier S.V., minutes. "Minutes of the general meeting of the founders of the Open Joint-Stock Company Premier S.V." (Moscow, December 29, 1994).

Ptichnikov interview. Igor Ptichnikov, executive director of the Foundation for Financial Support of the Reconstruction of the Cathedral of Christ the Savior, interviewed by the author in Moscow in September 1998.

Razbash interview. Andrei Razbash, general producer of TV Company VID, interviewed by the author in Moscow in November 1996.

Reardon interviews. David Reardon, one of the managers of British Petroleum's Moscow office, interviewed by the author in Moscow in December 1995 and January 1998.

Remnick, David. *Resurrection: The Struggle for a New Russia* (New York, 1997).

Remp interviews. Steve Remp, chairman of Ramco Energy, interviewed by the author in October 1994 and December 1995.

Reuben interview. David Reuben, chairman of Trans-World Metals Ltd., interviewed by the author by phone in December 1994.

Reymond interviews. Dominique Reymond, spokesman for the Federal Prosecutor-General of Switzerland, interviewed by the author by phone in January 2000.

Rich letter. A letter sent to the editors of *Izvestia* by metals trader Marc Rich in August 1992.

Rinaco-Plus analyst report. "Russian Industry Analysis: Automotive Production" (Moscow, January 1996).

RUOP memo. A "field intelligence" report entitled "Spravka po infor-matsii predstavlyaushei operativnyi interes po factu ubiistva..." ob-tained by the author in Moscow in October 1996 from one of the RUOP officers involved in the Listyev murder investigation.

———— source. A veteran officer of RUOP (the organized-crime squad) of the Moscow police, interviewed in Moscow in July–August 1993, September–November 1996, and January–March 1999.

———— spreadsheets. A spreadsheet compiled by Moscow RUOP officers investigating the Listyev murder, containing the names, addresses, passport numbers, previous criminal records, and suspected gangster ties of individuals involved with Premier S.V.

———— surveillance video. Filmed in July 1993 from a car parked across the street from the church. Viewed by the author in August 1993 in the presence of RUOP officers, who explained the identities of vari-ous individuals captured on the tape.

Rushailo interview. General Vladimir Rushailo, head of Moscow RUOP (the organized-crime squad of the Moscow police), interviewed by the au-thor in Moscow in August 1993.

Russian Petroleum Investor, Petroleum Intelligence Weekly, and Ernst & Young. *The Complete Guide to Russian Oil & Gas Exporters* (Los An-geles, 1993).

Ryzhkov interview. Nikolai Ryzhkov, prime minister of the Soviet Union, interviewed in *Forbes* (October 16, 1989).

Sakwa, Richard. "The Russian Elections of December 1993" in *Europe-Asia Studies* (March 1, 1995).

Salomon Smith Barney. "Aeroflot." (Research report dated August 11, 1998.)

Sayer interview. David Sayer, floor manager for the Cherry Casino, inter-viewed by the author in Moscow in July 1993.

Shelley interviews. Louise Shelley, head of the Center for Trans-National Crime at American University, interviewed by the author in Wash-ington, D.C., in March 1999 and by phone in December 1999.

Sibneft financial statements. "Related Party Transactions" addendum to the Sibneft financial statements, posted on *www.sibneft.ru* as of No-vember 23, 1999.

Sipachev interview. Vladimir Sipachev, president of Aeroflot Bank, inter-viewed by the author in Moscow in July 1993.

Skuratov, Yuri. Prosecutor general of the Russian Federation, 1995–2000, interviewed by the author in Moscow in May 2000.

———— letter. Official report addressed to G. N. Seleznyov, chairman of the Duma, from Prosecutor-General Yuri Skuratov, closing Criminal Case No. 18/221891-96, dated April 10, 1997.

———— interview with *The Moscow Times* (September 7, 1999).

Solzhenitsyn interview. Aleksandr Solzhenitsyn, Soviet dissident and Nobel Prize–winning author, interviewed by the author by phone in March 1994.

Soros, "Who Lost Russia?" An article by George Soros, head of Soros Fund Management and major investor in the Russian economy, printed in *The Moscow Times* (March 7, 2000).

Stacy interview. T. Don Stacy, president of Amoco-Eurasia, interviewed by the author by phone in December 1995.

Stepashin interview with *Nezavisimaya Gazeta* (January 14, 2000).

Sterligov interviews. Herman Sterligov, cofounder of Alisa trading company, interviewed by the author in New York in January 1992, in Moscow in May 1992, and by phone in September 1992 and July 1993.

Streletsky, Valery. *Mrakobesie* (Moscow, 1998).

———— interviews. Colonel Valery Streletsky, head of "Department P" (anticorruption) of the Presidential Security Service, interviewed by the author in Moscow in December 1998 and January–February 1999; witness statement signed February 1999.

———— analytical report. A thirty-four-page report entitled "Berezovsky, Boris Abramovich," compiled by former security service officers and signed by Colonel Valery Streletsky, obtained by the author in Moscow in February 1999.

Sysuyev interview. Oleg Sysuyev, former deputy prime minister, interviewed by the author in New York in February 2000.

Tarasov interviews. Artyom Tarasov, president of Istok trading company and cofounder of Alisa trading company, interviewed by the author in New York in January 1992 and by phone in May 1992.

Thompson interviews. Page Thompson, retired treasurer of Arco and director of International Trading Partners Ltd., interviewed by the author by phone in October 1996, December 1999, and January 2000.

Triboi interview. Pyotr Triboi, investigator for top-priority cases at the prosecutor-general's office, interviewed by the author in Moscow in January 1999.

Urazhtsev interview. Vitaly Urazhtsev, president of the military trade union, Scheet, interviewed by the author in Moscow in September 1990.

U.S. Department of Commerce. *Statistical Abstract of the United States 1998* (Washington, D.C., 1998).

U.S. State Department source, interviewed by the author by phone in September 1999.

U.S. Treasury Department source. A veteran law-enforcement agent charged with keeping track of the movements of American fugitive Marc Rich, interviewed by the author, on condition of anonymity, by phone in April 1992.

Usmanov interview. Lyoma Usmanov, Chechen president Aslan Maskhadov's envoy to the United States, interviewed by the author by phone in October 1999.

Vanous interview. Jan Vanous, president of PlanEcon Inc., interviewed by the author by phone in December 1989.

Vavilov interview. Andrei Vavilov, first deputy minister of finance, interviewed by the author in Moscow in May 1992.

Vecchio interview. Mark Vecchio, partner of Coudert Bros., interviewed by the author by phone in November 1996 and October 1999.

Veinshtock interview. Semyon Veinshtock, general director of Kogalymneftegaz, interviewed by the author in Kogalym, Russia, in December 1995.

Veselovsky memorandum. "Analytical memorandum of Colonel of the KGB of the U.S.S.R., V. G. Veselovsky, to chief of the Executive Administration of the Central Committee of the CPSU, N. V. Kruchina," excerpts of which were obtained by the author in New York in February 2000.

Volkov interviews. Nikolai Volkov, investigator for top-priority cases for the Prosecutor-General's Office of the Russian Federation, interviewed by the author in Moscow in December 1998 and February 1999, and by phone in February 2000.

Voronin interview. Oleg Voronin, Siberian coal miner trade union leader, interviewed by the author in New York in March 1991.

Waller, J. Michael, and Victor J. Yasmann. "Russia's Great Criminal Revolution: The Role of the Security Services" in *Journal of Contemporary Criminal Justice,* Vol. 11, No. 4 (December 1995).

Wanniski interview. Jude Wanniski, economic adviser to President Reagan and president of Polyeconomics Inc., interviewed by the author by phone in December 1989.

Wedel, Janine. *Collision and Collusion: The Strange Case of Western Aid to Eastern Europe 1989–1998* (New York, 1998).

Yasin interview. Yevgeny Yasin, economic adviser to President Yeltsin, interviewed by the author in Moscow in January 1993.

Yavlinsky interview. Grigory Yavlinsky, author of the 500-Day Plan and leader of the liberal Yabloko faction in parliament, interviewed by the author in Moscow in January 1998.

Yegorov press conference. A press conference in Moscow by General Mikhail Yegorov, head of the organized-crime division of the Russian Ministry of Internal Affairs, attended by the author in July 1993.

Yeltsin, Boris. *Zapiski Prezidenta* (Notes of a President) (Moscow, 1994).

Yergin interview. Daniel Yergin, historian and chairman of Cambridge Energy Research Associates, interviewed by the author in December 1995.

Newspapers and Periodicals cited

Agence France-Presse (Paris)
Argumenty i Fakty (Moscow)
BBC Worldwide Monitoring (London)
Boston Globe (Boston)
Chicago Sun-Times (Chicago)
The Christian Science Monitor (Boston)
The Daily Telegraph (London)
Delovoi Mir (Moscow)
Delovye Lyudi (Moscow)
Dengi (Moscow)
Dow Jones International News Services (New York)
Euromoney (London)
Le Figaro (Paris)
The Financial Post (Toronto)
Financial Times (London)
Forbes (New York)
Facts on File (New York)
The Globe and Mail (Toronto)
The Guardian (London)
L'Hebdo (Geneva)
The Independent (London)
Institutional Investor (New York)
Itar-Tass (Moscow)
Itar-Tass Express (Moscow)

Izvestia (Moscow)
Jane's Intelligence Review (London)
Jerusalem Post (Jerusalem)
Jewish Telegraphic Agency (New York)
Kommersant-Daily (Moscow)
Kommersant-Vlast (Moscow)
Kommersant-Weekly (Moscow)
Komsomolskaya Pravda (Moscow)
Kuranty (Moscow)
Los Angeles Times (Los Angeles)
Lyudi (Moscow)
Menatep Group News (Moscow)
Le Monde (Paris)
The Moscow Times (Moscow)
Moskovskie Novosti (Moscow)
Moskovsky Komsomolets (Moscow)
Newsweek (New York)
The New York Times (New York)
Nezavisimaya Gazeta (Moscow)
Novaya Gazeta (Moscow)
Novye Izvestia (Moscow)
The Observer (London)
Obshchaya Gazeta (Moscow)
Ogonyok (Moscow)
Rossiiskaya Gazeta (Moscow)
Russian Petroleum Investor (Los Angeles)
Sevodnya (Moscow)
Sovershenno Sekretno (Moscow)
Sunday Telegraph (London)
The Sunday Times of London (London)
Sydney Morning Herald (Sydney)
Time (New York)
Toronto Star (Toronto)
U.S. News & World Report (Washington, D.C.)
The Wall Street Journal (New York)
The Wall Street Journal—Europe (Brussels)
The Washington Post (Washington, D.C.)
The Washington Times (Washington, D.C.)

NOTES

Unless otherwise noted, the interviews cited in this book were conducted by the author. The simple notation "Berezovsky interview," for example, refers to the author's interview with Berezovsky. "Berezovsky interview with *Kommersant-Daily*," however, refers to his interview with the newspaper.

In cases of multiple citations for a single paragraph, the order in which the citations appear is deliberate. The first citation typically refers to the first significant fact or assertion mentioned in the paragraph, the second citation to the second assertion, and so on. If several citations refer to a single assertion in the paragraph, they are presented in order of importance, with the most important source coming first.

For a more extensive description of the sources cited in this book, see the List of Sources.

Introduction
1 *Forbes* (December 30, 1996).
2 Moody interviews; CSIS, pp. 23–24.
3 Streletsky interviews; Shelley interviews.

4 Nikolaev interview.

5 Berezovsky interview.

6 *Financial Times* (November 1, 1996).

7 Lebed interview with *Le Figaro* (February 17, 1997).

Chapter 1

1 RUOP surveillance video; RUOP source.

2 RUOP source; Streletsky interviews; Chebotarev interview.
  Berezovsky's close relationships with the Chechens would be revealed when he started dealing with the Chechen warlords and kidnapping gangs in 1996–97. (See Chapter 9 of this book.) Numerous Chechen officials confirmed that Berezovsky employed Chechens in his security service. (See Usmanov interviews, for instance.)
  With regard to the auto dealership market's being one of the most criminalized segments of the Russian economy, see: *Dengi* (March 3, 1999); *Sovershenno Sekretno* (February 1998); Cilluffo interview; *Sevodnya* (March 14, 1998).

3 RUOP source; *Kommersant-Daily* (July 21, 1993).

4 Rushailo interview.

5 Berezovsky interview.

6 RUOP source; Modestov, pp. 30–58; Maximov, pp. 161–63; Lieven, pp. 351–56; *Obshchaya Gazeta* (March 8, 1999); *Moskovsky Komsomolets* (September 29, 1999).

7 Konstantinov, pp. 342–44, 348–51; Lieven, p. 62; Moody interviews; Waller, "Russia's Great Criminal Revolution."

8 Modestov, pp. 30–58; Maximov, pp. 161–63; Konstantinov, pp. 348–51; *Obshchaya Gazeta* (March 18, 1999); RUOP source.

9 RUOP source; Konstantinov, pp. 377–81; Maximov, pp. 161–63; Kislinskaya interviews.

10 RUOP source; Kislinskaya interviews; *Moskovsky Komsomolets* (September 29, 1999); Maximov, pp. 161–63.

11 CSIS, pp. 27–28; Moody interviews; Handelman, *Comrade Criminal*; Modestov, pp. 76–92; RUOP source.

12 Skuratov interviews; Korzhakov interviews; Streletsky interviews; Onexim Bank analytical department; Barsukov memo. Badri's role as the intermediary between Berezovsky and various Chechen criminal organizations is also confirmed in Berezovsky-Udugov Conversation 1.

13 Korzhakov interviews.

14 All the information on Ivankov comes from RUOP source; Kislinskaya interviews; Moody interviews; Maximov, pp. 361–71; Modestov, pp. 258–75; Konstantinov, pp. 75–93.

15 RUOP source; Fyodorov interview; Konstantinov, pp. 81–83; Maximov, pp. 369–70; Kislinskaya interviews; Moody interviews; Kvantrishvili interview.

16 Fyodorov interview; RUOP source; Modestov, p. 269; Konstantinov, pp. 85–87; Moody interviews; Committee on International Relations, p. 22.

17 Moody interviews; RUOP source.

18 Kvantrishvili interview; RUOP source; Kislinskaya interviews; Modestov, pp. 276–89.

19 All the preceding information on Otarik is based on Kvantrishvili interview; RUOP source; Kislinskaya interviews; Modestov, pp. 276–89.

20 RUOP source; Modestov, pp. 35–36.

21 RUOP source; Maximov, pp. 296–343; Konstantinov, pp. 107–10; Moody interviews; CSIS, p. 42.

22 Streletsky analytical report; Chebotarev interview.

23 The following account of the Great Mob War comes from: RUOP source; Kislinskaya interviews; Moody interviews; Cilluffo interview; Konstantinov; Modestov; Maximov.

24 RUOP source; Modestov, pp. 118–24; Maximov, p. 79.

25 RUOP source; RUOP spreadsheet; Kvantrishvili interview; Maximov, pp. 79, 325, 342.

26 *Kommersant-Daily* (September 9, 1993); Ibid. (September 23, 1993); Ibid. (November 2, 1993).

   The *Kommersant* reporters noted that Logovaz did not have an officially registered security service. (See *Kommersant-Daily* [September 23, 1993].)

27 *Itar-Tass* (November 5, 1996); *Izvestia* (November 6, 1996). Berezovsky at first denied the fact that he had Israeli citizenship and threatened to sue *Izvestia*, but he later admitted it was true.

28 Berezovsky interview with *Sevodnya* (November 14, 1996).

29 *Delovoi Mir* (November 26, 1996); *Kuranty* (November 26, 1996).

30 Mayor interviews.

31 Sayer interview.

32 Aslund, p. 124; Yegorov press conference; Konstantinov, pp. 344–48; Lieven, p. 75.

33 RUOP source; Sayer interview.

34 Modestov, pp. 348, 350; Konstantinov, p. 565.

35 Gaidar interviews.

36 CSIS, p. 2; Rushailo interview.

   Approximately the same figures surfaced in an international conference on money laundering in Prague in February 1997 (see *Financial Times* [February 14, 1997]), as well as in the hearing on Russian organized crime before U.S. House of Representatives in 1996 (see Committee on International Relations).

37 Yegorov press conference; CSIS, p. 25; Freeh press conference.

38 Aslund, p. 195; Vavilov interview; Davydov interviews; Kharshan interviews; Gusinsky interview; Jordan interviews.

39 CSIS, p. 87.

40 Sipachev interview.

41 Modestov, p. 327; Dixon interview.

42 Konstantinov, pp. 552–53.

43 Khatsenkov interview.

44 Rushailo interview.

45 Pashkevich interview; RUOP source; Goskomstat figures cited in Aslund, p. 167; U.S. Department of Commerce, p. 210.

46 Maximov, p. 400.

47 RUOP source.

48 Shelley interviews.

49 Committee on International Relations, pp. 10–11, 38.

50 RUOP source.

51 Committee on International Relations, p. 35.

52 Lebed interviews.

53 Ibid.

54 Cockburn, p. 42; CSIS, pp. 61–63.

55 Korzhakov, p. 172.

56 Borovoi interviews.

57 Khatsenkov interview.

58 Borovoi interviews.

59 Yavlinsky interview.

The jewelry trader Georgi Khatsenkov, in our 1993 interview, reached the same conclusion (that the street-corner thugs acted on the orders of local government officials), based on his personal experience.

60 Berezovsky interview.

61 Moody interviews.

62 Korzhakov interviews; *Kommersant-Daily* (April 29, 1995).

63 Maximov, pp. 80, 302; Modestov, pp. 290–304.

64 Berezovsky interview; *Kommersant-Daily* (June 21, 1994); Berezovsky and Lesnevskaya appeal; Korzakov interviews.

65 Berezovsky interview.

66 Korzhakov interviews; Streletsky interviews; Onexim Bank analytical department, "Berezovsky"; *Moskovsky Komsomolets* (November 21, 1996).

67 *Christian Science Monitor* (June 14, 1994); Korzhakov interviews; Streletsky interviews.

68 Streletsky interviews.

The connection between the Logovaz bombing and Sylvester was also mentioned in *Moskovsky Komsomolets* (November 21, 1996) and *U.S. News & World Report* (January 13, 1997).

69 Lebed interviews; Lieven, pp. 62–66, 75–76, 79, 82–83, 351–56; Bekhoev interview; Davydov interviews; Konstantinov, pp. 351–58.
70 Davydov interviews.
71 Lebed interviews; Bekhoev interview; Basayev interview.
72 Lebed interviews.
73 Ibid.
74 Chubais interview with *Kommersant-Daily* (March 5, 1998). Chubais made the same claim about Big Business's having risen above the gangster world in his 1998 interview with me.
75 Berezovsky interview with *Kommersant-Daily* (March 13, 1998).
76 Chubais interview.

Chapter 2
1 Gorbachev interview.
2 American Petroleum Institute, section vi, table 14.
3 Chubais interview.
4 Davydov interviews.
5 Shelley interviews.
6 Ryzhkov interview.
7 Ibid.
8 Vanous interviews; Greenspan interview.
9 Greenspan interview.
10 Yavlinsky interview.
11 Gorbachev interview.
12 Berezovsky interview.
13 Ibid; Onexim Bank analytical department, "Berezovsky."
14 Berezovsky interview.
15 Brunswick analyst report; Rinaco Plus analyst report.
16 Berezovsky interview.
17 Ibid.; Berezovsky interview with *Kommersant-Daily* (October 5, 1995).
18 Kalugin interviews; Waller, "Russia's Great Criminal Revolution"; Shelley interviews; Ermarth, pp. 10–11; Borovoi interviews; Mavrodi interview.
19 Ibid.
20 Kalugin interviews.
21 Ibid.
22 Ibid.
23 Ibid.
24 Veselovsky memorandum.
25 Kalugin interviews; Waller, "Russia's Great Criminal Revolution"; Shelley interviews.
26 Kalugin interviews.
27 Ibid.

28 Ermarth, pp. 10–11.

29 RUOP source; Committee on International Relations, p. 135; *Facts on File* (February 21, 1991); Davydov interviews.

30 *The Moscow Times* (February 5, 1999). Excerpts of the PriceWaterhouse Coopers audit of Fimaco were reproduced in *The Moscow Times* (August 17, 1999).

31 U.S. Treasury Department source.

32 Boettcher interview.

33 Platts-Mills interview.

34 Medaas interview.

35 Boettcher interview; Medaas interview; Platts-Mills interview.

36 Davydov interviews; Aven interview; Tarasov interviews.

37 U.S. Treasury Department source.

38 Rich letter.

39 Aven interview; Davydov interviews.

40 Tarasov interviews.

41 Davydov interviews.

42 Ibid.

43 Russian Petroleum Investor, *The Complete Guide,* pp. 75–76; Davydov interviews.

    Berezovsky began his own international commodities trading career in 1992–93, at the tail end of Marc Rich's brief reign as the Soviet Union's dominant exporter of metals and oil. In 1993, for instance, Berezovsky's company, Logovaz, was registered as exporting 840,000 tons of aluminum—over a third of Russia's annual exports. It seemed to be a one-off deal, since Logovaz would not be a big player in metals trading in the future. Logovaz, it seemed, was providing its name and export license to some other player in the market. It is not impossible that Logovaz was fronting for Marc Rich, who dominated Russian aluminum exports at this time. Later, both Rich and Berezovsky would indirectly share an interest in the affairs of the Krasnoyarsk Aluminum smelter.

44 Davydov interviews.

45 Gaidar interview; Ermarth, pp. 10–11.

46 Kalugin interview.

47 Cuendet interviews; Maret interviews.

48 Mayor interviews.

49 Ibid.

50 Ibid.

51 Maret interviews.

52 Mayor interviews.

53 Ibid.

54 Cuendet interviews; Mayor interviews.

Since it did not participate in any of the recapitalizations, André's share of Anros shrank to 1 percent by May 1998. (See Anros S.A. registration documents.)

Berezovsky told me that Logovaz was owned by him and a few other individuals. (See Berezovsky interview.)

55 Berezovsky interview; Cuendet interviews; RUOP memo; Onexim Bank analytical service, "Berezovsky," pp. 45–46; Russian Petroleum Investor, *Complete Guide*, p. 75.

56 Berezovsky interview.

Berezovsky thus admitted that Anros S.A. (which held 50 percent of Logovaz shares) was owned by himself and his closest Russian collaborators; in 1998, André executives would tell me that Anros S.A. was almost completely owned by the Russian partners. (See Cuendet interviews.)

57 Cuendet interviews.

58 Luzhkov interview.

59 Voronin interview.

60 Korzhakov, pp. 83–86; Yeltsin, pp. 83–87.

61 Korzhakov, pp. 87–89; Yeltsin, pp. 99–133.

62 CSIS, pp. 51–56; Kalugin interviews; Waller, "Russia's Great Criminal Revolution."

63 Ibid.; Moody interviews.

64 Heffner interview; Russian Petroleum Investor, *Complete Guide*, pp. 75–76.

65 Kalugin interviews; Waller, "Russia's Great Criminal Revolution"; Ermarth, p. 11; Tarasov interviews; Khatsenkov interview; Borovoi interviews.

Chapter 3

1 *Facts on File* (April 25, 1991), p. 303.

2 Solzhenitsyn interview.

3 Yavlinsky interview.

4 Gaidar interviews.

5 Gaidar interviews; Yavlinsky interview; Vanous interview; Greenspan interview; Yasin interview; Goskomstat, *Rossiisky Statistichesky Ezhegodnik—1997*, p. 554.

6 Wanniski interview.

7 Yavlinsky interview.

8 Gaidar interviews.

9 Yavlinsky interview.

10 Ibid.

11 Sterligov interviews.

12 Mavrodi interview; Borovoi interviews; Yeltsin, p. 410; Sterligov interviews.

13 Sterligov interviews.

14 Tarasov interviews.

15 Ibid.; Streletsky interviews; Skuratov interviews; Onexim Bank analytical department, "Smolensky"; Korzhakov interviews; Kiriyenko interview.

16 Tarasov interviews.

17 Aven interview.

18 Tarasov interviews; Sterligov interviews.

19 Sterligov interviews.

20 Ibid.; Davydov interviews.

21 Borovoi interviews.

   Borovoi's investment banking arm—Rinaco Plus—went independent, with its own set of partners, and would become one of Russia's biggest securities brokerages before selling out to Nikoil (Lukoil's investment banking arm) during the financial crisis of 1998.

22 Khatsenkov interview.

23 Thompson interviews.

24 Forus Services S.A. registration documents; Kuppers interviews; Cuendet interviews; Mayor interviews; Maret interviews.

25 Kuppers interviews; Cuendet interviews; Mayor interviews; Forus Services S.A. registration documents.

26 Andava S.A. registration documents; Forus Services S.A. registration documents; Onexim Bank analytical department, "Berezovsky"; FOK registration documents; Grangeland Holdings registration documents.

27 Brunswick analyst report; Rinaco Plus analyst report; Mayor interviews; Nikolaev interview.

28 Nikolaev interview; Thompson interviews; Maret interviews; Mayor interviews; Brunswick analyst report; Rinaco Plus analyst report.

29 Nikolaev interview.

30 RUOP source; Cilluffo interviews; *Dengi* (March 3, 1999); *Sovershenno Sekretno* (February 1998).

   A special MVD anticrime operation at Avtovaz in 1997–98 uncovered evidence that gangsters connected to Avtovaz had carried out more than sixty-five murders of company managers and business rivals. (See *Sevodnya* [March 14, 1998].)

31 Thompson interviews.

32 Ibid.

33 Mayor interviews.

34 Maret interviews.

35 Thompson interviews.

36 Konstantinov, pp. 529–30; *Dengi* (March 3, 1999); *Sevodnya* (March 14, 1998).

37 *Kommersant-Daily* (September 23, 1991); Korzhakov interviews; Berezovsky interview; Nikolaev interview.

38 Nikolaev interview.
39 Ibid.; Brunswick analyst report; Rinaco Plus analyst report.
40 Nikolaev interview.
41 Korzhakov interviews; Davydov interviews.
42 Mayor interviews.
43 *Kommersant-Daily* (June 2, 1994); *Kommersant-Daily* (November 26, 1996); *Kommersant-Daily* (June 14, 1997); *Sevodnya* (August 1, 1997); *Kommersant-Vlast* (July 7, 1998); FOK 1996 annual report.
   AFK was reportedly owned by Avtovaz, Forus, Logovaz, Avva, Avtovaz Bank, Obyedinyonny Bank, and several individuals. The chairman was Vladimir Kadannikov. The first general director was Berezovsky's close partner, Nikolai Glushkov; he was later succeeded by Roman Sheinin, who was simultaneously general director of the Andava subsidiary, FOK.
44 Brunswick Brokerage, *Russian Equity Guide* (1997/98), "Avtovaz"; Brunswick analyst report; Rinaco Plus analyst report; Nikolaev interview.
45 Davydov interviews.
46 Aslund, p. 281; Davydov interviews; Aven interview.
47 Aven interview; Davydov interviews.
48 Davydov interviews.
49 Russian Petroleum Investor, *Complete Guide,* pp. 75–76.
50 Berezovsky interview.
51 Maret interviews.
52 Mordasov interview; Konstantinov, pp. 520, 531.
53 Davydov interviews.
54 Petrakov interview; Yavlinsky interview.
   By 1996, the total amount of Russian flight capital reached an estimated $150 billion (see Committee on International Relations, p. 34; CSIS, p. 4).
55 Yavlinsky interview.
56 Aven interview.
57 Gaidar interviews.
58 Vecchio interview.
59 Davydov interviews.
60 Mordasov interview; Davydov interviews; Glazyev interview; *Kommersant-Daily* (June 2, 1993).
61 Davydov interviews.
62 Mordasov interviews; Borovoi interviews; Konstantinov, pp. 517, 531.
63 Yasin interview.
64 Yavlinsky interview.
65 Yasin interview.
66 Eberstadt, pp. 4, 19.
67 Goskomstat, *Rossiisky Statistichesky Ezhegodnik—1997,* pp. 86, 686; Eberstadt, p. 7.

68 Eberstadt, p. 6.
69 Goldfarb and Becerra's article in *The Moscow Times* (January 29, 2000).
70 Urazhtsev interview; Aksyuchits interviews.
71 Goskomstat, *Rossiisky Statistichesky Ezhegodnik—1997*, p. 243; *The Moscow Times* (January 14, 2000).
72 *Novye Izvestia* (February 20, 1999).
73 Eberstadt, p. 10.
74 Goskomstat, *Rossiisky Statistichesky Ezhegodnik—1997*, p. 88; Goskomstat, *Rossiya v Tsyfrakh* (1998), p. 35.
75 Goskomstat, *Rossiisky Statistichesky Ezhegodnik—1997*, p. 236; Gray; Murray Feshbach writing in *The Moscow Times* (January 21, 1999).
76 Shelley interviews; *The New York Times* (January 11, 1998).
77 Goskomstat, *Rossiisky Statistichesky Ezhegodnik—1997*, pp. 59, 99; *The Moscow Times* (January 26, 2000).

The grim official statistics almost certainly underestimated the extent of Russia's population implosion. Russia's first post-Communist census was due to be conducted in 1999, but the government postponed it indefinitely—officially because it lacked $120 million in funding, but more likely because of a reluctance to expose the nation's demographic catastrophe. (See *The Moscow Times* [July 3, 1999].)

78 Goskomstat, *Rossiisky Statistichesky Ezhegodnik—1997*, p. 84; Agence France-Presse (November 18, 1998); Human Rights Watch, pp. 18–20.

Chapter 4

1 Korzhakov interviews.
2 Korzhakov, pp. 202–3.
3 Ibid., p. 205.
4 Ibid., pp. 213–20.
5 Ibid., pp. 303–4.
6 Ibid., pp. 140–50.
7 Ibid., p. 359.
8 Davydov interviews.
9 Maximov, pp. 110, 123; *Sydney Morning Herald* (March 4, 1995); *Interfax* report, reproduced by *BBC Worldwide Monitoring Service* (March 4, 1998).
10 Yeltsin, pp. 327–36.
11 Committee on International Relations, pp. 18–19.
12 Davydov interviews; *U.S. News & World Report* (August 3, 1998).
13 *Financial Times* (November 21, 1997); *Los Angeles Times* (December 3, 1997); *Izvestia* (July 1, 1997); *The Moscow Times* (July 2, 1997); *Institutional Investor* (April 30, 1997); Skuratov letter.
14 Wedel, pp. 121–63.
15 Waller, "Russia's Great Criminal Revolution"; Yeltsin, pp. 400, 406, 409.

16 Korzhakov, p. 10.
17 Ibid., pp. 136–37, 142.
18 Korzhakov interviews.
19 Ibid.; Berezovsky interview.
20 Korzhakov interviews; Yeltsin, *Zapiski Prezidenta.*
21 Nurnberg interviews.
22 Korzhakov interviews.
23 *U.S. News & World Report* (August 3, 1998).
24 Nurnberg interviews.
25 Korzhakov interviews.
   Korzhakov says he knows the sums involved because he occasionally handled Yeltsin's personal financial affairs and had special rights to use Yeltsin's safe. "The key to the safe was kept in a secret hiding place in Yeltsin's office. I was allowed to use this key and have access to the safe; when I carried out a personal mission for Yeltsin, I often had to open the safe to either deposit documents or remove them."
26 Ibid.
27 Korzhakov, p. 280.
28 Khatsenkov interview.
29 Korzhakov, pp. 158–60.
30 Yeltsin, pp. 378–87; Remnick, pp. 68–79.
31 Lebed interviews.
32 British political scientist Richard Sakwa produced an extensive analysis of the allegations of vote falsification and other illegalities in the 1993 election. (See Sakwa.)
33 Gaidar interviews.
34 Aslund, p. 235.
35 Yavlinsky interview.
36 Luzhkov interview.
37 Davydov interviews.
38 The $5 billion figure is derived in the following way: The street price of the voucher at the end of 1993 was about $7. A total of 151 million vouchers were distributed—slightly more than $1 billion worth. These were supposed to buy an average of 29 percent of the companies up for sale, which meant that these enterprises were being valued at $3 billion. In 1995, most of the rest of Russian industry was sold off for a further $1 billion. What remained—Svyazinvest, a few oil fields, and a big piece of the military-industrial complex—would have been valued (at 1993–95 prices) at another $1 billion.
39 Kharshan interviews.
   Another early millionaire in Russia, Oleg Boiko, also grew rich by using the post office network. He had started a check-cashing operation there.
40 Ibid.

41 Aleksandrov interview.
42 Ibid.
43 Ibid.
44 Orekhov interview; CSIS, p. 86.
45 Kharshan interviews.
46 Morgenstern interview.
47 CS First Boston, "Gazprom Voucher Auction Completed" (October 7, 1994).
48 Ibid.
49 Kharshan interviews.
50 CS First Boston, "Gazprom Voucher Auction Completed" (October 7, 1994); Kharshan interviews; Kokh interviews.
51 CS First Boston, *Russian Capital Markets* (June 1994); CS First Boston, "Gazprom Voucher Auction Completed" (October 7, 1994); MC Securities Ltd., *Russian Companies: Monthly Valuation Analysis* (August 5, 1997).

The voucher auction valuation for Lukoil is a combination of the company's biggest units, Lukoil Holding, Kogalymneftegaz, and Langepasneftegaz.
52 Kharshan interviews.
53 Mavrodi interview.
54 *The Moscow Times* (August 16, 1994); Ibid. (October 13, 1994).
55 Ibid. (November 1, 1994).
56 Berezovsky interview.
57 *Moskovskie Novosti* (April 27, 1994); *Kommersant-Daily* (September 9, 1993).
58 Berezovsky interview.
59 *Kommersant-Daily* (September 9, 1993).
60 Berezovsky interview.
61 Lou Hughes, head of international operations for GM, cited the "problems" with Avtovaz's dealership network as the primary reason why the American car company decided to minimize its investment in Russia (see Hughes interview).

Avva had promised to offer tens of thousands of cars as prizes in the lotteries (see *Kommersant-Daily* [September 9, 1993]), but only several hundred materialized (see *The Moscow Times* [February 18, 1994]).

Like MMM shares, Avva's shares went from being one of the country's most liquid securities to being worth zero. (See Kuppers interviews.)
62 Onexim Bank analytical department, "Berezovsky."
63 Kuppers interviews.
64 Davydov interviews.
65 *Sevodnya* (June 25, 1996).
66 Brunswick Brokerage, *Russian Equity Guide (1997/98)*, "Avtovaz." See also Appendix III below.

67 Korzhakov interviews.

At his October 11, 1996, press conference, when Korzhakov first accused Berezovsky of requesting the assassination of Gusinsky, he caused a sensation. He would repeat the accusation in numerous interviews and press conferences over the next several years (see *The New York Times* [August 13, 1997], for instance), though he never presented any concrete evidence to back it up. In November 1999, he claimed that he had taped conversations with Berezovsky proving that the tycoon had requested the assassination, but as of the publication of this book, the tapes have not been made public. (See Korzhakov interview with *Moskovsky Komsomolets* [October 30, 1999] and Korzhakov interview with *Moskovsky Komsomolets* [November 3, 1999].) However, Moscow mayor Yuri Luzhkov—another purported target of Berezovsky's assassination requests—told the press that he was "absolutely certain that Berezovsky had approached Korzhakov with a request to kill Gusinsky, Kobzon, and me." (See Luzhkov interview with *Moskovsky Komsomolets* [November 5, 1999].)

Chapter 5

1 Berezovsky interview; Berezovsky and Lesnevskaya appeal; *Kommersant-Daily* (April 29, 1995).

2 Berezovsky interview.

3 RUOP source; Premier S.V., "minutes"; Lis's registration documents; Kvantrishvili interview; Dickens interview.

4 RUOP source; RUOP spreadsheets; Maximov, pp. 79, 325, 342; Moody interviews.

5 Berezovsky interview; Streletsky interviews; Skuratov interviews; Korzhakov interviews; Barsukov memo.

6 Berezovsky interview.

7 Ibid.; Bull interview; Dickens interview; Razbash interview.

8 Berezovsky interview.

9 Korzhakov interviews.

10 Ibid.

11 Gusinsky interview with *Obshchaya Gazeta* (April 18, 1996); *Moskovskie Novosti,* No. 17 (March 12–19, 1995); Berezovsky interview; Korzhakov interviews; Berezovsky and Lesnevskaya appeal.

12 Gusinsky interview; Gazprom analytical department, "Gusinsky Vladimir Aleksandrovich"; Onexim Bank analytical department, "Gusinsky," pp. 14–15.

13 Gusinsky interview.

14 Gusinsky interview with *Obshchaya Gazeta* (April 18, 1996).

15 Korostelev interview.

16 Ibid.; Ptichnikov interview.

17 Gusinsky interview; Gazprom analytical department, "Gusinsky Vladimir Aleksandrovich"; Onexim Bank analytical department, "Gusinsky," pp. 14–15.

18 *The Washington Post* (February 20, 1994).

19 *The Observer* (December 11, 1994); Gusinsky interview with *The Boston Globe* (March 13, 1994); *The Guardian* (September 20, 1995); *Jane's Intelligence Review* (July 1, 1995).

20 *The Washington Post* (April 7, 1995).

21 Ibid.

22 Berezovsky and Lesnevskaya appeal.

23 *The Washington Post* (April 7, 1995); Gusinsky interview with *Obshchaya Gazeta* (April 18, 1996).

24 Korzhakov interviews.

25 Ibid.

The March 1995 videotaped complaint by Berezovsky to President Yeltsin confirmed Korzhakov's assertion. (See Berezovsky and Lesnevskaya appeal.)

26 *Moskovskie Novosti*, No. 17 (March 12–19, 1995).

27 Korzhakov interviews.

Though Korzhakov never produced any evidence that Berezovsky had asked him to kill Gusinsky, at a press conference in 1999, he claimed that he had tapes of his conversations with Berezovsky in which the tycoon made his request and that he was willing to give the tapes to the prosecutor-general's office for a criminal investigation. (See Korzhakov interview with *Moskovsky Komsomolets* [October 30, 1999] and Korzhakov interview with *Moskovsky Komsomolets* [November 3, 1999].) Luzhkov, however, said that he was "absolutely certain" that Berezovsky had requested the assassination of Gusinsky, Kobzon, and himself. (See Luzhkov interview with *Moskovsky Komsomolets* [November 5, 1999].)

28 Ibid.

29 Korzhakov interviews; Berezovsky and Lesnevskaya appeal.

30 Among the charges Yeltsin's cronies leveled against Gusinsky and Luzhkov was that Most Bank had helped cause the run on the ruble on "Black Tuesday." This event took place on October 11, 1994, when the Russian currency exchanges were hit by a massive wave of rubles in search of foreign currency. Russian banks unloaded huge holdings of the national currency in just a few hours. Before the panic ended and the Central Bank finally entered the fray, the ruble had lost 27 percent of its value. The Yeltsin clan claimed that Black Tuesday, a devastating blow to the credibility of Yeltsin's monetary stabilization policy, was the result of a conspiracy by Most and several other banks. Gusinsky denied the charges. (See *Argumenty i Fakty* [November 29, 1994]; Onexim Bank analytical service, "Gusinsky," pp. 15–16.)

Another incident making the Kremlin uncomfortable at this time was the investigation of a Luzhkov-allied newspaper, *Moskovsky Komsomolets*, into the corruption of Yeltsin's defense minister, Pavel Grachev. This investigation ended in October 1994, when the chief reporter on the job, Dimitri Kholodov, was blown up by a briefcase bomb.

31 Korzhakov, p. 285.
32 Ibid.
33 *Moskovskie Novosti* (May 28–June 4, 1995).
34 Ibid.
35 Korzhakov, p. 287.
36 *Moskovskie Novosti* (May 28–June 4, 1995).
37 Korzhakov, p. 286.
38 *Moskovskie Novosti* (May 28–June 4, 1995).
39 The British Broadcasting Corporation, *BBC Summary of World Broadcasts* (December 6, 1994); *Sevodnya* (December 8, 1994).
40 Korzhakov, p. 287; *Moskovskie Novosti* (May 28–June 4, 1995).
41 *Nezavisimaya Gazeta* (December 6, 1994).
42 Korzhakov, p. 297; Korzhakov interviews; Lebed interviews.
43 Remnick, pp. 212–13.
44 *Nezavisimaya Gazeta* (January 20, 1995).
45 *Euromoney* (January 1995).
46 Onexim Bank analytical service, "Gusinsky," p. 16.
47 Razbash interview.
48 Politkovsky interview; *Izvestia* (February 23, 1996); Berezovsky and Lesnevskaya appeal.
49 Korzhakov interviews.
50 Ibid.; *Kommersant-Daily* (April 29, 1995); Yavlinsky interview.
51 ORT registration documents.
52 Ibid.
53 Politkovsky interview.
54 Berezovsky interview; Korzhakov interviews.
55 Korzhakov interviews.
56 RUOP memorandum; RUOP source.

This information represented what the police termed "operational information"—intelligence detectives get from their informants in the Moscow underworld and elsewhere. It is not a version that is either proven or even double-checked—it is merely a lead that had to be followed up. This "operational information" on Vlad Listyev's murder is based on confidential police interviews with some of his closest friends and business associates.

57 Onexim Bank analytical department, "Berezovsky"; *Kommersant-Daily* (April 29, 1995).
58 Berezovsky and Lesnevskaya appeal.

59 Ibid.
60 Ibid.
61 Ibid; Korzhakov interviews.
62 Berezovsky and Lesnevskaya appeal; Korzhakov interviews.
    Berezovsky was apparently so concerned that he was about to be arrested that he spent two nights sleeping in Barsukov's office in the Kremlin. (See Korzhakov interview with *Moskovsky Komsomolets* [October 30, 1999].)
63 Berezovsky and Lesnevskaya appeal.
64 Ibid.
65 Ibid.
66 Korzhakov interviews.
67 Berezovsky interview.
    Two months after my interview with him, Berezovsky pointed the finger in yet another direction. He told the Russian press that he was convinced Listyev had been murdered by "the security services," seeking to impose their control over ORT. (See Berezovsky interview with *Itogi* and Berezovsky interview with *Obshchaya Gazeta* [December 3, 1998].)
68 *The Moscow Times* (March 7, 1995); Ibid. (August 9, 1995); Ibid. (November 17, 1995).
69 Maximov, p. 341.
70 Triboi interview.
71 Berezovsky interview.
72 Berezovsky interview with *Kommersant-Daily* (October 5, 1995).
73 Dimitri Lesnevsky, interview with *Radio Liberty* (September 27, 1995).
74 Berezovsky interview.
75 Ibid.

Chapter 6

1 Berezovsky interview.
2 Berezovsky interview with *Kommersant-Daily* (November 16, 1995).
3 SalomonSmithBarney, "Aeroflot"; Okulov interviews; Berezovsky interview with *Kommersant-Daily* (November 16, 1995).
4 Berezovsky interview with *Kommersant-Daily* (November 16, 1995).
5 Ibid.; Okulov interviews.
6 Korzhakov interviews; Berezovsky interview with *Kommersant-Daily* (November 16, 1995).
7 Korzhakov, p. 116.
8 Berezovsky interview with *Kommersant-Daily* (November 16, 1995).
9 Aeroflot, *1997 Annual Report* (English and Russian versions).
10 Andava S.A., *Procès-Verbal.*
11 Ferrero interviews.

12 Andava S.A., registration documents; Cuendet interviews; Maret interviews.
13 Maret interviews.
14 Mayor interviews.
15 Maret interviews.
16 Ibid.; Mayor interviews.
17 Ibid.
18 Nikolaev interview; Mayor interviews; *Moskovsky Komsomolets* (October 11, 1995); Onexim Bank analytical service, "Berezovsky"; Berezovsky interview with *Kommersant-Daily* (November 16, 1995).
    In the interview with *Kommersant*, Berezovsky admitted that he had taken out a big loan from Menatep Bank, but he didn't specify its purpose.
19 Meanwhile, the cofounder of Andava—the commodities trader André & Cie.—was on its way out. If Avtovaz was unhappy with the job Berezovsky was doing in his various financial operations, André & Cie. was unhappy too. Having seen most of its ventures with Berezovsky, chiefly the partnership with Avtovaz and the Avva investment scheme, end ignominiously, André decided to sell its stake in Andava as well. That, at least, is the version of André & Cie.
    "We simply didn't see big perspectives in the business; we wanted to deal with real industrial projects," says Yves Cuendet, the director authorized to speak for the secretive Swiss commodities trader.
    In fact, the promise of larger profits from servicing the new client, Aeroflot, may have impelled the Russian partners to recapitalize Andava and dilute the André stake. That at least is the implication of Andava's general director, William Ferrero. "Don't forget the Russians or the other shareholders," he says. "There are other people who have their say, to let or not let other shareholders participate [in the recapitalization]."
    The recapitalization occurred in February 1996 in Lausanne. Berezovsky emerged as the largest shareholder of Andava, with 37 percent; Glushkov came second, with 34 percent; André held 25 percent; the remaining 4 percent remained with the original Russian entity that had cofounded Andava, Avva International. (André would sell its 25 percent stake to the Russian partners a year later.)
    (See Cuendet interviews; Ferrero interviews; Andava S.A., *Procès-Verbal.*)
20 Okulov interview; *Obshchaya Gazeta* (April 17, 1997).
21 Korzhakov interviews; Barsukov memo.
22 Okulov interviews.
23 Maret interviews.
24 Ferrero interviews.
25 Volkov interviews; Griffiths interview.
26 Griffiths interview; Andava Group, "Contract"; Volkov interviews.

27 Aeroflot, *1997 Annual Report* (English version); Andava, "VEK filing."
28 Maret interviews.
29 Ferrero interviews.
30 Ibid.
31 *Forbes* (December 30, 1996); *L'Hebdo* (April 24, 1997); *Obshchaya Gazeta* (April 17, 1997).

    According to both Russian and Swiss prosecutors, as of the publication of this book, the investigation into the relationship between Berezovsky's companies and Aeroflot is continuing.
32 Aeroflot, *1996 Annual Report* (Russian version).
33 Korzhakov interviews.
34 The transcript of these tapes appeared in a magazine linked to Korzhakov: *Lyudi* (December 1998).
35 Ibid.
36 Ferrero interviews.
37 Davydov interviews.
38 Ferrero interviews.
39 Davydov interviews.
40 Aeroflot, *1997 Annual Report* (United States SEC filing), p. 8; Ferrero interviews.
41 Maret interviews; Cuendet interviews.
42 Aeroflot, *1997 Annual Report* (English version).
43 Volkov interviews; Maret interviews; Okulov interviews; Ferrero interviews; Kuppers interviews.
44 FOK registration documents; FOK, *1996 Annual Report.*
45 Aeroflot, "Contract."
46 Aeroflot, *1997 Annual Report* (English version).
47 Aeroflot—FOK—Grangeland payment invoices and receipts.
48 Grangeland Holdings registration documents.
49 Ibid.; Aeroflot—FOK—Grangeland payment invoices and receipts; Ferrero interviews; Volkov interviews.
50 FOK, *1996 Annual Report*; Aeroflot, *1997 Annual Report* (English version); Aeroflot, "Contract."
51 FOK, *1996 Annual Report.*
52 Volkov interviews; Okulov, press conference.
53 Okulov, press conference.
54 Aeroflot, *1997 Annual Report* (English version); Aeroflot, *1997 Annual Report* (United States SEC filing); Aeroflot list of shareholders; Aeroflot, "Contract."
55 Aeroflot, *1998 Annual Report* (Russian version); Aeroflot, *1997 Annual Report* (English version); Maret interviews.

56 Volkov interviews; Okulov press conference.
57 Okulov press conference.

Chapter 7

1 *The Guardian* (September 20, 1995). The same story is in Berezovsky's interview with *Obshchaya Gazeta* (December 3, 1998).
2 *Sevodnya* (July 26, 1995); *Kommersant-Daily* (July 27, 1995); *Kommersant-Daily* (March 1, 1996); Konstantinov, pp. 551–52.
  Gaft had quarreled with other Logovaz directors shortly before his death; he was fired on July 5. But the new general director of Logovaz, Yuly Dubov, told *Kommersant* that Gaft was rehired three days later. According to Dubov, Gaft fell to his death after he locked himself out of his apartment and attempted to lower himself onto his balcony from the floor above.
3 Stacy interview.
4 Yergin interview.
5 Alekperov interviews; Veinshtock interview.
6 Alekperov interviews.
7 Hamilton interviews.
8 Reardon interviews.
9 Alekperov interviews.
10 Ibid. The equity stake bought by Alekperov and other managers is my rough estimate based on interviews with the managers of Lukoil and its investment banking arm, Nikoil.
11 Remp interviews.
12 Gouliev interviews.
13 Stacy interview.
14 Alekperov interviews.
15 Hamilton interviews.
16 Alekperov interviews.
17 Derr interview; Alekperov interviews.
18 Ibid.
19 Hamilton interviews.
20 Alekperov interviews.
21 Kokh interviews; Korzhakov interviews; *The New York Times* (September 21, 1999); *The Wall Street Journal* (September 22, 1999); *The Moscow Times* (September 24, 1999).
22 Berezovsky interview.
23 Berezovsky interview with *Kommersant-Daily* (October 5, 1995).
  This was confirmed by General Korzhakov, who added the name of Tatyana Dyachenko as one of the key people Berezovsky lobbied to get Sibneft (see Korzhakov interviews).

24 Korzhakov interviews; Streletsky interviews; Berezovsky interview with *Kommersant-Daily* (October 5, 1995); *Itar-Tass* (May 31, 1995).

25 Korzhakov interviews.

26 Alekperov interviews.

27 MC Securities, "Siberian Oil Company," p. 13; Streletsky interviews; *Kommersant-Daily* (September 8, 1995).

28 Berezovsky interview with *Kommersant-Daily* (October 5, 1995).

29 *Russian Petroleum Investor* (February 1994) Appendix; Ibid. (March 1996) Appendix; Davydov interviews.

30 Berezovsky and Lesnevskaya, appeal; *The Moscow Times* (October 10, 1995); Ibid. (September 18, 1996).

31 Jordan interviews; *Kommersant-Daily* (November 16, 1995).

32 Khodorkovsky interviews.

33 Potanin interviews; *Kommersant-Daily* (November 16, 1995); Gazprom analytical department, "Potanin, Vladimir Olegovich."

34 MFK, "International Company for Finance and Investments—1997"; Potanin interviews; Jordan interviews; *Euromoney* (November 1997).

35 Potanin interviews.

36 Ibid.

37 Onexim Bank, "The United Export-Import Bank Open Joint-Stock Company" brochure; Potanin interviews; *Kommersant-Daily* (November 16, 1995).

38 Streletsky interviews.

39 Ibid. Korzhakov interviews.

40 Korzhakov interviews.

41 Ibid.; *The New York Times* (September 21, 1999); *The Wall Street Journal* (September 22, 1999); *The Moscow Times* (September 24, 1999).

42 Yeltsin, pp. 19–20.

43 Streletsky interviews.

44 Russian Petroleum Investor, *Complete Guide,* pp. 75–76.

Later Berezovsky would adopt his own pocket bank, Obyedinyonny Bank, for day-to-day operations. He would also use Avtovaz Bank until it went bankrupt in 1995. But Berezovsky would still turn to Menatep whenever major funds were required. It was Menatep who finally paid off Berezovsky's debts to Avtovaz in the summer of 1995. It was Menatep who guaranteed the financing on the Sibneft purchase several months later. It was with Menatep-controlled Yukos that Berezovsky tried to merge Sibneft in 1998.

45 Khodorkovsky interviews.

46 Ibid.; *Menatep Group News* (July 1, 1996).

47 Khodorkovsky interviews.

48 Ibid.; *The Washington Post* (October 7, 1996); Ibid. (February 17, 1998).

49 *The Moscow Times* (November 10, 1995).
50 Ibid. (December 6, 1995); *Rossiiskaya Gazeta* (December 29, 1995).
51 Brunswick Brokerage, *Russian Equity Guide,* "Norilsk Nickel."
52 Davydov interviews.
53 *The Moscow Times* (November 21, 1995).
54 Ibid. (December 6, 1995); *Rossiiskaya Gazeta* (December 29, 1995).
55 MC Securities, "Siberian Oil Company."
56 Kokh interviews; *Sevodnya* (December 29, 1995); Berezovsky interview.
57 Kokh interviews.
58 There was some kind of a "favor-for-a-favor" arrangement between Menatep and Stolichny Bank. While Menatep guaranteed the bid of NFK (Berezovsky, Stolichny Bank, and Abramovich) for Sibneft, Stolichny Bank guaranteed Menatep's bid for Yukos.
59 Korzhakov interviews.
60 Berezovsky interview.
61 Berezovsky interview; Kokh interviews.
62 Chubais interview.
63 Potanin interviews.
64 The official auction results as printed in *Rossiiskaya Gazeta* (December 29, 1995); MC Securities, "Russian Companies"; Brunswick Brokerage, *Russian Equity Guide.*
65 Streletsky interviews.
66 Berezovsky interview with *Kommersant-Daily* (November 16, 1995).
67 Davydov interviews.
68 Sysuyev interview.

Chapter 8

1 Korzhakov, p. 319.
2 Berezovsky interview.
3 Berezovsky interview with *Kommersant-Daily* (June 17, 1997); Korzhakov, p. 289.
4 Gusinsky interview with *Obshchaya Gazeta* (April 18, 1996).
5 Ibid.
6 Berezovsky interview with *Kommersant-Daily* (June 17, 1997).
7 Skuratov letter; *The Moscow Times* (July 2, 1997); Ibid. (July 25, 1997); *Institutional Investor* (April 30, 1997).
8 Berezovsky interview with *Kommersant-Daily* (March 13, 1998).
9 Korzhakov interviews.
10 Chubais interview with *Kommersant-Daily* (March 5, 1998).
11 Korzhakov, p. 381.
12 Ibid., p. 367.
13 Berezovsky, "Vyiti iz Tupika!"

14 Berezovsky interview with *Kommersant-Daily* (June 17, 1997).

15 Potanin interviews.

16 *The Moscow Times* (May 12, 1996).

17 Berezovsky interview.

18 Berezovsky interview with *Kommersant-Daily* (June 17, 1997).

19 Korzhakov, p. 324.

20 Berezovsky interview with *Kommersant-Daily* (June 17, 1997).

21 Korzhakov, p. 355.

22 Berezovsky interview with *Kommersant-Daily* (June 17, 1997).

23 Ibid.

24 Korzhakov, p. 355.

25 Streletsky interviews; CSIS, p. 21.

26 Streletsky interviews.

27 Chubais interview.

28 Korzhakov interviews.

29 Streletsky interviews.

30 Chernomyrdin transcript; Kiriyenko interview; Barsukov memo.

31 Streletsky interviews.

32 Berezovsky interview with *Kommersant-Daily* (March 13, 1998).

33 Streletsky interviews; Korzhakov interviews.

34 Ibid.

35 The capital gains calculations are the same as those in Note 64 and accompanying table in Chapter 7 of this book.

36 Chernomyrdin transcript.

37 Lebed interviews.

38 Committee on International Relations, p. 24.

39 Berezovsky interview.

40 *Chicago Sun-Times* (July 1, 1996).

    The only concrete accusation came nearly a year after the election, against the first major newspaper to break from the Yeltsin camp. In early 1997, one of Yeltsin's public-relations managers, Sergei Zverev, spoke out against *Moskovsky Komsomolets*, which had recently published the transcript of the scandalous Chubais cover-up of the Xerox box filled with cash. He accused the newspaper of failing to return $1.5 million (or "three Xerox boxes") it had received during the campaign. (See *The Moscow Times* [April 3, 1997].)

41 Streletsky interviews.

42 Berezovsky interview with *Moskovskie Novosti* (May 16, 1996).

43 *Los Angeles Times* (June 25, 1996).

44 *The Sunday Times of London* (April 14, 1996); *The Globe and Mail* (July 3, 1996).

45 *The Sunday Times of London* (April 14, 1996); *Time* (July 15, 1996); Korzhakov, p. 356.

46 *The Sunday Times of London* (April 14, 1996); *Time* (July 15, 1996); *The Globe and Mail* (July 9, 1996).

47 *Newsweek* (June 17, 1996).

48 *The Washington Post* (July 7, 1996).

49 *Toronto Star* (December 9, 1996); *The Wall Street Journal* (June 11, 1996).

50 In a war characterized by the Russian army's use of World War II–era technology and tactics, it was unclear how it had suddenly managed to employ such high-tech sophistication. This incident raised another question: If Russian army commanders truly possessed this kind of war-fighting technology all along, why did they never use it again—to kill the most fearsome Chechen warlord, Shamil Basayev, for instance?

51 Streletsky interviews.
Chubais would later confirm that the Yeltsin campaign deliberately began collecting compromising materials on Korzhakov in April. (See Chubais transcript.)

52 Streletsky interviews.

53 ORT registration documents; Sayer interview; Konstantinov, p. 533; Davydor interviews; Streletsky, p. 192; Streletsky interviews.

54 Konstantinov, p. 545; Streletsky interviews.

55 Streletsky interviews.

56 Ibid.

57 Ibid.; Fyodorov transcript.

58 Streletsky interviews.

59 Ibid.

60 Ibid.

61 *The Washington Post* (July 7, 1996).

62 *Dow Jones International News Services*, "Russian businessmen" (June 10, 1996).

63 Berezovsky interview with *Moskovskie Novosti* (May 16, 1996).

64 Chubais interview.

65 *The Moscow Times* (June 13, 1996).

66 Barsukov memo.

67 *The Moscow Times* (May 12, 1996).

68 Korzhakov interviews; Berezovsky interview with *Kommersant-Daily* (March 13, 1998).

69 Korzhakov interviews.

70 Skuratov letter; Streletsky interviews.

71 Streletsky interviews.

72 Ibid.

73 Ibid.

74 Skuratov letter.

75 Ibid.; Korzhakov, p. 456.

76 Skuratov letter.

77 *The Moscow Times* (June 21, 1996).

78 Ibid.

79 Streletsky interviews.

80 Ibid.

81 Fyodorov transcript.

Chubais was worried that Yeltsin, after having fired Korzhakov, would reappoint his old friend to the government. (See Chubais transcript.)

82 Fyodorov interview in *Komsomolskaya Pravda* (July 10, 1996).

83 Olga Fyodorov interview in *Komsomolskaya Pravda* (July 10, 1996).

84 Streletsky interviews.

Streletsky says he became convinced that the Fyodorov assassination attempt had been organized by members of the Yeltsin campaign team several months later, when both he and Korzhakov were already out of government and Korzhakov was running for a seat in parliament.

"During Korzhakov's election campaign, when he was running for a seat in the Duma for the region of Tula, these people around Yeltsin—Berezovsky, Chubais, Lisovsky, Gusinsky, etc.—used the same tactics," he says.

"In the course of the election campaign in Tula, Korzhakov had a real rival: Eduard Pashchenko. He was a local inhabitant and a deputy of the Duma from the previous session," Streletsky says. "The above-mentioned businessmen felt that Korzhakov was gaining momentum, that he was about to be elected; they started showing Pashchenko on Berezovsky's Channel 1 and on Gusinsky's Channel 4, where Pashchenko spoke about Korzhakov violating the election rules—he was supposedly distributing free vodka.

"Then, we—Korzhakov, I, and other members of Korzhakov's team—got a visit from some gangsters, who said: 'We don't want to meddle in politics, but we have been given a contract to blow Pashchenko up.' These gangsters told us who had taken out a contract to kill Pashchenko. The contract came from that small group of businessmen who had played the leading role in Yeltsin's election campaign. We had to provide Pashchenko with security ourselves, so that he wouldn't be assassinated. Because if Pashchenko were to be blown up, the suspicion would land first of all on Korzhakov."

85 Korzhakov, p. 12.

86 Korzhakov interviews.

87 Chubais transcript.

88 Korzhakov interviews.

89 Chubais transcript.

90 Ibid.

91 Ibid.

92 Ibid.
93 Ibid.
94 Ibid.
95 Ibid.
96 Skuratov letter.
97 Ibid.
98 Ibid.

Chapter 9

1 CSIS, p. 46.
2 Ibid., pp. 81, 85.
3 Ibid., p. 81; NTV and ORT news reporting, November 10–11, 1996.
4 House of Lords, pp. 7, 23; Berezovsky interview with *Sunday Telegraph* (March 11, 1999); *The Washington Times* (August 10, 1998); Ibid. (September 28, 1998); *The Sunday Times of London* (August 30, 1998).
5 Yavlinsky interview.

Yavlinsky explained what he meant: "Corporatist meant that the state and the government promoted the interests of individual corporations—oil, gas, energy, industrial, armaments, financial. Oligarchic meant that the main characteristic of this system was the simultaneous access to government financing (the budget and the treasury), the government and the press. In other words, a single individual could simultaneously be a government official, the owner of a TV channel, and a person with good access to [government] financing. Based on monopolized property rights meant that the people, the citizens of Russia, did not get access either to financing or to property. Everything was based on those same monopolies that were liberalized in 1992. Semicriminal relationships were the result of the fact that the work of the oligarchy was based on personal relationships—it functioned outside even the minimal legal code that we have."

6 Lebed interviews.
7 Korzhakov press conference; *Novaya Gazeta* (November 9, 1996); *Moskovsky Komsomolets* (November 15, 1996); *The Moscow Times* (April 3, 1997).
8 Kiriyenko interview; Korzhakov interviews; Berezovsky interview.
9 Streletsky interviews; *Jewish Telegraphic Agency* (November 11, 1996); *Agence France Presse* (November 14, 1996).
10 *Sevodnya* (November 14, 1996); *Agence France Presse* (November 14, 1996); *Komsomolskaya Pravda* (November 5, 1996); *Jewish Telegraphic Agency* (November 21, 1996).

In the 2000 presidential race, Berezovsky's TV network ORT was the primary vehicle fanning anti-Semitism in Russia; a week before the elections, ORT broadcast an analysis of presidential challenger Grigory Yavlinsky, labeling him as a Jew and noting that he was supported by Vladimir Gusinsky's

NTV, which it asserted was funded by "citizens of Israel." (See *The New York Times* [March 25, 2000].)

11 Berezovsky interview with *Kommersant-Weekly* (November 12, 1996); Lebed interview with *Le Figaro* (February 17, 1997); Lebed interviews.

12 Lebed press conference.

13 *Ogonyok* (May 5, 1997).

14 *Kommersant-Weekly* (November 12, 1996); Streletsky interviews; Maskhadov interview; Usmanov interview; Lebed interviews.

15 U.S. State Department source.

16 Maskhadov interview; *Los Angeles Times* (December 19, 1996).

17 Lebed interview with *Le Figaro* (February 17, 1997).

18 Maskhadov interview.

19 Berezovsky press conference (Moscow, May 13, 1997); Berezovsky press conference (Moscow, August 20, 1997); Maskhadov interview; U.S. State Department source.

20 Maskhadov interview.

21 Ibid.; Streletsky interviews.

22 Berezovsky-Udugov Conversation 2.

23 Korzhakov interviews; Streletsky interviews; RUOP memo; Berezovsky and Lesnevskaya appeal; Berezovsky-Udugov Conversations 1 and 2.

24 Berezovsky-Udugov Conversation 1.

25 Maskhadov interview; *The Daily Telegraph* (September 26, 1998); *The Guardian* (October 6, 1998); U.S. State Department source.

26 *Komsomolskaya Pravda* (November 3, 1998); *The Independent* (November 3, 1998).

27 Maskhadov interview.

28 *Lyudi*, pp. 12–13; Volkov interviews; *The Jerusalem Post* (March 24, 1998).

Since he was prohibited from advancing his private business interests while he held a government post, Berezovsky resigned his official posts at Logovaz, ORT, and Sibneft. Abroad, he resigned from the board of directors of his two Swiss companies, Forus and Andava.

At the same time his longtime Swiss partner, commodities trading firm André & Cie., sold off its last stake in Andava. Alain Mayor, the man who had been representing André's interests in its dealings with Berezovsky, decided to jump ship and join the Russian tycoon—exactly the way Nikolai Glushkov, head of finance for Avtovaz, decided in 1995 to resign and join Berezovsky after the auto company quarreled with the tycoon. Mayor resigned from André and joined the boards of directors of Andava and Forus.

"Sometimes it happens that people within a company develop business relations that they find so wonderful and so good that they ask themselves— hey, why shouldn't I profit from this myself, instead of making my boss rich," notes Yves Cuendet, one of André's top managers. "Nobody really wanted to see him stay," he adds. "Several times he was associated with things that weren't the activities we preferred to be in." (Cuendet interviews.)

29 Kokh interviews.

30 Ibid.

31 Khodorkovsky interviews.

32 Kokh interviews.

33 Ibid.; Lebed interview in *Le Figaro* (February 17, 1997).

34 Chubais interview.

35 Berezovsky interview with *Kommersant-Daily* (March 13, 1998).

36 Chubais interview with *Kommersant-Daily* (March 5, 1998).

37 Demtsov interview.

38 Kokh, pp. 129–31, 201–4; Kokh interviews; *The Financial Post* (November 26, 1996).

39 Berezovsky and Lesnevskaya appeal; Potanin interviews.

40 Potanin interviews; *Izvestia* (July 31, 1997); *Itar-Tass Express* (No. 31, 1997).

41 *Kommersant-Daily* (August 15, 1997).

42 Gusinsky interview with *Ekho Moskvy* (August 15, 1997).

43 *The Moscow Times* (August 14, 1997); Ibid. (October 2, 1997).

44 Chubais interview; Chubais interview with *Kommersant-Daily* (March 5, 1998); Gazprom analytical department, "Ekonomicheskie i Politicheskie Perspektivy Svyazki Chubais-Potanin."

45 Nemtsov interview; *Financial Times* (November 6, 1997); *The Washington Post* (September 26, 1997).

46 Nemtsov interview; *Financial Times* (November 21, 1997); *Los Angeles Times* (December 3, 1997); *Izvestia* (July 1, 1997); *The Moscow Times* (July 2, 1997); *Institutional Investor* (April 30, 1997); Skuratov letter.

47 Berezovsky interview with *Kommersant-Daily* (March 13, 1998).

48 Kokh interviews; Sibneft financial statements.

49 Kiriyenko interview.

50 Kokh interviews; Khodorkovsky interviews; Jordan interviews.

51 Berezovsky interview with *Kommersant-Daily* (March 13, 1998).

52 Kiriyenko interview.

53 Ibid.

54 Berezovsky interview with *Kommersant-Daily* (March 13, 1998).

55 Kiriyenko interview.

56 Chubais claims that the reason foreigners were not allowed to participate in the GKOs was that there was a fear that foreign capital would be more flighty than domestic Russian capital. (See Chubais interview.) But as August 1998 showed, the Russian banks were not the domestically rooted, patriotic institutions Chubais made them out to be. They were the champions of

"hot money" and fled the Russian market much faster than the foreign institutions. (See Kiriyenko interview.)

57 Kiriyenko interview.

58 *The Moscow Times* (September 10, 1998).

59 *The Moscow Times* (December 22, 1998); Soros, "Who Lost Russia?"

Chapter 10

1 Korzhakov interviews.

2 Kiriyenko interview.

3 Korzhakov interviews; Streletsky interviews; Volkov interviews.

4 Korzhakov interviews.

5 Volkov interviews.

6 Davydov interviews.

7 Volkov interviews; Okulov press conference.

8 Volkov interviews.

The fact that Berezovsky was a part owner of SBS-Agro was a widely held view in Russian government circles. Former Prime Minister Sergei Kiriyenko, for instance, claimed that Berezovsky owned 25 percent of SBS-Agro (see Kiriyenko interview). One fact is indisputable: SBS-Agro (known as Stolichny Bank until 1997) served as a vehicle for Berezovsky in his takeover of Sibneft, starting in the loans-for-shares auction of Sibneft in December 1995, when Stolichny Bank was declared to be the winner of a 51 percent stake in Sibneft (though, according to Berezovsky's own admission in my 1996 interview, these shares were bought by him). An SBS-Agro front company also bought that same 51 percent stake in the second round of the Sibneft auctions, in May 1997. Sibneft headquarters in Moscow were physically linked to a branch office of SBS-Agro.

9 Skuratov interview with *The Moscow Times* (September 7, 1999).

With Switzerland getting tough on money-laundering, many of these officials funneled their earnings through the Bank of New York to offshore tax havens in the Caribbean and the South Pacific. The use of obscure islands in the South Pacific, typically accessed via Australia or New Zealand, was a novel capital flight and money-laundering strategy in the late 1990s. In 1998–99, several top Russian government officials made short, unofficial visits to Australia and New Zealand.

10 *The New York Times* (February 17, 2000); Ibid. (April 21, 2000); *The Wall Street Journal* (February 18, 2000).

11 Volkov interviews; Reymond interviews; Skuratov interviews.

12 Volkov interviews.

13 Berezovsky interview with *Sunday Telegraph* (March 11, 1999); Volkov interviews.

14 *The Moscow Times* (January 29, 1999); Ibid. (March 3, 1999).

15 Berezovsky downplayed his role in the government shake-up in a press conference on June 4, 1999, though he did admit to having "normal relations" with Tanya Dyachenko and "good relations" with Yeltsin's chief of staff, Aleksandr Voloshin.

16 Berezovsky interview with *Le Figaro* (September 22, 1999).

17 *The Moscow Times* (May 18, 1999).

18 Kalugin interviews; *The Moscow Times* (July 24, 1999); *Moskovsky Komsomolets* (December 17, 1999); *Moskovskie Novosti* (April 27, 1994); Ibid. (June 22, 1994).

19 *The Moscow Times* (April 25, 2000).

20 Kalugin interviews.

21 Ibid.

22 *Nezavisimaya Gazeta* (September 13, 1997).

23 Streletsky interviews; Kalugin interviews; Kiriyenko interview.

24 Kiriyenko interview.

25 Kalugin interviews.

26 Ibid.

27 Berezovsky interview in *Kommersant-Daily* (March 13, 1998).

28 Stepashin interview with *Nezavisimaya Gazeta* (January 14, 2000).

29 Maskhadov interview; Lieven interview.

30 Maskhadov interview.

31 Ibid.; Berezovsky interview with *Le Figaro* (September 22, 1999); Berezovsky-Udugov Conversations 1 and 2; *Moskovsky Komsomolets* (September 14, 1999); Ibid. (September 22, 1999).

32 Maskhadov interview.

33 *Le Monde* (September 28, 1999); Ibid. (January 25, 2000); *The Moscow Times* (March 17, 2000); Lebed interview with *Le Figaro* (September 29, 1999); Kozlov interview.

34 Kozlov interview.

35 Maskhadov interview.

36 *The Moscow Times* (October 28, 1999); ibid. (September 17, 1999).

    George Soros hinted that he suspected the same thing. (See Soros, "Who Lost Russia?")

37 See, for example, the reporting of Carlotta Gall for *The New York Times* in October 1999.

38 Sysuyev interview.

39 Reuben interview.

40 Maximov, pp. 282–91; Konstantinov, pp. 298–310; *Wall Street Journal-Europe* (January 28, 1997); Cilluffo interview; Streletsky, p. 140.

    In the autumn of 1999, law-enforcement authorities in Switzerland, the U.S., and the U.K. opened investigations of Lev Chorny. (See *Financial Times* [October 12, 1999].)

41 Bykov interview; Konstantinov, pp. 298–310; Maximov, pp. 282–91; *Wall Street Journal-Europe* (January 28, 1997); Reuben interview; Polevanov interview; Kokh, pp. 66–70.

42 Konstantinov, pp. 298–310; *The Moscow Times* (April 15, 1995); CSIS, p. 86.

43 Clingman interview; CSIS, p. 85; *Financial Times* (May 2, 1996).

44 Kulikov letter; *The Moscow Times* (February 22, 1997).

45 Aluminum executive source.

46 Ibid.

47 Russian Petroleum Investor, *The Complete Guide*, p. 76; Davydov interviews; Aluminum executive source.

   The huge size of the aluminum shipment attributed to Logovaz is surprising, to say the least. The yearbook's information seems to be based on registrations with the Customs Service and the Ministry of Foreign Trade. The authors of the yearbook are reputable: a Los Angeles–based consultancy called Russian Petroleum Investor, industry newsletter *Petroleum Intelligence Weekly,* and the big accounting firm Ernst & Young. The yearbook is impressive in its level of detail. It lists the precise dates and registration numbers for foreign trade licenses, for instance. But I found no corroborating evidence that Logovaz handled a third of Russia's aluminum exports in 1992.

48 Sysuyev interview.

49 Maximov, pp. 18–35; CSIS, pp. 79–88; *The Daily Telegraph* (October 6, 1995).

50 *The Moscow Times* (September 7, 1999).

51 Shelley interviews.

52 *The Sunday Times of London* (December 8, 1996).

Epilogue

1 Committee on International Relations, pp. 1–2.

2 Ibid., p. 28.

3 *The New York Times* (November 23, 1998); Fritz Ermarth, "Seeing Russia Plain."

Appendix 2

1 Note the sudden use of the Communist term "comrade."

2 Again, a conscious use of an archaic Marxist term, "toilers." Is this sarcasm? Hypocrisy?

INDEX

Abramovich, Roman
  aluminum industry and, 307, 311
  banks and, 289
  oil industry and, 195, 196, 197, 201, 208,
    266, 275–277, 292
  Russian parliament and, 316
  Yeltsin family and, 201, 284–285
Aeroflot
  Andava and, 173–184, 186, 266, 287, 296
  control of, 189, 198, 230, 318
  FOK and, 183, 184, 185
  Forus Services and, 182–183, 186
  Glushkov and, 172, 176, 179, 183, 186,
    289–290
  Gusinsky and, 148
  Logovaz and, 172, 186
  NFQ and, 185
  Obyedinyonny Bank and, 89
  privatization of, 171–173, 187, 275
  Tarasov and, 85
Aeroflot Bank, 31
AFK (Automobile Finance Corporation),
    89, 94, 143, 176, 183, 184
Agroprom Bank, 253
Alekperov, Vagit, 190–195
Alfa Bank, 160, 201, 204, 239, 268, 271,
    279
Alisa, 82–87
aluminum industry
  assassinations in, 188
  Berezovsky and, 4, 307–313
  Bykov and, 288

commodities trading and, 96, 98
Logovaz and, 97, 307, 312
privatization and, 32, 129
Rich, Marc and, 63, 64, 85
Andava S. A., 89, 173–184, 186, 266, 274,
    287, 289, 296
André & Cie., 23–24, 68–72, 89, 92, 174,
    176–178, 182, 186
Anros S. A., 71, 89, 184
Atoll, 285, 286, 287
automobile industry. *See also* Avtovaz;
    Logovaz
  André & Cie. and, 68–72
  Chechen organized crime and, 15
  privatization and, 170–171
  Russian economy and, 90, 95
  Russian organized crime and, 11–15, 21,
    23, 24, 37, 90
  Thompson and, 87–95
Aven, Pyotr, 75, 85–86, 95, 96, 99, 131, 201,
    271
Avtovaz
  AFK and, 94, 183, 184
  Andava and, 175, 180
  André & Cie. and, 70, 92
  Avva and, 140–142, 143
  control of, 198
  establishment of, 53–54
  Forus Services and, 89
  Logovaz and, 55, 71–72, 93, 94, 143,
    175–176, 202
  Obyedinyonny Bank and, 89

Avtovaz (*continued*)
  privatization and, 136, 170–171, 275
  Russian organized crime and, 21, 90–95,
    141, 287
  Thompson and, 88, 91–92
  Yanchev and, 197
Avtovaz Bank, 90, 91, 92, 143, 148, 172
Avva, 39, 89, 140–143, 171, 175, 176, 293,
  318
Azerbaijan, 15, 73, 191–192

Baku consortium, 191–193
Baku-Novorossiisk pipeline, 258, 259
Balkar Trading, 196–197, 276
Bank of New York, 288, 289, 293
banks and banking. *See also* Central Bank;
  and specific banks
  Chechen organized crime and, 15
  Chechnya and, 288
  Chubais and, 200, 210, 214–215
  debt-for-equity deal and, 198
  election campaign of 1996 and, 213,
    221–222, 239
  Gaider and, 81
  Great Chechen Bank Fraud, 27–28
  Gusinsky and, 148–149
  KGB and, 60–61
  oil industry and, 198
  Potanin and, 199–200
  privatization and, 5, 200–201, 210–211,
    271
  Russian economy and, 279–282, 289
  Russian organized crime and, 31, 199
  Yeltsin regime and, 30
Barsukov, Mikhail
  election of 1996 and, 215, 217, 224, 243,
    244
  Fyodorov and, 232–233, 241
  Gusinsky and, 155
  Russian parliament and, 121–122
  Yeltsin and, 113, 237
Basayev, Shamil, 13, 41, 43, 255, 261,
  300–301, 303, 305, 306
Bauman Group, 22–23, 28
Berezovsky, Boris
  airline industry
    Aeroflot and, 171–173, 176, 187, 189,
      198, 230, 289, 290, 318
    Andava and, 173–184, 186, 266, 287,
      296
    privatization and, 171–173
  aluminum industry
    Chorny and, 311–313
    commodities trading and, 97, 171, 202,
      365*n*43

Logovaz and, 97, 307, 312
    Rich, Marc and, 64
  automobile industry
    André & Cie. and, 68–72
    Avtovaz and, 3–4, 21, 53–55, 68, 70, 71,
      90, 92–95, 136, 187, 198
    Avva and, 140–143
    Forus Services and, 89
    Logovaz and, 4, 12, 37–40, 55, 70, 71,
      75–76, 92–94, 96, 117, 319
    privatization and, 170–171
    Russian organized crime and, 11–15,
      21, 24, 37
  banks
    Avtovaz Bank, 90, 91, 92, 143, 148, 172
    commodities trading and, 68
    election of 1996 and, 214
    Gusinsky and, 148
    investigations of, 289
    Logovaz and, 92
    Obyedinyonny Bank, 89
    SBS-Agro and, 280, 288
    Stolichny Bank, 85
  oil industry
    commodities trading and, 97, 171, 202
    Logovaz and, 97–98
    privatization and, 209–210
    Rosneft and, 194–195, 196
    Sibneft and, 195–198, 205–208, 230,
      266, 267–269, 275–277, 307, 318
  politics
    Aeroflot and, 179
    Chechen hostages and, 259–266, 293,
      296, 301
    Chubais and, 44–45, 270, 275
    Communism's fall and, 75
    corruption and, 318–319, 325–326
    election campaign of 1996 and, 169,
      212–223, 234–235, 240–241, 243,
      246
    KGB and, 59
    Kiriyenko and, 277–282
    Lebed and, 304
    Primakov and, 284, 286–292
    privatization and, 2, 120, 270–271
    Putin and, 295–298, 305
    Russian parliament and, 314–316
    Second Chechen War and, 299,
      301–302
    as Security Council deputy secretary, 5,
      256–257, 269, 274
    television and, 146, 147, 153–155, 169,
      189, 196, 220, 221, 225, 226, 313
    Yeltsin and, 110, 116–120, 140, 143,
      144, 163–166, 169, 201, 211,

284–285, 289, 292, 293, 301, 317, 319
Russian organized crime
business empire and, 37
Chechen War and, 42
Great Mob War and, 23
Logovaz bombing and, 37–40
Mikhailov and, 21
television
Channel 1/ORT and, 4, 144–145, 146, 168–169, 189, 198, 207, 318
corruption and, 145
Gusinsky and, 148–155, 157–158, 256
Listyev and, 143, 158–167, 327, 331–335
privatization and, 171
Berezovsky, Galina, 23
Bin Laden, Osama, 300, 305
Bobkov, Philip, 58, 86, 151, 157, 219
Boiko, Oleg, 28–29, 231
Borodin, Pavel, 195, 219, 289, 293, 295
Borovoi, Konstantin, 36–37, 87
Brezhnev, Leonid, 53, 55, 302
Bykov, Anatoly, 288, 293, 309, 310, 311

Carnegie, Andrew, 324
Carr, Camilla, 264–265
Central Bank
Aeroflot and, 178, 179–181, 266
Fimaco and, 61, 239
Great Chechen Bank Fraud, 27
Inkom Bank and, 207
loans at negative interest rates, 5, 30, 210–211
Central Committee, 45, 55–56, 58–61, 63, 74–75, 76
Chechen organized crime
Berezovsky and, 13, 258–259, 312, 319, 361*n*2
Chechen government and, 14
Chechen War and, 40–44
Great Mob War and, 22–25
KGB and, 75
Logovaz and, 12
Moscow and, 13, 14–15
Russian organized crime and, 20
Slavic organized crime and, 37–40, 44
Chechnya
bank fraud and, 288
Berezovsky and, 257
Chechen rebels, 229
First Chechen War, 40–44, 188, 215, 254–255, 258, 292, 300, 306, 316
Great Chechen Bank Fraud, 27–28
hostages and, 4, 259–266, 293, 296

independence of, 14
Second Chechen War, 299–307, 313, 317
Chernomyrdin, Viktor
Avva and, 142
banks and, 200, 202
Berezovsky and, 162
Chechen War and, 43
Chechnya and, 255
election of 1996 and, 216, 218, 221, 223–224
firing of, 277–278
fortune of, 325
Gazprom and, 270
Luzhkov and, 158
as prime minister, 43, 101, 112, 113, 269, 282–283
privatization and, 114–115
Russian parliament and, 315
Sibneft and, 195–196, 268
Yuksi and, 276
Cherry Casino, 18, 25–29, 231
Chorny, Lev, 308–309, 310, 311, 312–313
Chorny, Mikhail, 308–309, 310, 311
Chubais, Anatoly
aluminum industry and, 312
assassination threats, 273–274
banks and, 200, 210, 214–215
Berezovsky and, 75, 256, 270, 299
book advance and, 274, 278
as chief of staff, 253
as deputy prime minister, 269, 274–275
economy and, 277
election campaign of 1996 and, 211, 213–215, 218–222, 224, 232, 235, 238, 241, 243–247, 256, 337–344
ethics and, 115–116
foreign trade and, 95
lifestyle of, 252
privatization and, 5, 121, 125–130, 133, 208, 214, 244, 269–273, 323
Putin and, 294–296
Russian economy and, 279, 281, 282, 288
Russian organized crime and, 44–45
Russia's decline and, 5
Soskovets and, 216
starvation and, 104
Clinton, William Jefferson, 111, 226, 229–230, 291, 325
commodities traders
Abramovich as, 195
Alisa and, 82–87
André & Cie. and, 68–72
ANT and, 65
Berezovsky as, 68, 70, 97, 171, 189, 202, 365*n*43

commodities traders (*continued*)
  Birstein and, 114
  Chorny as, 308
  democrats and, 5
  foreign trade and, 96–97
  Rich and, 61–65, 68, 85
  stock exchange and, 136–138
  taxation and, 98–100, 139–140
Communist Party
  banks and, 27
  Central Committee and, 45, 55–56,
    58–61, 63, 74–75, 76
  election of 1996 and, 212, 217–218,
    224–227, 234, 235, 237, 325
  elections of 1999 and, 316
  popular elections and, 48
  privatization and, 133
  Russian parliament and, 81, 121, 278, 282,
    284, 316
  Yeltsin and, 111
Communist Youth League, 59, 67, 149, 199,
  203
Credit Lyonnais, 88
CS First Boston, 132, 133, 136, 185, 198,
  205, 271, 281

Davydov, Oleg
  Aeroflot and, 180
  Avva and, 142
  Berezovsky and, 286
  foreign trade and, 64, 96, 97, 99–101
  oil industry and, 40
  privatization and, 129, 210
  ruble stabilization and, 181
democrats, 5, 6, 67–68, 74, 254, 320
Deutch, John, 33–34, 115
Dlugach, Valery ("Globus"), 22, 146
Dorenko, Sergei, 273, 314
drug trade, 14, 15, 22, 40, 107
Dubinin, Sergei, 179–180
Dudayev, Jokhar, 40, 41, 229, 254
Dyachenko, Leonid, 195, 201
Dyachenko, Tanya
  Berezovsky and, 154, 155, 201–202, 256,
    285–286
  Chernomyrdin and, 277
  election of 1996 and, 219–221, 227, 243
  Fyodor and, 230–233, 242
  Primakov and, 291
  Western education and, 252
  Yeltsin and, 113, 201–202

Fatherland-All Russia movement, 296, 298,
  316
FBI, 2–3, 75, 250, 288

Ferrero, William, 173–174, 178, 180–181
Filshin, Gennadi, 60–61, 65
Fimaco, 61, 65, 239
Finco, 69, 175
First Voucher Investment Fund, 130–131,
  134–135, 136, 137
Flamingo Bank, 289
FNK, 268
FOK (United Finance Company), 89, 183,
  184, 186, 287
Ford, Henry, 324
foreign trade
  banks and, 199
  Davydov and, 64, 96, 97, 99–101
  Gaider and, 95–101
  Gorbachev and, 61
Forus Services S. A., 88–89, 98, 174–175,
  176, 182–183, 186, 287, 289
Freeh, Louis, 320
FSB-KGB
  Gusinsky and, 157–158
  Patarkatsishvili and, 16
  Putin and, 296–297
  terrorism and, 303
Fyodorov, Boris, 115, 231–234, 241–243,
  246, 250, 256

Gaidar, Yegor
  banks and, 210
  Berezovsky and, 75
  Boiko and, 29, 231
  capital flight and, 65
  Great Chechen Bank Fraud, 27
  privatization and, 126–130
  reform of, 79–82, 102, 103, 104, 121,
    125
  Russia's decline and, 5
  Soviet Union dissolution and, 78
  state monopoly on foreign trade, 95–101
  taxation and, 100
Gamsakhurdia, Zviad, 77–78
Gavrilin, Lev, 231–232
Gazprom, 114, 134, 160, 180, 194, 270, 274
General Motors, 140–141, 142
Georgia, 16–17, 41, 77
GHK Corp., 97–98
Gilman, Benjamin, 34, 319
GKO market, 210, 279–281, 288
Glushkov, Nikolai
  Aeroflot and, 172, 176, 179, 183, 186,
    289–290
  Andava and, 173, 174–175, 179–180, 186
  André & Cie., 68, 70
  Avtovaz and, 72, 92, 143
  FOK and, 183

Forus Services and, 89
Okulov and, 287
Golden Ada, 115, 118
Goldfarb, Alexander, 106
Gorbachev, Mikhail
  anti-alcohol campaign, 48–50, 53
  Central Committee and, 55
  Communism and, 235
  democracy and, 320
  economy and, 103
  perestroika and, 2, 48, 69, 167
  Putsch, 72–76, 191, 294
  reform of, 46–48
  ruble/dollar exchange rate, 80
  ruble overhang and, 50–52, 53
  Russian organized crime and, 21, 45
  Soviet foreign trade practices and, 61
  Soviet Union dissolution and, 77
  weakening power of, 66
  Western support for, 67
Grachev, Pavel, 35, 41–42, 74, 113, 215,
  237
Grangeland Holdings Ltd., 183–185
Greenspan, Alan, 50–51
Gromov, General, 73, 157–158
Gurfinkel-Kagalovsky, Natasha, 203
Gusinsky, Vladimir
  Berezovsky and, 148, 153–155, 157, 158,
    257
  Chechen hostages and, 261
  Chubais and, 275
  election campaign of 1996 and, 213, 215,
    243, 253
  election campaign of 1999 and, 298
  Kiriyenko and, 278
  Korzhakov and, 143, 153–158, 230, 256,
    372n67, 373n26
  Lesnevskaya and, 152–153, 163, 164, 165
  lifestyle of, 252
  Listyev and, 163, 164, 165, 166
  loans-for-shares auctions and, 201
  Logovaz bombing and, 38
  Luzhkov and, 149–150, 154, 157, 158
  ORT and, 169
  privatization and, 271–273
  security service of, 151, 156–157
  Yuksi and, 276

Hamilton, Thomas, 191, 193, 194
Hussein, Saddam, 66
hyperinflation, 2, 5, 80–81, 102, 126

Ilyushenko, Alexei, 162–163, 167, 197
Ilyushin, Viktor, 215, 218, 244, 245, 246,
  247, 337–344

Inkom Bank, 201, 204, 207, 230
International Monetary Fund, 27, 31, 96,
  229, 239, 281
Iraq, 52, 62, 66
Itskov, Leonid, 185, 186
Ivankov, Vyacheslav ("Jap"), 17–18, 20, 22,
  38, 44, 75

James, Jon, 264–265
Jordan, Boris, 197–198

Kadannikov, Vladimir, 54, 55, 72, 118, 142,
  143, 171
Kagalovsky, Konstantin, 203, 204
Kalugin, Oleg, 57–60, 293, 297
Kantor, Oleg, 310
KGB
  activities of, 56–57
  Alisa and, 83, 86
  Berezovsky and, 290
  Central Committee and, 56, 58–61, 63,
    75, 76
  democracy and, 57–58
  Gusinsky and, 151
  Listyev and, 163, 164
  money laundering and, 100
  Pamyat and, 66–67
  privatization and, 133
  public discontent and, 50
  Putsch and, 73, 74
  Russian organized crime and, 17, 34, 35,
    36–37, 45, 75, 324
  Yeltsin and, 111
Kharshan, Mikhail, 130–131, 133, 134–135,
  136, 137
Khasbulatov, Ruslan, 122, 124
Khatsenkov, Georgi, 32, 36, 87, 120
Khattab, 300, 301, 305, 306
Khodorkovsky, Mikhail, 75, 176, 198, 200,
  202–204, 213, 267, 276
Khrushchev, Nikita, 55
Kiriyenko, Sergei, 276, 277–281, 285, 296
Kiselev, Yevgeny, 152, 243
Kobzon, Iosef, 17, 19
Kogan, Viktor ("the Kike"), 22–23
Kohl, Helmut, 112
Kokh, Alfred, 206, 266–269, 273
Korablinov, Valery, 250
Korzhakov, Aleksandr
  Abramovich and, 285
  Aeroflot and, 177
  Berezovsky and, 116–117, 120, 143, 147,
    148, 153–154, 163–164, 166, 201,
    230, 236, 257, 285, 305, 312, 327
  Channel 1 and, 159, 160

Korzhakov, Aleksandr (*continued*)
    election campaign of 1996 and, 215, 216,
        217, 219, 220, 223–224, 227, 232,
        237–238, 244, 246, 337
    First Chechen War and, 42
    Fyodorov and, 232, 241–243, 256
    Gusinsky and, 143, 153–158, 230, 256,
        372*n*67, 373*n*26
    Logovaz bombing and, 38, 154
    National Sports Fund and, 232–233
    Patarkatsishvili and, 16–17
    Putsch and, 73
    Russian organized crime and, 35–36
    Sibneft and, 195, 196
    Yeltsin and, 110–113, 116–119
Kovalyov, Nikolai, 296
Kozlyonok, Alexander, 115
Krasnoyarsk Aluminum Smelter, 309–311
Kruchina, Nikolai, 58, 76
Kulikov, Anatoly, 310–311
Kuppers, Rene, 142, 182
Kuwait, 52, 62, 66
Kuznetsov, German, 238, 240
Kvantrishvili, Otari ("Otarik"), 18–20, 22,
        24–25, 37–38, 75, 145, 231

Lebed, Aleksandr
    Berezovsky and, 257–258, 260, 269
    election of 1996 and, 224–225, 236–237
    First Chechen War and, 41, 42, 254–255
    as national security chief, 237, 238, 254
    ORT and, 169
    Putsch and, 74, 236
    Russian army and, 35
    Russian businessmen and, 6
    terrorism and, 304
    as trustworthy, 159
    Yeltsin and, 253–256
Lenin, Vladimir, 36
Lerner, Grigory, 266
Lesnevskaya, Irina, 152–153, 159, 163–165,
        168, 327–331, 334, 335–336
Lesnevsky, Dimitri, 152–153, 168–169
Liberal Democratic Party, 57
Likhodei, Mikhail, 251
Lisin, Vladimir, 308
Lisovsky, Sergei
    advertising business of, 145–147
    election of 1996 and, 228, 240, 243–246,
        337
    Listyev and, 160–161, 164, 168
    ORT and, 287
    Otarik and, 19, 22
Listyev, Vlad

Berezovsky and, 143, 158–168, 189,
        331–335
    Gusinsky and, 256
    Ilyushenko, 197
    Lesnevskaya and, 163–165, 168, 327–331,
        334, 335–336
Litvinenko, Aleksandr, 296
Lobov, Oleg, 42, 114
Logovaz
    Aeroflot and, 172, 186
    aluminum industry and, 97, 307, 312
    André & Cie. and, 70–71
    Avtovaz and, 55, 71–72, 93, 94, 143,
        175–176, 202
    Avva and, 140
    Berezovsky and, 4, 55, 189, 319
    bombing of, 37–40, 92, 164–165
    Chechen organized crime and, 12, 21
    Chechnya and, 261
    commodities trading and, 97
    Great Mob War and, 23
    Listyev and, 162
    Ministry of Foreign Trade and, 75–76
    Patarkatsishvili and, 16
    publishing and, 117–118
    television and, 146–147, 159
    wholesale/export trade and, 82, 87
Lukoil, 180, 190–191, 195, 196, 208
Luzhkov, Yuri
    Alekperov and, 194
    election of 1999 and, 298, 313, 314, 316
    election campaign of 1996 and, 213
    Gusinsky and, 149–150, 154, 157, 158
    Kovalyov and, 296
    Logovaz bombing and, 38
    ORT and, 169
    privatization and, 128
    Yeltsin and, 113, 154–155, 164, 165

Makarov, Andrei, 116
Manevich, Mikhail, 273–274
Maret, Christian, 70, 92, 175, 178
Maskhadov, Aslan, 259, 260, 265–266,
        300–301, 302, 304, 306
Mavrodi, Sergei, 138, 139, 140, 314
Mavrodi, Vyacheslav, 138–140
Mayor, Alain, 23–24, 69–71, 92, 94, 175
MC Securities, 271
MDM Bank, 289
Menatep Bank, 160, 176, 198, 202–204, 207,
        208, 211, 239, 267, 279
MFK, 199
Mikhailov, Sergei ("Mikhas"), 20, 44, 146
Milken, Michael, 252

Ministry of Foreign Trade, 21, 64, 65, 75–76, 96, 100
Ministry of Fuel and Energy, 100, 195, 196
Ministry of Internal Affairs (MVD), 17, 29–30, 123, 162, 287–288, 309, 314
Ministry of Posts and Communications, 131
MMM, 138–140, 143, 314
Mogilny, Aleksandr, 250
Moody, James, 37
Morgan, J. P., 324
Morgan Stanley Asset Management, 271
Morgenstern, Claudia, 133–134
mortality rates, 105–108, 291
Moscow Profbank, 31
Moscow RUOP, 12, 28, 32–33, 42, 156, 161
Most Bank, 123, 148, 149, 150–151, 201, 239
Mostorg Bank, 39
Murdoch, Rupert, 151, 252, 287
MVD. *See* Ministry of Internal Affairs (MVD)

National Sports Fund, 4, 160, 231–234, 241–242, 250
NATO, 188, 291
Natsionalny Kredit Bank, 28, 160, 231
Nemtsov, Boris, 243, 269, 270, 274, 277, 299, 315
NFK, 206, 207, 267–268
NFQ, 185, 186, 287
Nicholas II (czar of Russia), 49
Nikolaev, Alexei, 4, 90, 93, 94
Norilsk Nickel, 204–205, 209, 253
Noyabrskneftegaz, 195, 196, 197
NTV, 148, 152–153, 155, 253, 261, 271, 272, 298
Nurnberg, Andrew, 117, 118

Obyedinyonny Bank, 38, 89
oil industry
    Alekperov and, 190–194
    assassinations in, 188
    Berezovsky and, 189
    Chechnya and, 40–41
    commodities trading and, 96, 99, 100–101
    Cyprus and, 100
    foreign companies and, 189–190
    foreign trade and, 96–97
    government revenues and, 50
    Logovaz and, 97–98
    OPEC and, 47–48
    privatization and, 32, 129, 191, 266, 275–276
    Rich, Marc, and, 61, 62, 63, 85
    Russian organized crime and, 32, 101
    Tarasov and, 64

Okulov, Valery, 177, 178–179, 185, 186–187, 287
Olbi, 28, 231
oligarchy
    Berezovsky and, 169
    Chubais and, 270
    Dyachenko and, 201
    election campaign of 1996 and, 211, 213–214, 216–218, 220, 222–223, 234–235, 237, 253
    Kiriyenko and, 278–280
    Lebed and, 237
    privatization and, 210–211, 267
    Russian organized crime and, 37, 44
    Second Chechen War and, 301
    Western governments and, 288
OMON, 162
Omsk Oil Refinery, 4, 195, 196, 201, 276
Onexim Bank, 115, 199–200, 204, 205, 211, 213, 253, 268, 271, 273, 274, 279, 312
OPEC oil cartel, 47
Orekhov, Andrei, 132–133, 138
ORT (Russian Public Television)
    bankruptcy of, 287
    Chernomyrdin and, 196
    control of, 169, 198, 318
    election of 1999 and, 313–314
    election campaign of 1996 and, 220, 225
    Fyodorov and, 231
    hostages and, 260
    Inkom Bank and, 207
    Listyev and, 160–161, 168, 327–336
    privatization and, 159–160, 168–169, 171
    Yeltsin and, 148, 156
Otarik (Otari Kvantrishivili), 18–20, 22, 24–25, 37–38, 75, 145, 231
Ovchinnikov, Igor, 11–15, 146

Pamyat, 57, 66–67
Patarkatsishvili, Badri
    Chechnya and, 262–264
    Listyev and, 161, 162, 163, 164
    Mostorg Bank and, 39
    Russian organized crime and, 16–17, 146
Petrov, Aleksandr, 31
Pokrovksy, Vadim, 107
Potanin, Vladimir
    Berezovsky and, 169
    debt-for-equity deal and, 198
    election of 1996 and, 213, 215, 253
    loans-for-shares auctions and, 199–200, 204–205, 208, 267, 312
    telecommunications and, 271–273
    Yeltsin and, 269

Premier S. V., 146
Primakov, Yevgeny, 283, 284, 286, 290–292,
    298, 313, 314, 316
privatization. *See also* Russian economy
    Berezovsky and, 2, 120, 170–171,
        209–210, 270–271
    Channel 1 and, 147–148, 159–160
    Chernomyrdin and, 114
    Chubais and, 5, 121, 125–130, 133, 208,
        214, 244, 269–273, 323
    of debts, 170
    democrats and, 5
    discrediting of, 320
    election campaign of 1996 and, 223
    emerging markets and, 130–136
    failure of, 323
    502–Day Plan and, 51
    Greenspan and, 51
    hyperinflation and, 81
    loans-for-shares auctions and, 5, 197–211,
        214, 223, 230, 266–275
    oil industry and, 129, 191, 266, 275–276
    Putin and, 294
    Russian organized crime and, 32, 36
    voucher privatization, 5
Promstoi Bank, 266
Putilov, Aleksandr, 191, 193, 195, 275
Putin, Vladimir
    Berezovsky and, 295–296, 313, 314
    Chubais and, 294–295, 296
    FSB-KGB and, 296–297, 313
    KGB and, 57–58, 293–294
    as prime minister, 293, 297–298, 313
    Russia's rebuilding and, 326
    terrorism and, 302–305, 307
    Trans-World and, 311
    as Yeltsin's heir, 298–299, 316–317

Raduyev, Salman, 258, 259–260, 261, 265,
    305
Reagan, Ronald, 47
Reardon, David, 191
Renaissance Capital, 198, 271
Reuben, David, 308, 311
Rich, Marc, 61–65, 68, 84–87, 203
Rockefeller, John, 324
Rosneft, 193, 194–195, 196, 272, 275–276,
    277
Rosselkhoz Bank, 31
Rossiisky Kredit Bank, 201, 204, 205, 239
Runicom, 275–276
Rushailo, Vladimir, 12, 39, 42, 292–293
Russia
    Belarus and, 229

capital flight, 99–100, 178, 181
    Chechnya and, 14
    children and, 108–109
    decline of, 2, 5, 6
    democracy and, 320–321
    expatriate community in, 248–250
    foreign trade network, 95–101
    healthy society and, 322–323
    healthy state and, 321–322, 323
    mortality rates, 81, 105–108, 291
    NATO and, 188
    population decline, 108
    power structure of, 3
    public health, 106–107
    robber barons of, 1–2
    terrorism in, 302–305, 307
    values in, 321, 323
    Western democratic principles and, 6
Russian army
    back wages for, 277, 279
    corruption in, 35, 41–42
    First Chechen War and, 40–43
    Germany and, 112
    Putsch and, 73–74
    Russian organized crime and, 34–35
    withdrawal from Chechnya, 14
Russian economy. *See also* privatization
    automobile industry and, 90, 95
    constructive roles in, 148–149
    disintegration of, 113
    financial crash of, 277–283, 284, 288
    foreign trade and, 95–101
    hyperinflation and, 2, 5, 80–81, 102, 126
    as market economy, 2, 6, 45, 125–130,
        144, 323
    poverty and, 2, 103–105
    Russian organized crime and, 30, 87
Russian government
    banks and, 200
    Chechen War and, 40–44
    corruption and, 6, 30, 113–116, 122,
        144–148, 295
    crony capitalism and, 5–6, 232, 269, 274,
        284, 320, 324
    "Harvest 90" campaign, 64, 87
    Kvantrishvili on, 19
    Russian organized crime and, 2, 3, 4, 30,
        34, 36, 37, 319
    Soviet Union dissolution and, 78
Russian organized crime
    aluminum industry and, 307, 309, 311
    athletics and, 250
    automobile industry and, 11–15, 21, 23,
        24, 37, 90

banks and, 31, 199
charitable organizations and, 250–251
commodities trading and, 100–101
development of, 15–21
extent of, 29–37
government allies of, 3
Great Mob War and, 21–25, 113, 143,
146, 148, 188
investigations of, 289
Listyev and, 161, 164
Luzhkov and, 150
market economy and, 144
National Sports Fund and, 232
Ovchinnikov and, 11–12
Primakov and, 286
Russian government and, 2, 3, 4, 30, 34,
36, 37, 319
Russian parliament and, 315–316
Russia's decline and, 2
thieves-professing-the-code and, 15–21,
22
U.S. hearings on, 319–320
Russian parliament
capital flight and, 100
Communist Party and, 81, 121, 278, 282,
284, 316
immunity from prosecution and, 314–315,
316
Yeltsin and, 121–125, 290, 291, 316
Russia's Choice party, 29
Rutskoi, Aleksandr, 116, 120–122, 124
Ryzhkov, Nikolai, 50, 65

Sakharov, Andrei, 17, 58, 167, 188
Samaraneftegaz, 97–98, 204
Samara Refinery, 98, 204
Savostyanov, Yevgeni, 157, 158
Sayer, Dave, 26, 28
SBS-Agro, 253, 268–269, 280, 288, 289, 319
Shafranik, Yuri, 192, 193
Shaposhnikov, Yevgeny, 172, 176, 178
Shatalin, Stanislav, 51
Sheinin, Roman, 183
Shushkevich, Stanislav, 77
Sibneft, 195–198, 205–208, 230, 266–269,
275–277, 287, 292, 307, 318
Sidanco, 205
Simes, Dimitri, 151
Skorochkin, Sergei, 314
Skuratov, Yuri, 61, 245, 246, 288, 297
Slavic organized crime, 37–40, 44
Smolensky, Aleksandr
Berezovsky and, 84
Chubais and, 214

criminal proceedings against, 288
election of 1996 and, 85, 221–222, 224
loans-for-shares auctions and, 85, 200, 207
money-laundering affiliates and, 289
Sibneft and, 268
Yuksi and, 276
Sobchak, Anatoly, 58, 75, 294
Sobin Bank, 289
Solntsevo Brotherhood, 11, 20, 21, 22–25,
44, 75
Solzhenitsyn, Aleksandr, 58, 78
Soros, George, 271, 273, 282
Soskovets, Oleg
Berezovsky and, 120, 162, 172
Chorny and, 312
election of 1996 and, 215–216, 217, 220
First Chechen War and, 42
Lisin and, 308
oil industry and, 195
Yeltsin and, 238
Soviet Union
anti-alcohol campaign and, 49, 53
army, 46, 73
capital flight and, 64, 65–66
commodity trading and, 69
dissolution of, 14, 77–78, 103, 291, 301
economy of, 49, 50–52, 78
gold and foreign exchange reserves of,
65–66
Gorbachev's reform, 46–48
Putsch, 72–76
Russian businessmen and, 2
Special Export firms, 100–101
Stacy, T. Don, 189, 192
Stalin, Joseph, 6, 55, 105, 111, 316
Staravoitova, Galina, 286
State Property Committee (GKI), 125, 133,
206, 266
Stepashin, Sergei, 42, 290, 291, 292, 299,
306, 313, 315
Sterligov, German, 82–84, 86
Stolichny Bank, 84–85, 160, 200, 211, 214,
221–222, 239, 253, 274, 279
Stolypin, Pyotr, 2, 95
Streletsky, Valery, 200–202, 209, 221–223,
226, 230, 231–234, 238–242, 244, 246
Suleimanov, Khozha, 13, 15, 45
Svyazinvest, 271–273, 274, 282
Sylvester (Sergei Timofeyev), 22, 38, 39,
39–40, 146
Sysuyev, Oleg, 211, 307, 314

Tarasov, Artyom, 63–64, 84–87
Tarpishchev, Shamil, 113, 197, 231, 233

Tatum, Paul, 249–250
television. *See also* NTV; ORT (Russian
    Public Television)
    Channel 1 and, 144–145, 146, 168–169,
        189, 198, 207
    election campaign of 1996 and, 225–228,
        234
    election campaign of 1999 and, 298
    Lebed and, 237
    politics and, 146, 147, 153–155, 169, 189,
        196, 220, 221, 225, 226, 313
Thompson, Page, 87–95, 182
Tikhonov, Vladimir, 171–172
Timan Pechora project, 193
Timofeyev, Sergei ("Sylvester"), 22, 38, 39,
    39–40, 146
Trans-World Group, 308, 310, 311, 312
Triboi, Pyotr, 167–168
Turner, Ted, 149, 169

Udugov, Movladi, 261, 262–264, 300, 301
United States government, 52
Unity Party, 298, 299, 316
U.S. Eximbank, 172, 177

Vanner, 23, 28
VE Bank, 56
Veselovsky, Leonid, 58, 59
Vinogradov, Vladimir, 201, 207, 213
Voloshin, Aleksandr, 293, 303
Vyakhirev, Rem, 213

Wanniski, Jude, 80–81
Western Group of Forces, 41–42

Yabloko, 316
Yafyasov, Vadim, 309–310
Yakutian, Radik, 92
Yamani, Sheik Ahmed, 47
Yanchev, Pyotr, 197
Yasin, Yevgeny, 111–103
Yavlinsky, Grigory, 37, 51, 78, 81–82, 99,
    102, 127–129, 236, 253
Yeltsin, Boris
    absolute power and, 120
    army corruption and, 35
    Avtovaz and, 95
    banks and, 30
    Berezovsky and, 110, 116–120, 140, 143,
        144, 147, 163–166, 169, 201, 211,
        284, 289, 292, 293, 301, 317, 319, 327
    Chechnya and, 14, 42
    Chubais and, 214, 274, 281

counterdemonstration of, 67
crony capitalism and, 5–6, 232, 269, 274,
    284, 320, 324
democracy and, 2, 321, 322–323
election campaign of 1996 and, 177, 211,
    212, 215–216, 218, 226–229, 238,
    245–247, 252, 337–344
embezzlement charges and, 289
Gusinsky and, 157–158
health of, 215, 226–227, 252–253, 255,
    269
Korzhakov and, 110–113, 116–119, 244
Logovaz bombing and, 38, 39
Luzhkov and, 154–155
oil industry and, 194
oligarchy and, 210–211
Primakov and, 290–292
privatization and, 127, 269
Putin and, 297–298, 316–317
Putsch and, 73–74, 111, 117, 121, 123,
    172, 236
Russian economy and, 80, 277, 282–283,
    284
Russian organized crime and, 3, 21, 29,
    34, 320
Russian parliament and, 121–125, 290,
    291, 316
Skuratov and, 297
Smolensky and, 85
Soviet Union dissolution and, 77–78
Tarasov and, 64
television and, 144–145, 147–148
Yergin, Daniel, 189
Yerin, Viktor, 42, 162
Yevstafyev, Arkady, 240, 243–246, 337
Yfimov, Vitaly, 172
Yukos, 204, 207, 208, 276, 277
Yumashev, Valentin
    Berezovsky and, 165, 230, 232, 256
    Chernomyrdin and, 277
    as chief of staff, 269
    Dyachenko and, 201, 219
    election of 1996 and, 215
    Putin and, 295
    Yeltsin and, 113, 117–119, 165

Zhaboev, Samat, 72, 189
Zhirinovsky, Vladimir, 57, 124–125, 212,
    236, 292, 315
Zverev, Sergei, 158, 244, 337–344
Zyuganov, Gennady, 212–213, 218, 224–227,
    234, 235, 237, 325